MW00844547

OPEN-CHANNEL FLOW

OPEN-CHANNEL FLOW

SUBHASH C. JAIN

JOHN WILEY & SONS, INC.

New York • Chichester • Weinheim • Brisbane • Singapore • Toronto

Copyright © 2001 by John Wiley & Sons, Inc. All rights reserved.

Published simultaneously in Canada.

This publication is designed to provide accurate and authoritative information in regard to the subject matter covered. It is sold with the understanding that the publisher is not engaged in rendering professional services. If professional advice or other expert assistance is required, the services of a competent professional person should be sought.

Library of Congress Cataloging-in-Publication Data:

Jain, Subhash C. (Subhash Chandra), 1938–
    Open-channel flow / Subhash C. Jain
        p.   cm.
    Includes bibliographical references and index.
    ISBN 0-471-35641-7 (cloth: alk. paper)
    1. Channels (Hydraulic engineering)   2. Hydraulics.   I. Title.
    TC175 .J35   2000
    627'.23—dc21   99-089638

Printed in the United States of America.

10 9 8 7 6 5 4 3 2 1

CONTENTS

Contents   ix

PREFACE

A fundamental knowledge of flow in open channels, which is the subject matter of this book, is essential for the planning and design of systems to manage water resources. This book is intended to serve as a text for use by senior undergraduate or graduate students in civil engineering. Practicing engineers will find the book useful for updating their basic comprehension of flow in open channels. The book uses metric units.

The author believes that the basic principles of flow in open channels should be taught through practical problems that can be solved either analytically or by simple numerical methods that do not require computer software. Though there is a place for a text on computational hydraulics, it should be a separate book. Accordingly, the governing equations are simplified for particular flows and the analytical or numerical solutions of the simplified equations are described in this book. For many practical problems the solution of the simplified equations is more than adequate.

Though this textbook, like earlier textbooks, deals with one-dimensional equations of motion, it has several unique features. In this book the one-dimensional equations of motion are derived using two approaches: the simplified approach and the rigorous approach. The distinction between the momentum and mechanical-energy equations is explained. A great emphasis is placed on identifying the types and locations of the control sections that are essential in analyzing flow profiles, and a section on nonunique flow profiles that were recognized recently is also included. Analytical solutions of sluice-gate operations, which were either incorrect or inadequately explained in earlier books, are described herein. The book contains numerous worked examples that are helpful in understanding the basic principles and their practical applications.

The governing equations, both in differential and algebraic forms, are derived in Chapter 1. It is shown that the one-dimensional momentum and energy equations are very similar to each other for most flows, except for spatially varied flow. Solutions of the governing equations are presented in Chapters 2–8, starting with the simplest flow, that is, steady uniform flow, in Chapter 2. The concept of control sections is introduced in Chapter 3 through the theories of disturbance propagation and channel transitions. Chapter 4 deals with the analysis of gradually varied flow with emphasis on sketching flow profiles in channels with transitions. The analytical, numerical, and graphical techniques for computing flow profiles are included in Chapter 5, while Chapter 6 summarizes the analysis and computational methods for spatially varied flow. The problems of unsteady flow are divided into two types: one with negligible channel and energy slopes and the second with negligible local and connective accelerations. The unsteady flow is described in Chapters 7 and 8. Artificial channel controls, such as weirs, spillways, and gates, are included in Chapter 9 and the special topics are discussed in Chapter 10.

The author has drawn material from many sources and wishes to thank them all.

CHAPTER 1

BASIC EQUATIONS

1-1 Introduction

Liquids are conveyed from one location to another through conduits as closed-conduit flow (also known as pipe flow) or open-channel flow. While the conduit is completely full in the former case, it has a free surface in the latter case. Open-channel flow is, therefore, flow of a liquid in a conduit with a free surface subjected to atmospheric pressure. Flow of water in rivers, canals, and partially full sewers and drains and flow of water over land (i.e., surface runoff) are some examples of open-channel flow. Technical activities associated with the utilization of water for water supply, irrigation, hydropower, and navigation and the control of water for flood mitigation deal with open channels. Some of the practical applications of open-channel flow are the determination of—(i) flow depth in rivers, canals, and other conveyance conduits; (ii) changes in flow depth due to channel controls such as weirs, spillways, and gates and due to channel transitions such as expansions, contraction, and bends; (iii) changes in river stage during floods; and (iv) surface runoff from rainfall over land.

Problems of open-channel flow are relatively more difficult to solve than those of closed-conduit flow. While the flow area is known in the latter, it is unknown in the former as the location of the free surface is not known a priori. The shape and size of open channels vary over a wide range. The size varies from rivulets to mighty rivers. While the artificial (man-made) channels have regular geometric shapes, the cross-sectional shapes of natural channels are in general very irregular. The component of the weight of the liquid along the channel slope is the primary driving force, and the shear force on the channel boundaries is the main resisting force. Flow in open channels is mainly in the turbulent regime. Depending on the variation of flow depth and velocity with the length of the channel and time, different types of flow, classified as *uniform, nonuniform (varied), steady,* and *unsteady,* occur in open channels. The flow in a river during a flood is unsteady, varied flow, while the flow upstream from a spillway during steady operation is steady, varied flow.

Types of Channels

Channels are classified as rigid- and loose-boundary channels. In rigid-boundary channels the channel boundary is not deformable; consequently the channel cross section, profile, plan, and boundary roughness are not functions of discharge, and flow depth is the only channel parameter that changes with discharge. In loose-boundary channels the channel boundary undergoes deformation due to scour and deposition of sediment transported by the flow in the channel, and not only the flow depth but also the width, slope, and plan of the channel can change with discharge. Flow in loose-boundary channels usually is discussed in books on sediment transport and is not dealt with in this book. In most cases flow in channels can be analyzed assuming them to be rigid channels, unless their boundary undergo significant deformation.

An open channel in which the shape and size of the cross section and the slope of the bottom are constant is termed prismatic channel. Generally artificial channels are prismatic and natural channels are nonprismatic.

1-2 Governing Equations

Solutions of fluid-flow problems of the most general type involve prediction of distributions of fluid pressure, temperature, density, and three components of flow velocity. Six basic equations are required to solve for these six unknown quantities. The six equations are as follows: (i) the continuity equation based on the law of conservation of mass; (ii) the three momentum equations, along three orthogonal directions, derived from Newton's second law of motion; (iii) the thermal energy equation obtained from the first law of thermodynamics; and (iv) the equation of state, which is an empirical relation among fluid pressure, temperature, and density. Fluid problems do not require the last two equations, when the problems deal with (i) essentially homogeneous fluids; (ii) velocity changes that are small compared to the speed of sound; (iii) no exchange of thermal energy; and (iv) negligible changes in stored thermal energy. Density and temperature in such problems are considered as constant. Open-channel problems discussed in this book belong to this type and therefore can be solved by the continuity and momentum equations of motion. However, there is a class of problems (e.g., lateral outflow problems discussed in Chapter 6) to which the momentum equations are not readily applicable; such problems are solved using the mechanical-energy equation, which is derived from the momentum equations.

Most problems in open-channel flow are solved using the one-dimensional equations of motion. The solution of some problems requires the use of the differential forms of the equations of motion, while other problems are solved using the algebraic forms of the governing equations. The algebraic forms are obtained by integrating the differential forms; therefore the latter are derived first. The one-dimensional differential equations of motion are derived using three approaches: (i) elementary approach, (ii) simplified approach, and (iii) rigorous approach. In the elementary and simplified approaches, the assumptions are stated first and then the simplified governing equations are derived based on the stated assumptions. In the rigorous approach, the complete one-dimensional differential equations of motion are derived first by integrating the three-dimensional equations of motion over the channel cross section (Strelkoff, 1969; Yen, 1973) and then simplified after making certain assumptions. The rigorous approach, though it may seem tedious, bridges the gap between the exact three-dimensional equations of fluid mechanics and the approximate one-dimensional equations used in open-channel flow computations and clarifies the limitations of the one-dimensional equations. The derivation of the governing equations by the rigorous approach is presented in Appendix A of this chapter.

The difference between the elementary and simplified approaches is in one of the assumptions made in the two approaches. While the main component of velocity in the elementary approach is a function of only the longitudinal direction, the main component of velocity in the simplified approach is a function of the longitudinal direction as well as one of the lateral directions. Though this difference in the assumption makes the derivation by the elementary approach of the continuity and momentum equations simpler, the elementary approach has one deficiency. The mechanical-energy equation cannot be derived with the elementary approach.. The governing equations by the simplified approach are derived next. The derivations of the continuity and momentum equations by the elementary approach are presented in Examples 1-1 and 1-3, respectively

1-3 Basic Hypotheses

Most problems in open-channel flow deal with a channel that has depth dimension far smaller than the longitudinal and lateral dimensions. In addition, it may often be assumed that changes in cross section along the longitudinal direction are very gradual. For the channel geometry under consideration it is reasonable to assume that the main component of flow velocity is along the longitudinal direction and the components of velocity along the normal directions are negligible. Moreover, the one-dimensional method of approach neglects acceleration in the normal direction, which implies that the streamlines have small curvatures and the pressure distribution is hydrostatic. The cartesian coordinate system is used throughout the book. The x-axis along the longitudinal direction is parallel to the average bottom slope; the z-axis along the lateral direction is horizontal; and the y-axis along the depth direction is normal to the channel bottom, as shown in Figure 1-1. A horizontal datum and a G-axis along the vertical direction are also shown in the figure. The governing equations derived using the simplified approach are based on the following assumptions:

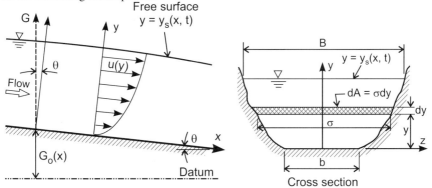

Figure 1-1. Definition sketch of channel geometry.

(1) The main component of velocity, u, is along the x-axis and is a function of x and y (see Figure 1-1). The velocity components, v and w, along the y- and z-axes are zero.

(2) The rate of change of shear stresses with x and z are small and assumed zero.

(3) The components of acceleration along the y- and z-axes are zero. This assumption leads to a pressure distribution that is hydrostatic.

(4) The density of water, ρ, is constant.

Hydrostatic Pressure Distribution

To derive an expression for hydrostatic pressure distribution, consider the forces in the y-direction acting on an element of fluid with an area ΔA normal to the y-axis and with side Δy parallel to the y-axis, as shown in Figure 1-2. The forces on the element include the body and surface forces. The body force is the weight of the element and acts vertically downward. Its component in the y-direction is equal to $\gamma \Delta y\, \Delta A \cos\theta$, where γ is the specific weight of fluid and θ is the angle between the channel bed and a horizontal plane. The surface forces include the pressure and shear forces. The pressure forces in the y-direction are shown in Figure 1-2. On the basis of the second assumption stated above, the net shear force in the y-direction is zero. From Newton's second law of motion, the resultant of the pressure and gravity forces must be equal to the product of the mass of the element and the y-component of the acceleration of the element, i.e.,

$$p\Delta A - \left(p + \frac{\partial p}{\partial y}\Delta y \right)\Delta A - \gamma \cos\theta\, \Delta y\, \Delta A = (\rho\, \Delta A\, \Delta y)\, a_y$$

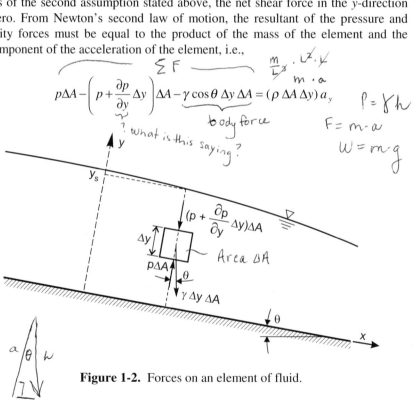

Figure 1-2. Forces on an element of fluid.

$$F \;=\; m \cdot a$$

or

$$\frac{\partial p}{\partial y} + \gamma \cos\theta = -\rho a_{y} \nearrow 0 \tag{1-1}$$

where p = pressure and a_y = fluid acceleration in the y-direction. Using the boundary condition that the pressure at the free surface ($y = y_s$) is zero and the flow condition that the acceleration in the y-direction is zero, integration of Eq. (1-1) gives

$$p = \gamma \cos\theta \,(y_s - y) \tag{1-2} \leftarrow$$

Equation (1-2) is referred to as the equation for hydrostatic pressure distribution in open-channel flow. For most open channels the bed slope is small and the $\cos\theta$ may be replaced by unity. Note that in Eq. (1-2) y_s is the depth of flow in the channel and it is measured normal to the channel bottom, not along the vertical direction.

PROBLEMS

1-1   Derive an expression for the pressure distribution in a steep channel in terms of channel slope θ and vertical distance ΔG from the free surface. The flow in the channel is uniform. Determine the percentage error in pressure for a channel slope of 0.01 if the pressure is assumed equal to $\gamma \Delta G$.

1-2   The depth of flow in a rectangular chute is 1.5 m and the chute is inclined at an angle of 30° with the horizontal. Determine the pressure force on the sidewalls of the chute.

1-4 Differential Continuity Equation

The continuity equation is based on the principle of conservation of mass, which states that the rate of increase of fluid mass within a given control volume must be equal to the difference between the mass influx into and mass efflux out of the control volume. Using this principle, the continuity equation is derived first for a differential volume and then integrated over the channel cross section to obtain the one-dimensional continuity equation. The differential volume is shown in Figure 1-3 and is located at a distance y from the channel bottom. The dimensions of the differential volume along the x-, y-, and z-directions are dx, dy, and $\sigma\,(x,y)$, respectively; where $\sigma\,(x,y)$ = channel width. For the fixed differential volume and the constant-density fluid, the rate of change of fluid mass within the differential volume is zero. The mass influx into and mass efflux out of the differential volume must be equal: influx - must equal -   outflux

rainfall
term $+\;\; \rho u \sigma \, dy = \rho \left[u\sigma + \frac{\partial(u\sigma)}{\partial x}\,dx \right] dy + \rho V_n \, dP \, ds \quad\leftarrow$

where $V_n = V_{n\ell} + V_{nr}$ = flow velocity normal to the banks of the channel and is

BASIC EQUATIONS

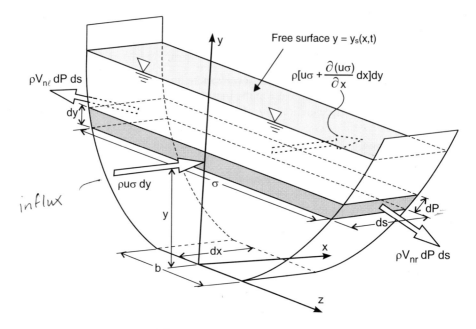

Figure 1-3. Differential volume.

positive for outflow; $dPds$ is the surface area of the lateral surface on the right and left banks of the channel; and the subscripts r and ℓ respectively denote the right and left banks (looking toward the flow direction). The above equation can be written as

$$\frac{\partial(u\sigma)}{\partial x}+V_n\frac{dP}{dy}\frac{ds}{dx}=0 \qquad (1\text{-}3)$$

Equation (1-3) will be integrated over the cross section of the channel to obtain one-dimensional continuity equation. The integration of Eq. (1-3) over the cross-sectional area can be written as

$$\int_0^{y_s}\left[\frac{\partial(u\sigma)}{\partial x}+V_n\frac{dP}{dy}\frac{ds}{dx}\right]dy=0 \qquad (1\text{-}4)$$

depth integration ?

Note that y_s is a function of x and t. Applying the Leibnitz rule of integration[*] (Hildebrand, 1966) to the first term in Eq. (1-4) and noting that the lower limit of integration is constant, one obtains

$$[*]\int_{A(x,t)}^{B(x,t)}\frac{\partial f(x,y,t)}{\partial x}dy=\frac{\partial}{\partial x}\int_{A(x,t)}^{B(x,t)}f(x,y,t)\,dy-f(B)\frac{\partial B}{\partial x}+f(A)\frac{\partial A}{\partial x} \qquad ?$$

$f = u\sigma$
$A = 0$
$B = y_s$

$f(A) = 0$

$f(B)$

$$\int_0^{y_s} \frac{\partial(u\sigma)}{\partial x}\,dy = \frac{\partial}{\partial x}\int_0^{y_s} u\sigma\,dy - (u\sigma)\Big|_{y=y_s}\frac{\partial y_s}{\partial x} \quad \text{evaluate @ } y_s \quad (1\text{-}5)$$

The second term in Eq. (1-4) represents the net volume flux out of the channel bed and banks and can be written as

Distortion

$$\int_0^{y_s} V_n\frac{dP}{dy}\frac{ds}{dx}\,dy = q_\ell \tag{1-6}$$

where q_ℓ is the lateral discharge per unit length of the channel and is positive for lateral outflow. Note that q_ℓ in Eq. (1-6) includes the outflow through the channel bed; the integrand at $y = 0$ becomes $V_n\,b\,dx$, where $b =$ bottom width of the channel cross section. Substitution of Eqs. (1-5) and (1-6) into Eq. (1-4) yields   *what is this saying?*

$$\frac{\partial}{\partial x}\int_0^{y_s} u\sigma\,dy - (u\sigma)\Big|_{y=y_s}\frac{\partial y_s}{\partial x} + q_\ell = 0 \tag{1-7}$$

at surface

Equation (1-7) can be simplified by introducing a boundary condition, known as the kinematic boundary condition, which has to be satisfied by the flow. The kinematic condition in physical terms states that the velocity of a fluid particle relative to the surface in which it lies must be either wholly tangential or zero; otherwise there is a finite flow of fluid across it. As shown in Appendix A, the mathematical expression for the kinematic boundary condition at the free surface $F = y - y_s(x,t) = 0$, with no flow across it, is

Boundary Condition $\longrightarrow$

$$u_s\frac{\partial y_s}{\partial x} + \frac{\partial y_s}{\partial t} = 0 \quad\Rightarrow\quad \frac{\partial y_s}{\partial x} = -\frac{1}{u_s}\frac{\partial y_s}{\partial t} \tag{1-8}$$

where the subscript s denotes the conditions at the free surface. Placing Eq. (1-8) into Eq. (1-7) and noting that the integral term in Eq. (1-7) is discharge Q in the channel, Eq. (1-7) can be written as

$$\frac{\partial Q}{\partial x} + B\frac{\partial y_s}{\partial t} + q_\ell = \frac{\partial Q}{\partial x} + \frac{\partial A}{\partial t} + q_\ell = 0 \tag{1-9}$$

where $B = \sigma_s =$ channel width at the free surface; $dA = B\,dy_s$; and $A =$ cross-sectional area of flow. Equation (1-9) is the one-dimensional differential continuity equation. It is an exact equation for incompressible fluids. This equation without the lateral flow is derived by the elementary approach in Example 1-1.

Example 1-1. Applying the law of conservation of mass to a control volume between two cross sections an infinitesimal distance apart, derive the continuity equation for one-

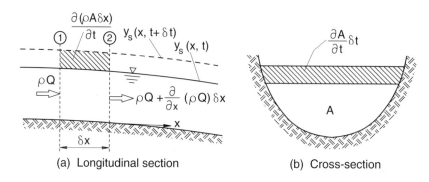

(a) Longitudinal section   (b) Cross-section

Example 1-1

dimensional flow in open channels. Assume no lateral flow across the channel boundaries.

Consider a control volume between two cross sections, distance δx apart, of a channel, as shown in the figure above. The net rate of mass inflow into the control volume can be expressed as

$$\rho Q - \left(\rho Q + \frac{\partial(\rho Q)}{\partial x} \delta x \right) = -\frac{\partial(\rho Q)}{\partial x} \delta x$$

and this term must be equal to the temporal rate of change of the mass contained within the control volume, i.e.,

$$-\frac{\partial(\rho Q)}{\partial x} \delta x = \frac{\partial(\rho A \, \delta x)}{\partial t}$$

In the limit as $\delta x \to 0$,

$$\frac{\partial(\rho A)}{\partial t} + \frac{\partial(\rho Q)}{\partial x} = 0$$

With the assumption of an incompressible liquid, this equation reduces to

$$\frac{\partial A}{\partial t} + \frac{\partial Q}{\partial x} = 0$$

which is the one-dimensional continuity equation for flow of an incompressible liquid in open channels and could be obtained from Eq. (1-9) by setting $q_\ell = 0$.

Example 1-2. Water level in a river reach is increasing due to a flood. Estimate the rate of rising of the river stage if the inflow rate into and outflow rate from the river reach are 100 and 50 m³/s, respectively. The width of the river at the free surface is 250 m and the length of the river reach is 5 km.

The continuity equation (1-9) without lateral flow can be written as

$$\frac{\partial A}{\partial t} = -\frac{\partial Q}{\partial x} = -\frac{Q_2 - Q_1}{\Delta x} = -\frac{50 - 100}{5 \times 1000} = 0.01 \ \text{m}^3/\text{s/m}$$

Noting that $dA = B\,dy_s$, the above equation can be written as

$$B\frac{\partial y_s}{\partial t} = 0.01 \quad \text{or} \quad \frac{\partial y_s}{\partial t} = \frac{0.01}{250}\times 3600 = 0.144 \text{ m/h}$$

The water level in the river is rising at a rate of 14.4 cm/h.

PROBLEMS

1-3   Modify the one-dimensional continuity equation (1-9) to include rainfall over the free surface.

1-4   Water level in a river is rising due to a flood. At a river cross section it is estimated that the river discharge at a particular time is about 2200 m^3/s and that the river stage is rising at a rate of 0.5 m/h. Estimate the river discharge 4 km upstream if the surface width of the river in this reach is 800 m.

1-5   Rework Example 1-1 if there is a net outflow q_l across the lateral boundaries of the channel.

1-6   The velocities measured at the centerline of a wide rectangular channel of depth 3 m are given below. Find the average velocity.

y (m)	0	0.5	1.0	1.5	2.0	2.5	3.0
u (m/s)	0	1.7	2.5	3.0	3.2	3.4	3.5

1-5 Differential Momentum Equation

The momentum equations are based on Newton's second law of motion, which states that the sum of all external forces acting on a system is equal to the product of the mass and acceleration of the system. Note that it is vector law that implies the three scalar equations along the x-, y-, and z-directions. As the acceleration components in the y- and z-directions are considered to be negligible, the momentum equation along the x-direction only is derived. It is assumed that the only important external forces acting on the control volume are pressure, gravity, and shear forces. The scalar equation along the x-direction for the differential volume shown in Figure 1-4 is

$$\sum F_x = f_p + f_g + f_s = ma_x \tag{1-10}$$

where $\sum F_x$ = sum of the forces acting on the differential volume in the x-direction; f_p, f_g, and f_s = pressure, gravity, and shear forces in the x-direction on the differential volume, respectively; $m = \rho\sigma\,dx\,dy$ = mass of fluid within the differential volume; and a_x = acceleration in the x-direction of fluid in the differential volume and is

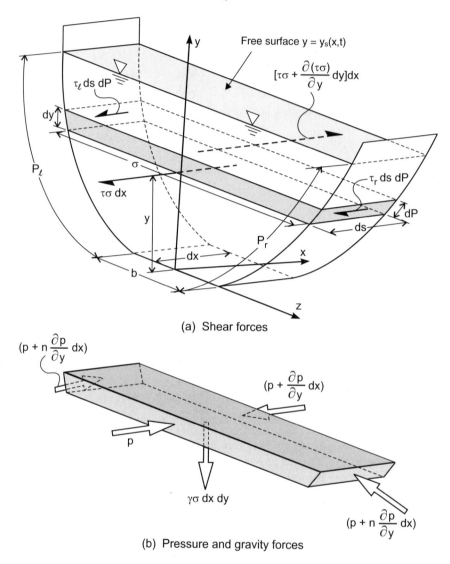

(a) Shear forces

(b) Pressure and gravity forces

Figure 1-4. Forces on the differential volume.

$$a_x = \frac{\partial u}{\partial t} + u\,\frac{\partial u}{\partial x} \qquad (1\text{-}11)$$

In general Eq. (1-11) includes two additional terms, namely, $v(\partial u/\partial y)$ and $w(\partial u/\partial z)$, that are zero because v and w are assumed negligible. The pressures on the upstream, downstream, and lateral faces of the differential volume (see Fig. 1-4b), respectively, are p, $p+p'$, and $p+np'$, where $p' = (\partial p/\partial x)\,dx$ and n is a weighting factor having a value between 0 and 1, i.e., $1>n>0$. Though the pressure forces on the upstream and downstream faces are in the x-direction, the pressure forces on the lateral surfaces are not in the x-direction. The problem of determining the x-component of the pressure forces can be simplified by superimposing a uniform pressure of $-(p+np')$ on the differential volume. The net force due to superimposition of a uniform pressure all around the surface is zero. After superimposing the indicated pressure, the pressures on the upstream, downstream, and lateral faces become $-np'$, $(1-n)p'$, and zero, respectively. The x-component of the pressure force on the differential volume then can be written as

$$f_p = -n\frac{\partial p}{\partial x}\,dx\sigma(x,y)\,dy - (1-n)\frac{\partial p}{\partial x}\,dx\left[\sigma(x,y)+\frac{\partial \sigma}{\partial x}\,dx\right]dy = -\sigma(x,y)\frac{\partial p}{\partial x}\,dy\,dx$$

where the second-order term has been neglected. Substitution for the pressure from Eq. (1-2) into the above equation yields

$$f_p = -\sigma(x,y)\frac{\partial}{\partial x}\left[\gamma\cos\theta(y_s-y)\right]dy\,dx = -\gamma\cos\theta\,\sigma(x,y)\frac{\partial y_s}{\partial x}\,dy\,dx \qquad (1\text{-}12)$$

The component in the x-direction of weight of water in the differential volume in Figure 1-4b is

$$f_g = \gamma\,S_0\sigma(x,y)\,dy\,dx \qquad (1\text{-}13)$$

where $S_0 = \sin\theta =$ the channel bottom slope and $\theta =$ angle between channel bed and a horizontal plane. The shear forces on the four faces of the differential volume are shown in Figure 1-4a. The shear force on the right face is $\tau_r\,ds\,dP$, where $ds\,dP$ is the surface area of the face and τ_r is the shear stress on the right face. The x-component of the shear force is $-\tau_r\,ds\,dP\cos\phi = -\tau_r\,dx\,dP$, where ϕ is the angle that the shear force on the right face makes with the x-direction and $ds\cos\phi = dx$. Similarly the x-component of the shear force on the left face is $-\tau_\ell\,dx\,dP$, where τ_ℓ is the shear stress on the left face. The x-component of the shear force on the lower face is $-\tau\sigma\,dx$, where τ is the shear stress on the lower face and $\sigma\,dx$ is the area of the lower face. The x-component of the shear force on the upper face can be written as $\{\tau\sigma+[\partial(\tau\sigma)/\partial y]dy\}dx$. The net x-component of the shear force on the upper and lower faces is then $[\partial(\tau\sigma)/\partial y]\,dy\,dx$. The component in the x-direction of the shear force on the differential volume is

$$f_s = -\left[\tau_r\,dP + \tau_\ell\,dP - \frac{\partial(\tau\sigma)}{\partial y}\,dy\right]dx \qquad (1\text{-}14)$$

Substitution of the expressions for forces from Eqs. (1-12), (1-13), and (1-14) and for the acceleration from Eq. (1-11) into Eq. (1-10) yields

$$
-\gamma\cos\theta\,\frac{\partial y_s}{\partial x}\sigma(x, y)\,dy\,dx + \gamma\,S_0\sigma(x, y)\,dy\,dx - \left[\tau_r\,dP + \tau_\ell\,dP - \frac{\partial(\tau\sigma)}{\partial y}\,dy\right]dx
$$

$$
= (\rho\sigma\,dy\,dx)\left(\frac{\partial u}{\partial t} + u\,\frac{\partial u}{\partial x}\right) \tag{1-15}
$$

Division of Eq. (1-15) by dx and integration over the cross-sectional area of the flow give

$$
-\gamma\cos\theta\,\frac{\partial y_s}{\partial x}\int_0^{y_s}\sigma(x, y)\,dy + \gamma\,S_0\int_0^{y_s}\sigma(x, y)\,dy - \left[\int_0^{P_r}\tau_r\,dP + \int_0^{P_\ell}\tau_\ell\,dP - \int_0^{y_s}\frac{\partial(\tau\sigma)}{\partial y}\,dy\right]
$$

$$
= \rho\int_0^{y_s}\sigma(x, y)\left(\frac{\partial u}{\partial t} + u\,\frac{\partial u}{\partial x}\right)dy \tag{1-16}
$$

where P_r and P_ℓ are the wetted perimeters of the right and left banks, respectively. Noting that

$$
\int_0^{y_s}\sigma(x, y)\,dy = A
$$

the first two terms on the left-hand side of Eq. (1-16) become

$$
-\gamma\cos\theta\,\frac{\partial y_s}{\partial x}\int_0^{y_s}\sigma(x, y)\,dy + \gamma\,S_0\int_0^{y_s}\sigma(x, y)\,dy = -\gamma\cos\theta\,\frac{\partial y_s}{\partial x}A + \gamma\,S_0 A \tag{1-17}
$$

Integration of the fifth term on the left-hand side of Eq. (1-16) gives

$$
\int_0^{y_s}\frac{\partial(\tau\sigma)}{\partial y}\,dy = (\tau\sigma)\Big|_0^{y_s} = -\tau_b b
$$

where the shear stress at the free surface ($y = y_s$) is assumed to be zero and $\tau_b =$ shear stress at the channel bottom ($y = 0$). Using the above expression, the shear force term in Eq. (1-16) becomes

$$
\int_0^{P_r}\tau_r\,dP + \int_0^{P_\ell}\tau_\ell\,dP + \tau_b b = \tau_0 P \tag{1-18}
$$

where $P = P_r + P_\ell + b =$ wetted perimeter of the channel cross section, and τ_0 is the average wall shear stress given by

$$\tau_0 = \frac{1}{P}\left[\int_0^{P_r} \tau_r\, dP + \int_0^{P_\ell} \tau_\ell\, dP + \tau_b b\right]$$

Addition of a term $u\{\partial(u\sigma)/\partial x + V_n(dP/dy)(ds/dx)\}$, which is zero by virtue of Eq. (1-3), to the right-hand side of Eq. (1-16) yields

$$\rho\int_0^{y_s} \sigma\left(\frac{\partial u}{\partial t} + u\frac{\partial u}{\partial x}\right)dy + \rho\int_0^{y_s} u\left\{\frac{\partial(u\sigma)}{\partial x} + V_n\frac{dP}{dy}\frac{ds}{dx}\right\}dy$$

$$= \rho\int_0^{y_s}\left[\frac{\partial(u\sigma)}{\partial t} + \frac{\partial(\sigma u^2)}{\partial x} + uV_n\frac{dP}{dy}\frac{ds}{dx}\right]dy \qquad (1\text{-}19)$$

in which the use is made of the fact that σ is not a function of t. Applying the Leibnitz rule of integration to the first two terms on the right-hand side of Eq. (1-19) and noting that the lower limit of integration is constant, one obtains

$$\rho\int_0^{y_s}\left[\frac{\partial(u\sigma)}{\partial t} + \frac{\partial(\sigma u^2)}{\partial x}\right]dy$$

$$= \rho\left[\frac{\partial}{\partial t}\int_0^{y_s} u\sigma\, dy - (u\sigma)\Big|_{y=y_s}\frac{\partial y_s}{\partial t}\right] + \rho\left[\frac{\partial}{\partial x}\int_0^{y_s} \sigma u^2\, dy - (\sigma u^2)\Big|_{y=y_s}\frac{\partial y_s}{\partial x}\right]$$

$$= \rho\frac{\partial}{\partial t}\int_0^{y_s} u\sigma\, dy + \rho\frac{\partial}{\partial x}\int_0^{y_s} \sigma u^2\, dy - (u\sigma)\Big|_{y=y_s}\left(\frac{\partial y_s}{\partial t} + u_s\frac{\partial y_s}{\partial x}\right) \qquad 0 \ast 0$$

On the right-hand side of the above equation, the first term can be written as $\rho\,(\partial Q/\partial t)$; the second term is the spatial rate of momentum flux across the channel cross section; and the third term by virtue of Eq. (1-8) is zero. The above equation then becomes

$$\rho\int_0^{y_s}\left[\frac{\partial(u\sigma)}{\partial t} + \frac{\partial(\sigma u^2)}{\partial x}\right]dy = \rho\frac{\partial Q}{\partial t} + \rho\frac{\partial}{\partial x}\left(\beta V^2 A\right) \qquad (1\text{-}20)$$

where V = average flow velocity and β = momentum coefficient and is defined as

$$\beta = \frac{\int_0^{y_s} \sigma u^2 \, dy}{V^2 A} \tag{1-21}$$

The third term on the right-hand-side of Eq. (1-19) is the momentum flux of lateral outflow in the x-direction and can be written as

$$\rho \int_0^{y_s} uV_n \frac{dP}{dy}\frac{ds}{dx} dy = \rho q_\ell V_\ell \tag{1-22}$$

where V_ℓ = average longitudinal velocity of lateral flow. Combining Eqs. (1-17), (1-18), (1-20), and (1-22) with Eq. (1-16), one obtains

$$-\gamma \cos\theta \frac{\partial y_s}{\partial x} A + \gamma S_0 A - \tau_0 \overset{+\tau_s \beta}{\overbrace{P}} = \rho \left[\frac{\partial Q}{\partial t} + \frac{\partial(\beta V^2 A)}{\partial x} + V_\ell q_\ell \right] \tag{1-23}$$

The above equation is the preferred form of the one-dimensional momentum equation commonly used in mathematical modeling. This equation can be written in another form that is more amicable to analytical solutions. Using $Q = VA$, the first and second terms on the right-hand side of Eq. (1-23) can be expanded as

$$\frac{\partial Q}{\partial t} + \frac{\partial(\beta V^2 A)}{\partial x} = \left\{ V\frac{\partial A}{\partial t} + A\frac{\partial V}{\partial t} \right\} + \left\{ \beta V\frac{\partial(VA)}{\partial x} + \beta VA\frac{\partial V}{\partial x} + V^2 A\frac{\partial\beta}{\partial x} \right\} \tag{1-24}$$

Some manipulations on the last two terms within the second pair of brackets in Eq. (1-24) and the use of the continuity equation (1-9) yield

$$\frac{\partial Q}{\partial t} + \frac{\partial(\beta V^2 A)}{\partial x} = \left\{ V\frac{\partial A}{\partial t} + A\frac{\partial V}{\partial t} \right\}$$
$$+ \left\{ -\beta V\left(\frac{\partial A}{\partial t} + q_\ell \right) + A\frac{\partial(\beta V^2/2)}{\partial x} + \frac{V^2 A}{2}\frac{\partial\beta}{\partial x} \right\} \tag{1-25}$$

Combining Eqs. (1-23) and (1-25), dividing by γA, and rearranging the terms, one obtains

$$-\cos\theta \frac{\partial y_s}{\partial x} + S_0 - \frac{\tau_0 P}{\gamma A} + \frac{\tau_s \beta}{\gamma A}$$
$$= \frac{1}{g}\frac{\partial V}{\partial t} + \frac{\partial(\beta V^2/2g)}{\partial x} + \frac{V_\ell - \beta V}{gA}q_\ell + \frac{V^2}{2g}\frac{\partial\beta}{\partial x} - \frac{(\beta - 1)V}{gA}\frac{\partial A}{\partial t} \tag{1-26}$$

Noting that $S_0 = -dG_0/dx$, where G_0 = elevation of the channel bed (see Figure 1-1), and transposing some terms from one side to the other, Eq. (1-26) becomes

$$\frac{1}{g}\frac{\partial V}{\partial t}+\frac{\partial H_\beta}{\partial x}+\frac{V_\ell-\beta V}{gA}q_\ell = -\frac{\tau_0}{\gamma R}+\frac{(\beta-1)V}{gA}\frac{\partial A}{\partial t}-\frac{V^2}{2g}\frac{\partial \beta}{\partial x} \tag{1-27}$$

where
$$H_\beta = \beta\frac{V^2}{2g}+y_s\cos\theta+G_0 = \text{total head} \tag{1-28}$$

and $R = A/P$ and is termed hydraulic radius. Equation (1-27) is the one-dimensional momentum equation. This equation for steady flow and without the lateral flow is derived by the elementary approach in Example 1-3. Though most problems in open-channel flow are solved using Eq. (1-27), it is not readily applicable to problems with lateral outflow, such as flow through a side weir or a bottom rack, because the longitudinal velocity of lateral flow, V_ℓ, cannot be reasonably estimated. For such flow a reasonable assumption is that the total velocity (not the longitudinal velocity) of the lateral outflow is equal to the total velocity of the main flow. Such problems are solved using the mechanical-energy equation as it deals with the total velocity. The mechanical energy equation is derived in the next section.

Example 1-3. Applying Newton's second law of motion to a control volume between two cross sections an infinitesimal distance apart, derive the momentum equation for one-dimensional steady flow in open channels. Assume that there is no lateral flow across channel boundaries and the pressure distribution is hydrostatic.

Consider the forces acting in the flow direction on an element of the channel section shown in the accompanying figure. The pressure, gravity, and shear forces acting on the element can be written as follows:

Net pressure force:
$$\gamma A\bar{y}-\left(\gamma A\bar{y}+\frac{d(\gamma A\bar{y})}{dx}\delta x\right)=-\frac{d(\gamma A\bar{y})}{dx}\delta x$$

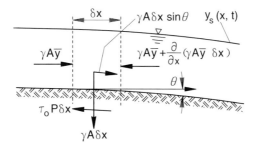

Example 1-3

Weight component: $\gamma A\, \delta x \sin\theta$

Shear force: $-\tau_0 P\, \delta x$

where $\bar{y}$ is the distance along the vertical direction from the free surface to the centroid of the cross section. From Newton's second law of motion (i.e., the net force on an element is equal to the product of the mass and acceleration of the element), the momentum equation in the x-direction is

$$\rho V \frac{dV}{dx}\, \delta x\, A = -\left(\frac{d(\gamma A\bar{y})}{dx}\delta x\right) + \gamma A\, \delta x \sin\theta - \tau_0 P\, \delta x$$

It can be shown that $d(A\bar{y})/dx = A\{d(y_s \cos\theta)/dx\}$(Prob. 1-7). Substituting this in the above equation, dividing it by $\gamma A\, \delta x$, and transposing some terms from one side to the other, one obtains the one-dimensional momentum equation as

$$\frac{d}{dx}\left(\frac{V^2}{2g} + y_s \cos\theta + G_0\right) = -\frac{\tau_0}{\gamma R}$$

where use has been made of the relations $-dG_0/dx = \sin\theta$ and $R = A/P$. Introducing the momentum coefficient β [Eq. (1-21)], the above equation can be modified to include the effect of nonuniform velocity distribution over the channel cross section (see Section 1-7) as

$$\frac{d}{dx}\left(\beta\frac{V^2}{2g} + y_s \cos\theta + G_0\right) = -\frac{\tau_0}{\gamma R}$$

PROBLEMS

1-7   Show that $d(A\bar{y})/dx = A\{d(y_s \cos\theta)/dx\}$; $\bar{y}$ is defined in Example 1-3, in which this identity has been used.

1-8   Rework Example 1-3 for the case of a lateral flow q_ℓ.

1-9   Modify the one-dimensional momentum equation (1-27) to include the effect of wind forces on the free surface.

1-6 Differential Mechanical-Energy Equation

It should be mentioned that the mechanical-energy equation is not an independent equation as it is derived from the momentum equation. However, unlike the momentum equations that deal with vector quantities in terms of force and momentum flux, the mechanical-energy equation handles the scalar quantities in terms of work and energy. Multiplication of Eq. (1-15) by u, division by dx, and integration of the resulting equation over the cross section of the flow yield

$$-\gamma\cos\theta\,\frac{\partial y_s}{\partial x}\int_0^{y_s}u\sigma(x,y)dy+\gamma\,S_0\int_0^{y_s}u\sigma(x,y)dy-\int_0^{P_r}u\tau_r dP$$

$$-\int_0^{P_\ell}u\tau_\ell dP+\int_0^{y_s}u\frac{\partial(\tau\sigma)}{\partial y}dy=\rho\int_0^{y_s}u\sigma\left(\frac{\partial u}{\partial t}+u\frac{\partial u}{\partial x}\right)dy \qquad (1\text{-}29)$$

The terms on the left-hand side of Eq. (1-29) represents the rates of work done by the pressure, gravity, and shear forces on the control volume. As

$$Q=\int_0^{y_s}u\sigma(x,y)dy$$

the first two terms on the left-hand side of Eq. (1-29) become

$$-\gamma\cos\theta\,\frac{\partial y_s}{\partial x}\int_0^{y_s}u\sigma(x,y)dy+\gamma\,S_0\int_0^{y_s}u\sigma(x,y)dy=-\gamma\cos\theta\,\frac{\partial y_s}{\partial x}Q+\gamma\,S_0 Q \quad (1\text{-}30)$$

The velocity is zero along the banks of the channel. Consequently, the third and fourth terms on the left-hand side of Eq. (1-29) vanish. The fifth term on the left-hand side of Eq. (1-29) can be written as

$$\int_0^{y_s}u\frac{\partial(\tau\sigma)}{\partial y}dy=\int_0^{y_s}\left[\frac{\partial(u\tau\sigma)}{\partial y}-\tau\sigma\frac{\partial u}{\partial y}\right]dy$$

$$=(u\tau\sigma)\big|_{y=y_s}-(u\tau\sigma)\big|_{y=0}-\int_0^{y_s}\tau\sigma\frac{\partial u}{\partial y}dy \qquad (1\text{-}31)$$

The velocity is zero at the channel bottom ($y = 0$) and the shear stress is zero at the free surface ($y = y_s$), unless strong wind acts on the water surface. Hence, the first two terms on the right-hand side of Eq. (1-31) vanish. The last term on the right-hand side is called the *dissipation integral*. The rate of work done by the shear forces becomes

$$-\int_0^{y_s}\tau\sigma\frac{\partial u}{\partial y}dy=-\overline{\varepsilon} \qquad (1\text{-}32)$$

in which $\overline{\varepsilon}$ is the rate of work done by the shear force per unit length of channel. This rate of work cannot be converted back into the mechanical energy and is considered as loss of energy. Therefore $\overline{\varepsilon}$ is termed the energy loss per unit length of channel.

The addition of $(u^2/2)\{\partial(u\sigma)/\partial x +V_n(dP/dy)(ds/dx)\}$, which is zero by virtue of Eq. (1-3), to the right-hand side of Eq. (1-29) yields

$$\rho \int_0^{y_s} u\sigma \left(\frac{\partial u}{\partial t} + u\frac{\partial u}{\partial x} \right) dy + \rho \int_0^{y_s} \frac{u^2}{2} \left\{ \frac{\partial (u\sigma)}{\partial x} + V_n \frac{dP}{dy}\frac{ds}{dx} \right\} dy$$

$$= \rho \int_0^{y_s} \left[\frac{\partial (u^2\sigma/2)}{\partial t} + \frac{\partial (\sigma u^3/2)}{\partial x} + V_n \frac{u^2}{2}\frac{dP}{dy}\frac{ds}{dx} \right] dy \qquad (1\text{-}33)$$

in which the use is made of the fact that σ is not a function of t. Applying the Leibnitz rule of integration to the first two terms on the right-hand side of Eq. (1-33) and noting that the lower limit of integration is constant, one obtains

$$\rho \int_0^{y_s} \left[\frac{\partial (u^2\sigma/2)}{\partial t} + \frac{\partial (\sigma u^3/2)}{\partial x} \right] dy = \rho \left[\frac{1}{2}\frac{\partial}{\partial t}\int_0^{y_s} u^2\sigma\, dy - \frac{1}{2}\left(u^2\sigma\right)\Big|_{y=y_s} \frac{\partial y_s}{\partial t} \right]$$

$$+ \rho \left[\frac{1}{2}\frac{\partial}{\partial x}\int_0^{y_s} \sigma u^3\, dy - \frac{1}{2}\left(\sigma u^3\right)\Big|_{y=y_s} \frac{\partial y_s}{\partial x} \right]$$

$$= \frac{\rho}{2}\frac{\partial}{\partial t}\int_0^{y_s} u^2\sigma\, dy + \frac{\rho}{2}\frac{\partial}{\partial x}\int_0^{y_s} \sigma u^3\, dy - \frac{1}{2}\left(u^2\sigma\right)\Big|_{y=y_s} \left(\frac{\partial y_s}{\partial t} + u_s\frac{\partial y_s}{\partial x} \right)$$

On the right-hand side of the above equation, the first term upon using Eq. (1-21) can be written as $(\rho/2)\{\partial(\beta V^2 A)/\partial t\}$; the second term is the spatial rate of the kinetic-energy flux across the channel cross section; and the third term is zero by virtue of Eq. (1-8). The above equation then becomes

$$\frac{\rho}{2}\int_0^{y_s} \left[\frac{\partial (u^2\sigma)}{\partial t} + \frac{\partial (\sigma u^3)}{\partial x} \right] dy = \frac{\rho}{2}\frac{\partial(\beta V^2 A)}{\partial t} + \frac{\rho}{2}\frac{\partial}{\partial x}\left(\alpha V^3 A\right) \qquad (1\text{-}34)$$

where α is called the energy coefficient and is defined as

$$\alpha = \frac{\int_0^{y_s} \sigma u^3\, dy}{V^3 A} \qquad (1\text{-}35)$$

The third term on the right-hand-side of Eq. (1-33) is the kinetic-energy flux of lateral outflow and can be written as

$$\frac{\rho}{2}\int_0^{y_s} u^2 V_n \frac{dP}{dy}\frac{ds}{dx}\, dy = \frac{\rho}{2}\overline{U}_\ell^2 q_\ell \qquad (1\text{-}36)$$

where $\overline{U}_\ell^2$ = average of the square of the total velocity of lateral flow. Combining Eqs. (1-30), (1-32), (1-34), and (1-36) with Eq. (1-29), one gets

$$-\gamma \cos\theta \frac{\partial y_s}{\partial x} Q + \gamma S_0 Q - \bar{\varepsilon} = \frac{\rho}{2}\left[\frac{\partial}{\partial t}\left(\beta V^2 A\right) + \frac{\partial}{\partial x}\left(\alpha V^3 A\right) + \bar{U}_\ell^2 q_\ell\right] \qquad (1\text{-}37)$$

The first and second terms on the right-hand side of Eq. (1-37) can be expanded as

$$\frac{\partial}{\partial t}\left(\beta V^2 A\right) + \frac{\partial}{\partial x}\left(\alpha V^3 A\right)$$

$$= 2\beta VA\frac{\partial V}{\partial t} + \beta V^2\frac{\partial A}{\partial t} + V^2 A\frac{\partial \beta}{\partial t} + VA\frac{\partial}{\partial x}\left(\alpha V^2\right) + \alpha V^2\frac{\partial(VA)}{\partial x} \qquad (1\text{-}38)$$

Combining Eqs. (1-37) and (1-38), making use of the continuity equation (1-9), dividing by γQ, and performing some mathematical manipulations, one obtains (Prob. 1-10)

$$-\cos\theta \frac{\partial y_s}{\partial x} + S_0 - \frac{\bar{\varepsilon}}{\gamma Q} = \frac{\beta}{g}\frac{\partial V}{\partial t} + \frac{V\beta}{2gA}\frac{\partial A}{\partial t} + \frac{V}{2g}\frac{\partial \beta}{\partial t} + \frac{\partial}{\partial x}\left(\frac{\alpha V^2}{2g}\right)$$

$$+ \frac{\alpha V^2}{2gQ}\left(-\frac{\partial A}{\partial t} - q_\ell\right) + \frac{\bar{U}_\ell^2}{2gQ}q_\ell \qquad (1\text{-}39)$$

Noting that $S_0 = -dG_0/dx$ and transposing some terms from one side to the other, Eq. (1-39) becomes

$$\frac{\beta}{g}\frac{\partial V}{\partial t} + \frac{\partial H_\alpha}{\partial x} + \frac{\bar{U}_\ell^2 - \alpha V^2}{2gQ}q_\ell = -\frac{\bar{\varepsilon}}{\gamma Q} + (\alpha - \beta)\frac{V}{2gA}\frac{\partial A}{\partial t} - \frac{V}{2g}\frac{\partial \beta}{\partial t} \qquad (1\text{-}40)$$

in which
$$H_\alpha = \alpha\frac{V^2}{2g} + y_s\cos\theta + G_0 = \text{total head} \qquad (1\text{-}41)$$

Equation (1-40) is the one-dimensional mechanical-energy equation. A comparison of the momentum and mechanical-energy equations (1-27) and (1-40) shows that the two equations are similar to each other. The similarity between the two equations should not be a surprise, as the mechanical-energy equation is derived from the momentum equation. There are some differences among the various terms in the two equations; these differences are discussed in Section 1-8, where the two equations are compared for specific flows.

PROBLEMS

1-10   Using Eqs. (1-9), (1-37), and (1-38), derive Eq. (1-39).

1-11   Simplify the one-dimensional momentum and mechanical-energy equations (1-27)
        and (1-40) for a steady, uniform flow and compare the two equations.

1-7 Momentum and Energy Coefficients

The momentum and energy coefficients for one-dimensional flow are defined in
Eqs. (1-21) and (1-35), respectively. Due to the velocity variation in a channel
section, the true momentum flux through the section, $\int_A \rho u^2 dA$, is not equal to the
momentum flux based on mean flow velocity, ρQV. A momentum coefficient, β, is
defined so that β times the momentum flux based on mean velocity is equal to true
momentum flux, i.e.,

$$\beta \rho QV = \int_A \rho u^2 dA$$

Upon using $Q = VA$, an expression for β given can be obtained from the above
equation as

$$\beta = \frac{\int_A u^2 dA}{V^2 A} \tag{1-42}$$

Equation (1-42) for β is more general than Eq. (1-21). The velocity u in Eq. (1-42)
is considered to be a function of both y and z.

The true kinetic-energy flux through the section, $\int_A (\rho u^3/2) \, dA$, is not equal to
the kinetic-energy flux based on mean velocity, $\rho QV^2/2$. An energy coefficient, α,
is defined so that α times the kinetic-energy flux based on average velocity equals
the true kinetic-energy flux, i.e.,

$$\alpha \rho Q \frac{V^2}{2} = \int_A \frac{\rho u^3}{2} dA$$

Upon using $Q = VA$, an expression for α can be obtained from the above equation
as

$$\alpha = \frac{\int_A u^3 dA}{V^3 A} \tag{1-43}$$

Equation (1-43) for α is more general than Eq. (1-35). The values of the
coefficients are greater than unity, except when the velocity is uniform across the
section. The energy coefficient is greater than the momentum coefficient due to the
fact that u in the expression for the former occurs in the third power and for the
latter in the second power. The ranges of values of α and β, respectively are
$1.10 - 1.20$ and $1.03 - 1.07$ for regular channels and $1.15 - 1.50$ and $1.05 - 1.17$
for natural channels (Chaudhry, 1993). Unless α and β are known to be

significantly greater than unity as in compound channels, they are generally taken as unity.

Example 1-4. The velocity distribution in a wide rectangular channel may be approximated by the equation $u = 0.6 + 0.4\,y$, where u is in meters per second and y is in meters. Find the mean velocity and the momentum coefficient if the flow depth is 1.2 m.
Consider a unit width of the channel, and let q be the discharge per unit width:

$$V = \frac{q}{y_s} = \frac{\int_0^{y_s} u\,dy}{y_s} = \frac{\int_0^{1.2} (0.6+0.4\,y)\,dy}{1.2} = \frac{\left(0.6+0.4\,y^2/2\right)\Big|_0^{1.2}}{1.2} = 0.84 \text{ m/s.}$$

From Eq. (1-42),

$$\beta = \frac{\int_0^{1.2} u^2\,dy}{V^2 y_s} = \frac{\int_0^{1.2} (0.6+0.4\,y)^2\,dy}{0.84^2 \times 1.2} = \frac{\int_0^{1.2} \left(0.36+0.48\,y+0.16\,y^2\right)dy}{0.84^2 \times 1.2}$$

$$= \frac{\left(0.36\,y+0.48\,y^2/2+0.16\,y^3/3\right)\Big|_0^{1.2}}{0.84^2 \times 1.2} = 1.03.$$

PROBLEMS

1-12 Derive an expression for the momentum coefficient β for the velocity distribution $u/y_s = (y/y_s)^n$ in a wide channel. What is the value of β for $n = 1/7$?

1-13 Derive an expression for the energy correction factor α for the velocity distribution given in Problem 1-12 and determine its value for $n = 1/7$.

1-14 Find values of α and β for the velocity distribution given in Problem 1-6.

1-8 Governing Equations for Specific Flows

The complete one-dimensional differential equations of motion—the continuity equation and the momentum or mechanical-energy equation—can be solved only numerically by using sophisticated computer programs for the two unknown flow quantities, Q or V and A or y (the flow depth hereinafter is denoted by y instead of y_s). Such numerical methods are beyond the scope of this book and are described in books on computational hydraulics (Abbott, 1979; Cunge et al., 1980). However, these equations can be simplified for particular flows, and the simplified equations can be solved by either analytical or simple numerical methods. For many practical problems the solution of the simplified one-dimensional differential equations is more than adequate.

Flow can be classified into steady or unsteady and uniform or nonuniform (varied) flow depending on the variation with x and t of the two unknown flow quantities.

Steady Uniform Flow

Flow in open channels is steady if the two unknown flow quantities, V and y, do not change with time. Flow is uniform if V and y do not vary with the longitudinal distance. Deleting the temporal and spatial terms in Eq. (1-9), the continuity equation for such flows is

$$q_\ell = 0 \quad \text{and} \quad Q = VA = \text{constant} \tag{1-44}$$

Deletion of the temporal and spatial terms in the momentum equation along with q_ℓ reduces Eq. (1-27) to a trivial form that states that the wall shear stress be zero. In physical terms it means that steady uniform flow (referred to hereinafter simply as uniform flow) in open channels without a force other than the resisting force can occur only if the latter is zero. In the case of a finite resisting force the uniform flow is possible only if on a flow element the resisting force is balanced by a driving force, so that the net force on the element is zero. Such a driving force is due to gravity and is represented by the term $-dG_0/dx$ in Eq. (1-27). Thus the momentum equation for uniform flow becomes

$$S_0 \equiv -\frac{dG_0}{dx} = \frac{\tau_0}{\gamma R} \tag{1-45}$$

Theoretically the channel slopes need not to be constant to satisfy Eq. (1-45) because the wall shear stress for a given flow velocity and depth can be varied, as shown in Chapter 2, by changing the wall roughness. In reality uniform flow exists only in prismatic channels, though the uniform-flow condition in rivers is frequently assumed in flow computations.

Similarly, deletion of temporal and spatial terms, except the gravity term, in the mechanical-energy equation (1-40) yields

$$-\frac{\partial G_0}{\partial x} = \frac{\bar{\varepsilon}}{\gamma Q} \tag{1-46}$$

Equating the right-hand sides of Eqs. (1-45) and (1-46) and using $Q = VA$ and $R = A/P$, one obtains

$$(\tau_0 P)V = \bar{\varepsilon} \tag{1-47}$$

In other words, for steady uniform flow, the rate of work done by the shear force on unit length of the channel boundary is equal to the rate of energy dissipation per unit channel length. When the friction slope S_f is defined as

$$S_f \equiv \frac{\tau_0}{\gamma R} \tag{1-48}$$

and the energy slope S_e is defined as

$$S_e \equiv \frac{\bar{\varepsilon}}{\gamma Q} \qquad (1\text{-}49)$$

the momentum and mechanical-energy equations for uniform flow reduce to the single equation

$$S_0 = S_f = S_e \qquad (1\text{-}50)$$

The three slopes are equal in uniform flow.

Steady Varied Flow

In varied flow the two flow quantities, V and y, vary with x. Varied flow can be classified as gradually and rapidly varied flow. Gradually varied flow can occur without or with lateral flow. The latter flow is termed spatially varied flow.

Gradually Varied Flow. In gradually varied flow the rate of change of flow depth with distance is small. Therefore, the streamline curvature is negligible; the pressure distribution is hydrostatic; and Eqs. (1-27) and (1-40) apply. The backing up of water in a river due to a dam is an example of gradually varied flow. Deletion of the temporal terms and the lateral-flow term in the governing equations (1-9), (1-27), and (1-40) yields

$$Q = VA = \text{constant} \quad [\text{Eq. } (1\text{-}44)]$$

$$\frac{d}{dx}\left(\beta \frac{V^2}{2g} + y\cos\theta + G_0\right) = -S_f - \frac{V^2}{2g}\frac{\partial\beta}{\partial x} \qquad (1\text{-}51)$$

$$\frac{d}{dx}\left(\alpha \frac{V^2}{2g} + y\cos\theta + G_0\right) = -S_e \qquad (1\text{-}52)$$

where the use of Eqs. (1-48) and (1-49) has been made. It is tempting to drop the last term in Eq. (1-51) in an attempt to make the momentum and mechanical-energy equations similar to each other. However, doing so is not consistent with the retention of an identical term on the left-hand side of the equation. If the value of unity is assumed for both coefficients α and β, and S_f and S_e are assumed equal, then the momentum and mechanical-energy equations become identical. Arguably, these assumptions are justified because of the inexact empirical methods involved in evaluating the slopes (Farell, 1966). For example, the uniform-flow formula, such as the Manning equation discussed in Chapter 2, is used to evaluate the slopes in a nonuniform flow. With these approximations the momentum and mechanical-energy equations reduce to the single equation

$$\frac{dH}{dx} = -S_f \equiv -S_e \tag{1-53}$$

where

$$H = \frac{V^2}{2g} + y\cos\theta + G_0 \tag{1-54}$$

In most engineering problems the mechanical-energy equation (1-52) is commonly used as it is easy to include into the energy slope term additional energy-loss terms due to local features such as bends, bridge piers, and other structures that produce change in velocity. The energy slope term in Eq. (1-52) is replaced with the friction slope term. The governing equation for gradually varied flow used in common practice is

$$\frac{dH_\alpha}{dx} = -S_f \tag{1-55}$$

where

$$H_\alpha = \alpha\frac{V^2}{2g} + y\cos\theta + G_0$$

Spatially Varied Flow. Examples of spatially varied flow are flow in roof gutters; flow in side-channel spillways for dams; and flow in channels with withdrawal of flow through a side-discharge weir or a bottom rack. Deletion of the temporal terms and the assumption that $\alpha = \beta = 1$ in the governing equations (1-9), (1-27), and (1-40) yield

$$\frac{\partial Q}{\partial x} + q_\ell = 0 \tag{1-56}$$

$$\frac{\partial H}{\partial x} + \frac{V_\ell - V}{gA}q_\ell = -\frac{\tau_0}{\gamma R} \tag{1-57}$$

$$\frac{\partial H}{\partial x} + \frac{\overline{U}_\ell^2 - V^2}{2gQ}q_\ell = -\frac{\overline{\varepsilon}}{\gamma Q} \tag{1-58}$$

The lateral-flow term in Eq. (1-58) upon dividing the numerator and the denominator by V can be restated as

$$\frac{\overline{U}_\ell^2 - V^2}{2gQ}q_\ell = \frac{\overline{U}_\ell^2/V - V}{2gQ/V}q_\ell = \frac{\overline{U}_\ell^2/V - V}{2gA}q_\ell \tag{1-59}$$

A comparison of the lateral-flow terms in Eqs. (1-57) and (1-59) shows that they are similar except that the factor 2 does not appear in the denominator of Eq. (1-57). The magnitude of the lateral-flow term in the momentum equation is about twice that of the lateral-flow term in the energy equation. In order that Eq. (1-57) is compatible with Eq. (1-58), the right-hand terms in these equations can no longer be assumed equal. The momentum and mechanical-energy equations for spatially varied flow are different; a choice between them must be made to solve flow problems.

There are two additional unknown quantities, V_ℓ (longitudinal velocity of the lateral flow) in the momentum equation and $\overline{U}_\ell$ (total velocity of the lateral flow) in the mechanical energy equation. Estimates of these quantities are needed to solve the equations. In most problems, the direction of the lateral outflow is not known; consequently, the longitudinal velocity of lateral outflow V_ℓ cannot be reasonably estimated. However, it is reasonable to assume that the total velocities of the lateral outflow and the main flow are equal. This makes the momentum equation unsuitable for lateral outflow.

Furthermore the right-hand terms in the momentum and mechanical-energy equations are estimated from empirical expressions developed for uniform flows. The empirical expressions cannot be applicable to both terms, as they are no longer equal. The appropriate term for lateral outflow and inflow is determined as follows. The momentum and mechanical-energy equations (1-57) and (1-58) remain compatible if

$$\frac{\overline{\varepsilon}}{\gamma Q} = \frac{\tau_0}{\gamma R} - \frac{V}{2gA}\left(1+\frac{\overline{U}_\ell^{\,2}}{V^2}-2\frac{V_\ell}{V}\right)q_\ell \qquad (1\text{-}60)$$

Equation (1-60) is obtained by equating $\partial H/\partial x$ in Eqs. (1-57) and (1-58) and rearranging the terms. It can be shown that the quantity inside the parentheses in Eq. (1-60) is positive for $\overline{U}_\ell > V_\ell$ (Prob. 1-16). For lateral outflow q_ℓ is positive; hence, from Eq. (1-60)

$$\frac{\tau_0}{\gamma R} > \frac{\overline{\varepsilon}}{\gamma Q} \qquad (1\text{-}61)$$

Equation (1-61) implies that either the rate of work done by the shear force should increase if the rate of energy dissipation assumed constant or the rate of energy dissipation should decrease if the shear stress is assumed unaffected. It is reasonable to assume that the main body of the flow is relatively unaffected and the energy dissipation is the same as without lateral outflow. The increase in shear stress cannot be easily estimated; therefore the energy equation is more appropriate for lateral outflow. For lateral inflow, q_ℓ is negative; consequently, from Eq. (1-60)

$$\frac{\tau_0}{\gamma R} < \frac{\overline{\varepsilon}}{\gamma Q} \qquad (1\text{-}62)$$

Equation (1-62) implies that either the rate of work done by the shear force should decrease if the rate of energy dissipation is assumed unaffected or the rate of energy dissipation should increase if the shear stress is assumed unchanged. It is an observed fact that the lateral inflow generates increased turbulence resulting in additional energy dissipation. Though the increased turbulence affects the shear stress τ_0, this is probably a secondary effect as τ_0 depends on the velocity profile near the boundary (Strelkoff, 1969). The momentum equation is, therefore, more appropriate for lateral inflow.

Rapidly Varied Flow.   In rapidly varied flow the streamlines have pronounced curvature. Therefore the pressure distribution is nonhydrostatic, and Eqs. (1-27) and (1-40) do not apply. The flow profile may be continuous, such as flow over spillway crests and buckets, or it may be discontinuous such as in a hydraulic jump. Traditionally such flows have been investigated by conducting laboratory experiments and the experimental results have been used empirically (Chapter 9). The theoretical solution to most problems of rapidly varied flow requires the use of two- or three-dimensional governing equations that are solved numerically. A discussion of such computational methods is beyond the scope of this book. Approximate solutions of some problems can be obtained if the curvature of the streamlines can be presumed (Section 1-10).

Unsteady Uniform Flow

The governing equations for unsteady uniform flow can be obtained by deleting the spatial terms in Eqs. (1-9), (1-27), and (1-40). However, these equations are not of much interest as unsteady uniform flow is rare.

Unsteady Varied Flow

Unsteady varied flow, similar to steady varied flow, can be classified as gradually and rapidly varied flow. The full equations (1-9) and (1-27) or (1-40) are applicable to unsteady varied flow, except in limited number of short reaches wherein the flow profiles have pronounced curvature of the streamlines. The lateral-flow terms in these equations can be deleted as lateral flow in most cases are represented by point inflow and outflow, which in turn are represented by the special flow conditions at the sections where lateral flows occur. Furthermore, if the coefficients α and β are taken as unity and the rate of work done by the shear force is assumed equal to the rate of energy dissipation, the momentum and mechanical-energy equations become identical. Based on these simplifications, the governing equations for unsteady flow are

$$\frac{\partial Q}{\partial x} + \frac{\partial A}{\partial t} = 0 \qquad (1\text{-}63)$$

$$\frac{1}{g}\frac{\partial V}{\partial t} + \frac{\partial H}{\partial x} = -S_f \qquad (1\text{-}64)$$

where use has been made of Eq. (1-48) and H is defined in Eq. (1-54). Equations (1-63) and (1-64) are referred to as de Saint-Venant equations. In mathematical modeling of flow in rivers (Cunge et al., 1980; Abbott, 1979) the following form of the momentum equation, which is obtained from Eq. (1-23), is commonly used:

$$\frac{\partial Q}{\partial t} + \frac{\partial}{\partial x}\left(\beta\frac{Q^2}{A}\right) + gA\left(\frac{\partial y}{\partial x} - S_0\right) + gAS_f + V_\ell q_\ell = 0 \qquad (1\text{-}65)$$

PROBLEMS

1-15   Using time and space criteria, classify the following flow cases:
    (a) River flow around a bridge pier.
    (b) Flow over a rooftop due to steady rainfall.
    (c) Flow in a gutter resulting from (b).
    (d) Flood flow in a river.

1-16   Show that for $\overline{U}_\ell > V_\ell$ the quantity inside the parentheses in Eq. (1-60) is positive.

1-17   Make the necessary assumptions and derive the following equations from Eqs. (1-27) and (1-40):

    (i) $$Q = CA\sqrt{RS_0}$$

    (ii) $$\frac{dE}{dx} = S_0 - S_f$$

    where S_0 = slope of channel bottom; C = constant; $E = y + V^2/2g$; and $S_f = V^2/C^2R$.

1-9 Algebraic Equations of Motion

The algebraic equations of motion are commonly used for steady flow and for an unsteady flow that can be transformed into a steady flow. These equations are derived by integrating the differential equations of motion from $x = x_1$ to $x = x_2$, where x_1 and x_2 are the x-coordinates of sections 1 and 2 of the control volume in Figure 1-5. These equations are generally used to relate flow conditions upstream and downstream of the control volume, i.e., at $x = x_1$ and $x = x_2$. The upstream and downstream sections are usually selected in the regions of gradually varied flow where the pressure distribution can be assumed hydrostatic. As there is no necessity of knowledge of the details of the flow between the two sections, these equations are especially useful where the flow in the interior of the control volume is fairly complex.

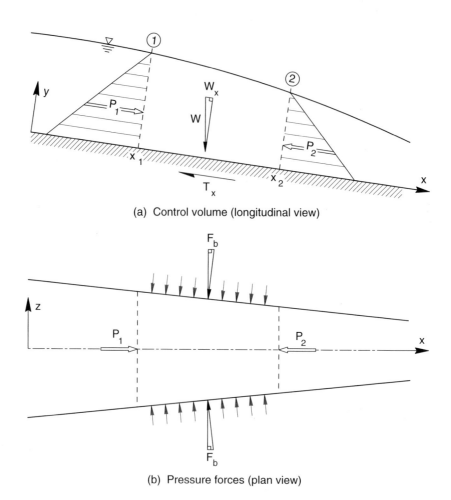

Figure 1-5. Definition sketch of the control volume.

Continuity Equation

Deletion of the unsteady term in Eq. (1-9) and the integration of the resulting equation from $x = x_1$ to $x = x_2$ yield

$$\int_{x_1}^{x_2} \frac{dQ}{dx}\, dx + \int_{x_1}^{x_2} q_\ell\, dx = 0 \qquad (1\text{-}66)$$

or

$$Q_2 - Q_1 + Q_\ell = 0 \qquad (1\text{-}67)$$

where Q_ℓ = rate of lateral outflow between sections 1 and 2 and the subscripts 1 and 2 denote the sections. If there is no lateral flow, Eq. (1-67) becomes

$$Q_1 = Q_2 \quad \text{or} \quad V_1 A_1 = V_2 A_2 \tag{1-68}$$

Momentum Equation

The differential momentum equation (1-27) was derived from Eq. (1-23) after carrying out some mathematical manipulations. The manipulations were made to obtain the one-dimensional momentum equation in a particular form and are not needed in deriving the algebraic momentum equation. Therefore, Eq. (1-23) is used to derive the algebraic momentum equation. In deriving Eq. (1-23), the assumption of hydrostatic pressure distribution was invoked in Eq. (1-12). This assumption also is not needed in deriving the algebraic momentum equation. Without this assumption the pressure force term in Eq. (1-23) is

$$-\int_0^{y_s} \sigma \frac{\partial p}{\partial x} dy = -\int_0^{y_s} \frac{\partial(\sigma p)}{\partial x} dy + \int_0^{y_s} p \frac{\partial \sigma}{\partial x} dy$$

$$= -\frac{\partial}{\partial x} \int_0^{y_s} (\sigma p) dy + \int_0^{y_s} p \frac{\partial \sigma}{\partial x} dy$$

$$= -\frac{\partial P}{\partial x} + \int_0^{y_s} p \frac{\partial \sigma}{\partial x} dy \tag{1-69}$$

where P = pressure force on a cross section of the channel. The second term represents the pressure force, P_b, on the fluid in the x-direction per unit length, and it is due to a variation of channel width. Deletion of the unsteady term in Eq. (1-23), introduction of Eq. (1-69) for the pressure force term into Eq. (1-23), and integration of the resulting equation from $x = x_1$ to $x = x_2$ yield

$$-\int_{x_1}^{x_2} \frac{\partial P}{\partial x} dx + \int_{x_1}^{x_2} P_b dx + \int_{x_1}^{x_2} \gamma S_0 A dx - \int_{x_1}^{x_2} \tau_0 P dx$$

$$= \rho \int_{x_1}^{x_2} \frac{\partial(\beta V^2 A)}{\partial x} dx + \rho \int_{x_1}^{x_2} V_\ell q_\ell dx \tag{1-70}$$

Equation (1-70) can be written as

$$\rho(\beta_2 Q_2 V_2 - \beta_1 Q_1 V_1) + \rho Q_\ell \overline{V}_\ell = P_1 - P_2 + F_b + W_x - T_x \tag{1-71}$$

where

$$\bar{V}_\ell = \frac{\int_{x_1}^{x_2} q_\ell V_\ell \, dx}{Q_\ell} = \text{average longitudinal velocity of lateral flow} \qquad (1\text{-}72)$$

$$F_b = \int_{x_1}^{x_2} P_b \, dx = x\text{-component of pressure force on channel boundary} \qquad (1\text{-}73)$$

$$W_x = -\int_{x_1}^{x_2} \gamma S_0 A \, dx = x\text{-component of weight of control volume} \qquad (1\text{-}74)$$

$$T_x = \int_{x_1}^{x_2} \tau_0 P \, dx = x\text{-component of shear force on control surface} \qquad (1\text{-}75)$$

and P_1 and P_2 are the pressure forces on sections 1 and 2, respectively. Equation (1-71) states that the rate of change of x-momentum within a control volume is equal to the sum of the x-components of the forces acting on the control volume. If there is no lateral flow, Eq. (1-71) reduces to

$$\rho Q(\beta_2 V_2 - \beta_1 V_1) = P_1 - P_2 + F_b + W_x - T_x \qquad (1\text{-}76)$$

Energy Equation

Deletion of the unsteady term in the one-dimensional mechanical-energy equation (1-40) yields

$$\frac{dH_\alpha}{dx} + \frac{\bar{U}_\ell^2 - \alpha V^2}{2gQ} q_\ell = -\frac{\bar{\varepsilon}}{\gamma Q} \qquad (1\text{-}77)$$

Equation (1-77) is based on the assumption that the pressure distribution is hydrostatic and the velocity components in the y- and z-directions are zero. By including certain correction terms, this equation can be modified for flow where these assumptions are not valid in the interior of the control volume. The correction terms can be obtained upon deriving the one-dimensional mechanical-energy equation by the rigorous approach, as shown in Appendix A, and can be included in the energy dissipation term on the right-hand side of Eq. (1-77). Let the term on the right-hand side, including the correction terms, be denoted by h_f, the head loss per unit channel length. The algebraic mechanical-energy equation is then obtained by integrating Eq. (1-77) from $x = x_1$ to $x = x_2$ as

$$\int_{x_1}^{x_2} \frac{dH_\alpha}{dx} \, dx + \int_{x_1}^{x_2} \frac{\bar{U}_\ell^2 - \alpha V^2}{2gQ} q_\ell \, dx = -\int_{x_1}^{x_2} h_f \, dx \qquad (1\text{-}78)$$

Equation (1-78) can be simplified for flows either without lateral flow or with lateral outflow. For flows with lateral outflow one can assume that $\overline{U}_\ell^2 = \alpha V^2$ in the second term on the left-hand side of Eq. (1-78). For such flows integration of Eq. (1-78) gives

$$\int_{x_1}^{x_2} \frac{dH_\alpha}{dx}\, dx = -\int_{x_1}^{x_2} h_f\, dx$$

or
$$H_{\alpha_2} - H_{\alpha_1} = H_f \qquad\qquad (1\text{-}79)$$

where H_f = head loss between the two sections. Equation (1-79) is applicable to a flow reach wherein the pressure distribution is not hydrostatic, provided sections 1 and 2 are located in regions of hydrostatic pressure distribution.

Because Eq. (1-79) contains new information which is used to evaluate the head-loss term, the momentum and mechanical-energy equations in the algebraic form, Eqs. (1-71) and (1-79), are considered as independent equations. A similar approach is employed in the boundary-layer integral methods where the momentum integral relation is supplemented by the mechanical-energy integral relation (Schlichting, 1968; Kline et al., 1968). Furthermore, Eq. (1-79) can also be derived from the first law of thermodynamics that is independent of Newton's second law of motion. Equation (1-79) is simply referred to as the energy equation. An example of hydraulic jump where both the momentum and energy equations are used as independent equations in analyzing the flow conditions is given in Chapter 3. The momentum equation is used to relate the flow conditions upstream and downstream from the jump, and the energy equation is applied to evaluate the energy loss in the jump. Another illustration is presented below.

Application

The practical use of the algebraic equations of motion is illustrated for the flow under a sluice gate shown in Figure 1-6. Assuming that the head loss and the boundary shear stress between sections 1 and 2 are negligible, the algebraic equations of motion for a unit width of the channel can be written as follows:

Continuity equation: $\qquad\qquad q = V_1 y_1 = V_2 y_2 \qquad\qquad (1\text{-}80)$

Energy equation: $\qquad\qquad y_1 + \frac{V_1^2}{2g} = y_2 + \frac{V_2^2}{2g} \qquad\qquad (1\text{-}81)$

Momentum equation: $\quad \rho q (V_2 - V_1) = \gamma \left(\frac{y_1^2}{2} - \frac{y_2^2}{2} \right) - F_b \qquad\qquad (1\text{-}82)$

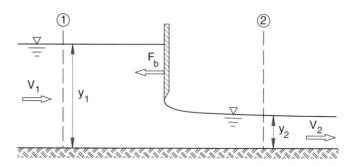

Figure 1-6. Flow under a sluice gate.

where q is the discharge per unit width and F_b is the force per unit width acting on the sluice gate. Expressing V_1 and V_2 in terms of q, y_1, and y_2 from Eq. (1-80) into Eq. (1-81), one obtains

$$y_1 + \frac{q^2}{2gy_1^2} = y_2 + \frac{q^2}{2gy_2^2} \tag{1-83}$$

The unit discharge from Eq. (1-83) is

$$q = \frac{\sqrt{2g}\, y_1 y_2}{\sqrt{y_1 + y_2}} \tag{1-84}$$

The force F_b in terms of q, y_1, and y_2 from Eq. (1-82) is

$$F_b = \gamma \left(\frac{y_1^2}{2} - \frac{y_2^2}{2} \right) - \rho q^2 \left(\frac{1}{y_2} - \frac{1}{y_1} \right)$$

Substitution for q from Eq. (1-84) into the above equation and simplification yield

$$F_b = \frac{\gamma}{2} \frac{(y_1 - y_2)^3}{(y_1 + y_2)}$$

PROBLEMS

1-18 Using the continuity and the energy equations, determine the unit discharge, q, under a sluice gate for depths short distances upstream and downstream of the sluice gate as 2 and 0.5 m, respectively, Neglect the head loss across the gate.

1-19 Determine the force on the gate in Problem 1-18.

1-20 The width of a rectangular channel is contracted from 4.7 to 3.8 m. Calculate the discharge in the channel if the flow depths upstream and downstream of the contraction are 1.9 and 1.8 m, respectively. The head loss in the contraction is one-tenth of the upstream velocity head.

1-21   The flow depths upstream and downstream of a submerged weir in a rectangular channel are 2.2 and 1.9 m, respectively. The discharge in the channel is 4.5 m³/s per meter width of the channel. Calculate the force on the weir and the energy loss due to the weir.

1-22   A rectangular channel carries wastewater at flow depth of 2.0 m and flow velocity of 2.5 m/s to a waste treatment plant. The channel has two identical gates: one normal to the flow and the second parallel to the flow in one of the sidewalls. Due to repairs in the treatment plant the gate normal to the flow is closed, and the wastewater is diverted through the gate in the sidewall. Calculate the height of the closed gate so that the wastewater would not spill over it using (a) the momentum equation and (b) the energy equation. Which answer is correct and why? Neglect the energy loss.

1-10 Pressure Distribution in Curvilinear Flow

In rapidly varied flow the streamlines have substantial curvature and the pressure distribution is nonhydrostatic. Streamline curvature produces appreciable acceleration components normal to the flow direction. The pressure distribution deviates from the hydrostatic if curvilinear flow occurs in a vertical plane. Such curvilinear flow may be either convex flow such as over a spillway crest, shown in Figure 1-7a, or concave flow such as over a spillway bucket, shown in Figure 1-7b. In convex flow the vertical component of the normal acceleration is opposite to the gravitational acceleration; hence the curve for the actual pressure distribution lies below the curve for the hydrostatic pressure distribution. In concave flow the vertical component of the normal acceleration and the gravitational acceleration act in the same direction; consequently, the curve for the actual pressure distribution

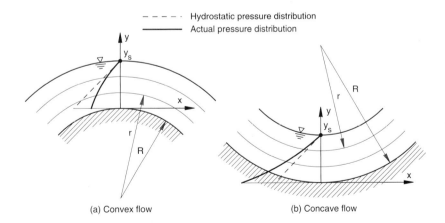

Figure 1-7. Pressure distribution in curvilinear flow.

lies above the curve for the hydrostatic pressure distribution. It can be shown (Rouse, 1946) that the normal acceleration at any point in a curvilinear flow is given by u^2/r, where u = velocity and r = radius of curvature of the streamline at that point. The normal acceleration, also called centrifugal acceleration, is positive for convex flow and negative for concave flow. For such curvilinear flow, Eq.(1-1)modifies as

$$\frac{\partial p}{\partial y} + \gamma \cos \theta = \pm \rho \frac{u^2}{r}$$ (1-85)

Integration of Eq. (1-85) with the boundary condition that $p = 0$ at $y = y_s$ yields

$$p = \gamma \cos \theta (y_s - y) \pm \rho \int_{y_s}^{y} \frac{u^2}{r} \, dy$$ (1-86)

The determination of pressure distribution from Eq. (1-86) requires knowledge of the velocity distribution and the radius of curvature of streamlines. For $y_s \ll r$, it can be assumed that $u \cong V$ and $r \cong R$, and Eq. (1-86) can be written as

$$p = \rho \left(g \cos \theta \pm \frac{V^2}{R} \right)(y_s - y)$$ (1-87)

where R is defined in Figure 1-7. A better approximation for the velocity distribution is that it is the same as in free-vortex flow, i.e.,

$$u = \frac{C}{r}$$ (1-88)

where C is a constant and can be determined from the continuity equation. The pressure distribution can be determined from Eqs. (1-86) and (1-88), as shown below in Example 1-5.

Example 1-5. Assuming that the flow in a two-dimensional circular bucket is irrotational, show that the maximum pressure head, h, in the bucket is

$$h = y_s + \frac{V_s^2}{2g} \left(1 - \frac{(R - y_s)^2}{R^2} \right)$$

where y_s is the flow depth in the bucket, V_s is the surface velocity in the bucket, and R is the radius of the bucket.

From Eq. (1-86) the maximum pressure head in the bucket, where $y = 0$ and $\cos \theta = 1$, is

$$h = \frac{p}{\rho g} = y_s - \frac{1}{g} \int_{y_s}^0 \frac{u^2}{r} dy \qquad (1)$$

For an irrotational flow the velocity distribution is given by

$$u = \frac{C}{r} \qquad (2)$$

where C is a constant and can be evaluated from the condition that $u = V_s$ at $r = (R - y_s)$ as

$$C = V_s(R - y_s) \qquad (3)$$

Substituting into Eq. (1) for u from Eq. (2), for C from Eq. (3), and $r = (R - y)$, and integrating the resulting equation, one obtains

$$h = y_s - \frac{C^2}{g} \int_{y_s}^0 (R - y) dy = y_s - \frac{C^2}{2g}(R - y)^{-2}\Big|_{y=y_s}^{y=0} = y_s + \frac{V_s^2}{2g}\left[1 - \frac{(R - y_s)^2}{R^2}\right]$$

PROBLEMS

1-23   Compute the pressure exerted upon the floor surface of a spillway bucket if the radius of the bucket is 30 m and the flow velocity and depth are 40.0 m/s and 3.0 m, respectively.

1-24   Solve Problem 1-23 assuming that the streamlines are concentric circles and the velocity distribution in the bucket is the same as in free-vortex flow, i.e., $u = C/r$, where C is a constant and r is the radius of streamlines.

1-25   The shape of the crest of a spillway can be approximated by a circular arc of radius 4.0 m. The discharge over the spillway is 6.0 m³/s/m. Calculate the pressure on the face of the spillway at a point that makes an angle of 30° with the vertical direction. The flow depth at that point is 2.0 m. Assume that the velocity distribution is $u = C/r$, where C is a constant and r is the radius of the streamlines, which can be assumed concentric.

APPENDIX A

RIGOROUS APPROACH

This appendix deals with the derivation of the one-dimensional differential equations of motion by the rigorous approach. The complete one-dimensional equations are derived first by integrating the three-dimensional equations and then simplified after making certain assumptions.

A-1 Differential Continuity Equation

The continuity equation is based on the principle of conservation of mass, which states that the rate of increase of fluid mass within a given volume must be equal to the difference between the mass influx into and the mass efflux out of the control volume. Using this principle, the three-dimensional continuity equation is obtained first. The one-dimensional continuity equation is then derived from the three-dimensional continuity equation.

Three-Dimensional Equation

To derive the three-dimensional continuity equation, consider an element of space with sides $\delta x, \delta y, and \, \delta z$ parallel to axes $x, y,$ and z, respectively (see Figure A-1). Let the components of velocity in the $x, y,$ and z directions be u, v, and w. The net

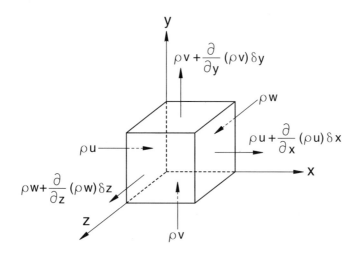

Figure A-1. Three-dimensional rectangular control volume.

mass influx in the x-direction is equal to the rate of mass inflow minus the rate of mass outflow and can be expressed as

$$\rho u \delta y \delta z - \left(\rho u + \frac{\partial(\rho u)}{\partial x} \delta x \right) \delta y \delta z = -\frac{\partial(\rho u)}{\partial x} \delta \forall$$

where $\delta \forall = \delta x \delta y \delta z$ = volume of the element and ρ = density of fluid. Similar expressions are obtained for the net mass influx in the other two directions. The total excess of mass passing into the element per unit time, therefore, is the sum of the three, which must be equal to the temporal rate of change of the mass contained within the control volume:

$$-\left(\frac{\partial(\rho u)}{\partial x} + \frac{\partial(\rho v)}{\partial y} + \frac{\partial(\rho w)}{\partial z} \right) \delta \forall = \frac{\partial(\rho \delta \forall)}{\partial t}$$

In the limit as $\delta \forall \to 0$ one obtains the exact expression for the continuity equation as

$$\frac{\partial(\rho u)}{\partial x} + \frac{\partial(\rho v)}{\partial y} + \frac{\partial(\rho w)}{\partial z} = -\frac{\partial \rho}{\partial t} \tag{A-1}$$

Equation (A-1) can be expanded as

$$-\rho \left(\frac{\partial u}{\partial x} + \frac{\partial v}{\partial y} + \frac{\partial w}{\partial z} \right) = \frac{\partial \rho}{\partial t} + u \frac{\partial \rho}{\partial x} + v \frac{\partial \rho}{\partial y} + w \frac{\partial \rho}{\partial z} \tag{A-2}$$

The right-hand side of Eq. (A-2) is the total derivative of the density with respect to time, i.e., $D\rho / Dt$ which is zero for an incompressible flow. The continuity equation for an incompressible flow from Eq. (A-2) is

$$\frac{\partial u}{\partial x} + \frac{\partial v}{\partial y} + \frac{\partial w}{\partial z} = 0 \tag{A-3}$$

One-Dimensional Equation

The one dimensional continuity equation is derived by integrating Eq. (A-3) over the cross section of the channel shown in Figure A-2. The integration of Eq. (A-3) over the cross-sectional area A of the flow can be written as

$$\int_A \left(\frac{\partial u}{\partial x} + \frac{\partial v}{\partial y} + \frac{\partial w}{\partial z} \right) dA = \int_{z_\ell}^{z_r} dz \int_{y_b}^{y_s} \left(\frac{\partial u}{\partial x} + \frac{\partial v}{\partial y} + \frac{\partial w}{\partial z} \right) = 0 \tag{A-4}$$

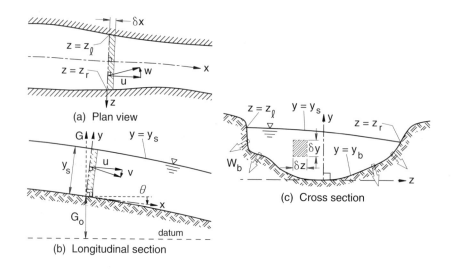

Figure A-2. Channel geometry.

where

$$A = \int_{z_\ell}^{z_r} (y_s - y_b)\,dz \tag{A-5}$$

$z_\ell(x,t)$ and $z_r(x,t)$ are the z-coordinates of the intersections of the free surface with the left and right banks, respectively; and $y_s(s,z,t)$ and $y_b(x,z,t)$ are the y-coordinates of the free surface and channel bottom, respectively. Applying the Leibnitz rule of integration to the first term on the right-hand side of Eq. (A-4), on obtains

$$\int_{z_\ell}^{z_r} dz \int_{y_b}^{y_s} \frac{\partial u}{\partial x}\,dy = \int_{z_\ell}^{z_r} dz \left\{ \frac{\partial}{\partial x} \int_{y_b}^{y_s} u\,dy - u_s \frac{\partial y_s}{\partial x} + u_b \frac{\partial y_b}{\partial x} \right\}$$

Application of the Leibnitz rule again to the first term on the right-hand side of the above equation yields

$$\int_{z_\ell}^{z_r} dz \int_{y_b}^{y_s} \frac{\partial u}{\partial x}\,dy = \int_{z_\ell}^{z_r} dz \int_{y_b}^{y_s} u\,dy - \int_{z_\ell}^{z_r} \left(u_s \frac{\partial y_s}{\partial x} - u_b \frac{\partial y_b}{\partial x} \right) dz \tag{A-6}$$

in which the use is made of the condition that $y_s = y_b$ at $z = z_\ell$ and $z = z_r$. The second term on the right-hand side of Eq. (A-4) can be written as

$$\int_{z_\ell}^{z_r} dz \int_{y_b}^{y_s} \frac{\partial v}{\partial y}\,dy = \int_{z_\ell}^{z_r} (v_s - v_b)\,dz \tag{A-7}$$

The third term, similar to the first term, on the right-hand side of Eq. (A-4) can be integrated using the Leibnitz rule, and the result is

$$\int_{z_\ell}^{z_r} dz \int_{y_b}^{y_s} \frac{\partial w}{\partial z}\, dy = \int_{z_\ell}^{z_r} dz \left\{ \frac{\partial}{\partial z} \int_{y_b}^{y_s} w\, dy - w_s\, \frac{\partial y_s}{\partial z} + w_b\, \frac{\partial y_b}{\partial z} \right\}$$

$$= -\int_{z_\ell}^{z_r} \left(w_s\, \frac{\partial y_s}{\partial z} - w_b\, \frac{\partial y_b}{\partial z} \right) dz \qquad (A-8)$$

Once again the condition that $y_s = y_b$ at $z = z_\ell$ and $z = z_r$. is applied to Eq. (A-8). Substitution of Eqs. (A-6), (A-7), and (A-8) into Eq. (A-4) yields

$$\frac{\partial}{\partial x} \int_{z_\ell}^{z_r} dz \int_{y_b}^{y_s} u\, dy - \int_{z_\ell}^{z_r} \left(u_s\, \frac{\partial y_s}{\partial x} - v_s + w_s\, \frac{\partial y_s}{\partial z} \right) dz$$

$$+ \int_{z_\ell}^{z_r} \left(u_b\, \frac{\partial y_b}{\partial x} - v_b + w_b\, \frac{\partial y_b}{\partial z} \right) dz = 0 \qquad (A-9)$$

The simplification of Eq. (A-9) requires boundary conditions, known as kinematic boundary conditions, which have to be satisfied by the flow. The general kinematic condition in physical terms states that the velocity of a fluid particle relative to the surface in which it lies must be wholly tangential (or zero), otherwise there is a finite flow of fluid across it. The mathematical expression for the kinematic boundary condition can be derived as follows.

Kinematic Boundary Condition

Let the shape of a bounding surface shown in Figure A-3 be given by

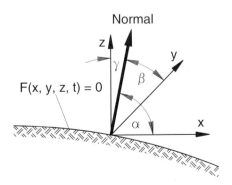

Figure A-3. Definition sketch of a surface.

$$F(x, y, z, t) = 0 \tag{A-10}$$

and v_{bs} be the rate of motion of the surface along the normal direction. In time δt, the surface move by $v_{bs}\delta t$. Equation (A-10) can be written as

$$F(x + lv_{bs}\delta t, \, y + mv_{bs}\delta t, \, z + nv_{bs}\delta t, \, t + \delta t) = 0 \tag{A-11}$$

where l, m, and n are the directional cosines of the normal at (x, y, z),

$$l = \cos\beta = (\partial F/\partial x)/R, \quad m = \cos\beta = (\partial F/\partial y)/R, \quad n = \cos\gamma = (\partial F/\partial z)/R \tag{A-12}$$

and

$$R = \sqrt{\left(\frac{\partial F}{\partial x}\right)^2 + \left(\frac{\partial F}{\partial y}\right)^2 + \left(\frac{\partial F}{\partial z}\right)^2} \tag{A-13}$$

Expanding Eq. (A-11) by using the Taylor series and substituting Eq. (A-10) into the resulting equation, one obtains

$$v_{bs}\left(\ell\frac{\partial F}{\partial x} + m\frac{\partial F}{\partial y} + n\frac{\partial F}{\partial z}\right) + \frac{\partial F}{\partial t} = 0$$

or

$$v_{bs} = -\frac{1}{R}\frac{\partial F}{\partial t} \tag{A-14}$$

If a fluid particle on the surface has velocity components on u, v, and w, then the velocity, v_f, of the fluid particle normal to the surface is

$$v_f = \ell u + mv + nw \tag{A-15}$$

If there is a flow of fluid across the surface, then

$$v_f - v_{bs} = W \tag{A-16}$$

where W is the net outflow velocity normal to the surface. Using Eqs. (A-12), (A-14), and (A-15), Eq. (A-16) can be written as

$$\frac{1}{R}\left(u\frac{\partial F}{\partial x} + v\frac{\partial F}{\partial y} + w\frac{\partial F}{\partial z}\right) + \frac{1}{R}\frac{\partial F}{\partial t} = W \tag{A-17}$$

If the fluid particle remains on the surface, there is no flow across the surface, i.e., $W = 0$, and Eq. (A-17) becomes

$$u\frac{\partial F}{\partial x} + v\frac{\partial F}{\partial y} + w\frac{\partial F}{\partial z} + \frac{\partial F}{\partial t} = 0 \tag{A-18}$$

The kinematic boundary condition at the free surface $F = y - y_s(x,z,t) = 0$ with no flow across it (i.e., $W = 0$) is obtained from Eq. (A-18) as

$$u_s \frac{\partial y_s}{\partial x} - v_s + w_s \frac{\partial y_s}{\partial z} + \frac{\partial y_s}{\partial t} = 0 \qquad \text{(A-19)}$$

where the subscript s denotes the conditions at the free surface. Assuming a lateral flow through the channel bottom as shown in Figure A-2, the kinematic boundary condition at the bottom $F = y - y_b(x,z,t) = 0$ is obtained from Eq. (A-17) as

$$u_b \frac{\partial y_b}{\partial x} - v_b + w_b \frac{\partial y_b}{\partial z} + \frac{\partial y_b}{\partial t} = W_b R_b \qquad \text{(A-20)}$$

where the subscript b denotes the conditions at the channel bottom and R_b from Eq. (A-13) is

$$R_b = \sqrt{1 + \left(\frac{\partial y_b}{\partial x}\right)^2 + \left(\frac{\partial y_b}{\partial z}\right)^2}$$

Using Eqs. (A-19) and (A-20) along with the Leibnitz rule, Eq. (A-9) becomes

$$\frac{\partial}{\partial x} \int_{z_\ell}^{z_r} dz \int_{y_b}^{y_s} u\, dy + \frac{\partial}{\partial t} \int_{z_\ell}^{z_r} (y_s - y_b)\, dz + \int_{z_\ell}^{z_r} W_b R_b\, dz = 0 \qquad \text{(A-21)}$$

Using Eq. (A-5), Eq. (A-21) can be written as

$$\frac{\partial Q}{\partial x} + \frac{\partial A}{\partial t} + q_\ell = 0 \qquad \text{(A-22)}$$

where

$$Q = \int_{z_\ell}^{z_r} dz \int_{y_b}^{y_s} u\, dy = VA = \text{discharge in the channel} \qquad \text{(A-23)}$$

$$V = \text{average flow velocity in the channel}$$

$$q_\ell = \int_{z_\ell}^{z_r} W_b R_b\, dz = \text{lateral outflow per unit of channel} \qquad \text{(A-24)}$$

Equation (A-22) is the one-dimensional differential continuity equation and is identical to Eq. (1-9). It is an exact equation for incompressible flow.

A-2 Differential Momentum Equations

Similar to the continuity equation, the three-dimensional momentum equations will be derived first, which then will be integrated over the cross section to obtain the one-dimensional momentun equation.

Three-Dimensional Equations

To derive the three-dimensional differential momentum equations, consider the forces acting on an element shown in Figure A-4. The forces on the element include the body and surface forces. The body force due to gravitational attraction, i.e., the weight of the element, is the only body force to be considered. The gravity force on the element is equal to $\rho g \delta \forall$ that acts vertically downward.

The gravity force can be expressed in terms of a potential, G (the potential energy per unit weight), given by

$$G = G_0(x) + y \cos \theta \tag{A-25}$$

such that the components of the gravity force are:

$$\text{In the } x\text{-direction:} \quad -\rho g \delta \forall \, \frac{\partial G}{\partial x} = \rho g \delta \forall \sin \theta$$

$$\text{In the } y\text{-direction:} \quad -\rho g \delta \forall \, \frac{\partial G}{\partial y} = \rho g \delta \forall \cos \theta \tag{A-26}$$

$$\text{In the } z\text{-direction:} \quad -\rho g \delta \forall \, \frac{\partial G}{\partial z} = 0$$

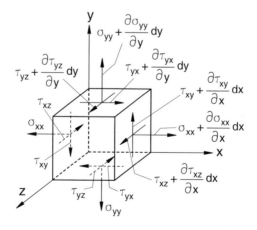

Figure A-4. Stress components.

where
$$\frac{\partial G}{\partial x} = \frac{\partial G_0}{\partial x} = -\sin\theta; \quad \frac{\partial G}{\partial y} = \cos\theta; \quad \frac{\partial G}{\partial z} = 0 \tag{A-27}$$

G_0 is the elevation of the channel bed, i.e., the lowest point in the cross section, measured from a reference horizontal datum along the gravitational direction, and θ is the angle between the channel bed and a horizontal plane (see Figure A-2). The surface forces include the normal forces and the tangential forces exerted on the element by the fluid in contact with it. These forces are expressed in terms of stresses, i.e., force per unit area. The normal and tangential stresses are designated by the symbols σ and τ, respectively. Both symbols are provided with two subscripts; the first subscript indicates the direction of the normal to the plane of action, and the second subscript indicates the direction of action. With this notation the net surface forces on the element in the x-direction (see Figure A-4) are

$$\left(\frac{\partial \sigma_{xx}}{\partial x} + \frac{\partial \tau_{yx}}{\partial y} + \frac{\partial \tau_{zx}}{\partial z} \right) \delta \forall \tag{A-28}$$

The normal stresses include in part the pressure p and are expressed as

$$\sigma_{xx} = -p + \sigma'_{xx}; \quad \sigma_{yy} = -p + \sigma'_{yy}; \quad \sigma_{zz} = -p + \sigma'_{zz} \tag{A-29}$$

where $\sigma'_{xx}, \sigma'_{yy}$, and σ'_{zz} are the deviatoric normal stresses. The momentum equation in the x-direction can be written as

$$ma_x = \sum F_x \tag{A-30}$$

where
$$m = \text{mass of fluid within the element} = \rho\,\delta\forall$$

$$a_x = \text{fluid acceleration in the } x\text{-direction}$$

$$= \frac{Du}{Dt} = \frac{\partial u}{\partial t} + u\frac{\partial u}{\partial x} + v\frac{\partial u}{\partial y} + w\frac{\partial u}{\partial z} \tag{A-31}$$

$$\sum F_x = \text{sum of the forces acting on the element in the } x\text{-direction}$$

Introducing Eq. (A-29) in Eq. (A-28) and returning to Eq. (A-30), the momentum

$$\rho\left(\frac{\partial u}{\partial t} + u\frac{\partial u}{\partial x} + v\frac{\partial u}{\partial y} + w\frac{\partial u}{\partial z} \right) = -\frac{\partial p}{\partial x} - \rho g\frac{\partial G}{\partial x} + \frac{\partial \sigma'_{xx}}{\partial x} + \frac{\partial \tau_{yx}}{\partial y} + \frac{\partial \tau_{zx}}{\partial z} \tag{A-32}$$

The momentum equations in the y- and z-directions are similar to that in the x-direction, and these can be written as

$$\rho\left(\frac{\partial v}{\partial t}+u\frac{\partial v}{\partial x}+v\frac{\partial v}{\partial y}+w\frac{\partial v}{\partial z}\right)=-\frac{\partial p}{\partial y}-\rho g\frac{\partial G}{\partial y}+\frac{\partial \tau_{xy}}{\partial x}+\frac{\partial \sigma'_{yy}}{\partial y}+\frac{\partial \tau_{zy}}{\partial z} \qquad (A\text{-}33)$$

$$\rho\left(\frac{\partial w}{\partial t}+u\frac{\partial w}{\partial x}+v\frac{\partial w}{\partial y}+w\frac{\partial w}{\partial z}\right)=-\frac{\partial p}{\partial z}-\rho g\frac{\partial G}{\partial z}+\frac{\partial \tau_{xz}}{\partial x}+\frac{\partial \tau_{yz}}{\partial y}+\frac{\partial \sigma'_{zz}}{\partial z} \qquad (A\text{-}34)$$

It should be pointed out that the components of velocity, u, v, and w, are time averaged velocity components over a short time interval. This means that the stresses include both the viscous stresses and the Reynolds' stresses. A detailed description about these stresses can be found in books on fluid mechanics (Dailey and Harleman, 1966; White, 1999).

One-Dimensional Equation

The one-dimensional momentum equation along the main flow direction (i.e., x-direction) can be derived by integrating Eq. (A-32) over the channel cross section after substituting for pressure p. An expression for pressure is obtained by integrating Eqs. (A-33) and (A-34) with respect to y and z, respectively. Using Eq. (A-27) and the boundary condition that the pressure p is zero at $(y = y_{st}, z = z_t)$, where y_{st} is the value of y_s at $z = z_t$, the integration of Eq. (A-33) yields

$$p(x,y,z,t)-\rho g\cos\theta\{y_{st}(x,z_\ell,t)-y\}=\int_{y_{st}}^{y}\left(\frac{\partial \tau_{xy}}{\partial x}+\frac{\partial \sigma'_{yy}}{\partial y}+\frac{\partial \tau_{zy}}{\partial z}\right)_{z=z_\ell}dy$$

$$-\int_{y_{st}}^{y}\rho\left(\frac{\partial v}{\partial t}+u\frac{\partial v}{\partial x}+v\frac{\partial v}{\partial y}+w\frac{\partial v}{\partial z}\right)_{z=z_\ell}dy \qquad (A\text{-}35)$$

Using Eq. (A-27) and the condition that the pressure at $(y = y_{st}, z = z_t)$ is zero, the integration with z of Eq. (A-34) gives

$$p(x,y,z,t)=\int_{z_\ell}^{z}\left(\frac{\partial \tau_{xz}}{\partial x}+\frac{\partial \tau_{yz}}{\partial y}+\frac{\partial \sigma'_{zz}}{\partial z}\right)_{y=y_{s\ell}}dz$$

$$-\int_{z_\ell}^{z}\rho\left(\frac{\partial w}{\partial t}+u\frac{\partial w}{\partial x}+v\frac{\partial w}{\partial y}+w\frac{\partial w}{\partial z}\right)_{y=y_{s\ell}}dz \qquad (A\text{-}36)$$

From Eqs. (A-35) and (A-36) the pressure distribution is given as

$$p(x,y,z,t)=\rho g\cos\theta\{y_{st}(x,z_\ell,t)-y\}+C_s(x,y,z,t)+C_l(x,y,z,t) \qquad (A\text{-}37)$$

where

$$C_s = \text{correction term due to stresses}$$

$$= \int_{y_{s\ell}}^{y} \left(\frac{\partial \tau_{xy}}{\partial x} + \frac{\partial \sigma'_{yy}}{\partial y} + \frac{\partial \tau_{zy}}{\partial z} \right)_{z=z_\ell} dy$$

$$+ \int_{z_\ell}^{z} \left(\frac{\partial \tau_{xz}}{\partial x} + \frac{\partial \tau_{yz}}{\partial y} + \frac{\partial \sigma'_{zz}}{\partial z} \right)_{y=y_{s\ell}} dz \qquad \text{(A-38)}$$

C_l = correction due to nonzero normal acceleration

$$= -\int_{y_{s\ell}}^{y} \rho \left(\frac{\partial v}{\partial t} + u\frac{\partial v}{\partial x} + v\frac{\partial v}{\partial y} + w\frac{\partial v}{\partial z} \right)_{z=z_\ell} dy$$

$$- \int_{z_\ell}^{z} \rho \left(\frac{\partial w}{\partial t} + u\frac{\partial w}{\partial x} + v\frac{\partial w}{\partial y} + w\frac{\partial w}{\partial z} \right)_{y=y_{s\ell}} dz \qquad \text{(A-39)}$$

Addition of a term, $u(\partial u/\partial x + \partial v/\partial y + \partial w/\partial z)$, which is zero by virtue of Eq. (A-3), to the left-hand side of Eq. (A-32) yields

$$\rho\left(\frac{\partial u}{\partial t} + \frac{\partial u^2}{\partial x} + \frac{\partial u v}{\partial y} + \frac{\partial u w}{\partial z} \right) = -\frac{\partial p}{\partial x} - \rho g \frac{\partial G}{\partial x} + \frac{\partial \sigma'_{xx}}{\partial x} + \frac{\partial \tau_{yx}}{\partial y} + \frac{\partial \tau_{zx}}{\partial z} \qquad \text{(A-40)}$$

Integration of Eq. (A-40) over the cross-sectional area of the flow gives

$$\rho \int_A \left(\frac{\partial u}{\partial t} + \frac{\partial u^2}{\partial x} + \frac{\partial u v}{\partial y} + \frac{\partial u w}{\partial z} \right) dA$$

$$= \int_A \left(-\frac{\partial p}{\partial x} - \rho g \frac{\partial G}{\partial x} + \frac{\partial \sigma'_{xx}}{\partial x} + \frac{\partial \tau_{yx}}{\partial y} + \frac{\partial \tau_{zx}}{\partial z} \right) dA \qquad \text{(A-41)}$$

Equation (A-41) will be integrated over the cross section term by term. The integration procedure is similar to that used for the continuity equation. The integration of the first through the fourth term on the left-hand side of Eq. (A-41) can be written as

$$\rho \int_{z_\ell}^{z_r} dz \int_{y_b}^{y_s} \frac{\partial u}{\partial t} dy = \rho \left\{ \frac{\partial}{\partial t} \int_{z_\ell}^{z_r} dz \int_{y_b}^{y_s} u \, dy - \int_{z_\ell}^{z_r} \left(u_s \frac{\partial y_s}{\partial t} - u_b \frac{\partial y_b}{\partial t} \right) dz \right\} \qquad \text{(A-42)}$$

$$\rho \int_{z_\ell}^{z_r} dz \int_{y_b}^{y_s} \frac{\partial u^2}{\partial t} dy = \rho \left\{ \frac{\partial}{\partial x} \int_{z_\ell}^{z_r} dz \int_{y_b}^{y_s} u^2 \, dy - \int_{z_\ell}^{z_r} \left(u_s^2 \frac{\partial y_s}{\partial x} - u_b^2 \frac{\partial y_b}{\partial x} \right) dz \right\} \qquad \text{(A-43)}$$

$$\rho \int_{z_\ell}^{z_r} dz \int_{y_b}^{y_s} \frac{\partial u\,v}{\partial y}\,dy = \rho \int_{z_\ell}^{z_r} \left(u_s v_s - u_b v_b \right) dz \tag{A-44}$$

$$\rho \int_{z_\ell}^{z_r} dz \int_{y_b}^{y_s} \frac{\partial u\,w}{\partial z}\,dy = -\rho \int_{z_\ell}^{z_r} \left(u_s w_s \frac{\partial y_s}{\partial z} - u_b w_b \frac{\partial y_b}{\partial z} \right) dz \tag{A-45}$$

Combining Eqs. (A-42)–(A-45), one obtains

$$\rho \int_{z_\ell}^{z_r} dz \int_{y_b}^{y_s} \left(\frac{\partial u}{\partial t} + \frac{\partial u^2}{\partial x} + \frac{\partial uv}{\partial y} + \frac{\partial u\,w}{\partial z} \right) dy$$

$$= \rho \frac{\partial}{\partial t} \int_{z_\ell}^{z_r} dz \int_{y_b}^{y_s} u\,dy + \rho \frac{\partial}{\partial x} \int_{z_\ell}^{z_r} dz \int_{y_b}^{y_s} u^2\,dy$$

$$- \rho \int_{z_\ell}^{z_r} u_s \left(\frac{\partial y_s}{\partial t} + u_s \frac{\partial y_s}{\partial x} - v_s + w_s \frac{\partial y_s}{\partial z} \right) dz$$

$$+ \rho \int_{z_\ell}^{z_r} u_b \left(\frac{\partial y_b}{\partial t} + u_b \frac{\partial y_b}{\partial x} - v_b + w_b \frac{\partial y_b}{\partial z} \right) dz \tag{A-46}$$

Consider the terms on the right-hand side of Eq. (A-46). Using Eq. (A-23), the first term can be written as $\rho(\partial Q/\partial t)$. The second term is the spatial rate of momentum flux across the channel cross section. The third term by virtue of Eq. (A-19) is zero, and the expression inside the parentheses of the fourth term by virtue of Eq. (A-20) is equal to $W_b R_b$. Equation (A-46) then becomes

$$\rho \int_{z_\ell}^{z_r} dz \int_{y_b}^{y_s} \left(\frac{\partial u}{\partial t} + \frac{\partial u^2}{\partial x} + \frac{\partial uv}{\partial y} + \frac{\partial u\,w}{\partial z} \right) dy = \rho \frac{\partial Q}{\partial t} + \rho \frac{\partial}{\partial x} \left(\beta V^2 A \right) + \rho q_\ell V_\ell \tag{A-47}$$

where

$$\beta = \frac{\int_A u^2\,dA}{V^2 A} = \text{momentum coefficient} \tag{A-48}$$

$$V_\ell = \frac{\int_{z_\ell}^{z_r} u_b W_b R_b\,dz}{q_\ell} = \text{average longitudinal velocity of lateral flow} \tag{A-49}$$

Now the terms on the right-hand side of Eq. (A-41) will be integrated over the cross section. Using Eq. (A-37), the first term can be written as

$$-\int_A \frac{\partial p}{\partial x}\,dA = -\rho g\,\cos\theta\,\frac{\partial y_{s\ell}}{\partial x} \int_A dA - \int_A \frac{\partial C_s}{\partial x}\,dA - \int_A \frac{\partial C_l}{\partial x}\,dA$$

$$= -\rho g \cos\theta \frac{\partial y_{s\ell}}{\partial x} A - \int_A \frac{\partial C_s}{\partial x} dA - \int_A \frac{\partial C_I}{\partial x} dA \tag{A-50}$$

The integration of the second term is straightforward and is given by

$$\int_A \rho g \frac{\partial G}{\partial x} dA = \rho g A \frac{\partial G_0}{\partial x} \tag{A-51}$$

because $\partial G/\partial x = \partial G_0/\partial x$ from Eq. (A-25). The third term $\partial\sigma'_{xx}$ can be either neglected assuming a slow variation in the longitudinal direction or included in the correction term C_s. In the latter case, the modified correction term becomes

$$C'_s = C_s - \sigma'_{xx} \tag{A-52}$$

The fourth and fifth terms can be integrated over the cross section as

$$\int_A \left(\frac{\partial \tau_{yx}}{\partial y} + \frac{\partial \tau_{zx}}{\partial z} \right) dA = \int_{z_\ell}^{z_r} \left\{ (\tau_{yx})_s - (\tau_{yx})_b \right\} dz - \int_{z_\ell}^{z_r} \left\{ (\tau_{zx})_s \frac{\partial y_s}{\partial z} - (\tau_{zx})_b \frac{\partial y_b}{\partial z} \right\} dz$$

$$= \int_{z_\ell}^{z_r} \left\{ (\tau_{yx})_s - (\tau_{zx})_s \frac{\partial y_s}{\partial z} \right\} dz - \int_{z_\ell}^{z_r} \left\{ (\tau_{yx})_b - (\tau_{zx})_b \frac{\partial y_b}{\partial z} \right\} dz \tag{A-53}$$

It will be shown that the second integral term on the right-hand side of Eq. (A-53) is equal to the wall shear stress in the x-direction. Applying Eq. (A-30) to a differential element close to the channel bottom as shown in Figure (A-5), one obtains

$$\rho \delta A a_x = \rho g \delta A + \frac{\partial \sigma_{xx}}{\partial x} \delta A + \tau_{yx} \delta z + \tau_{zx} \delta y - \tau_b \frac{\delta z}{\cos\phi} \tag{A-54}$$

where the symbols are defined in Figure A-5. In the limit as δA approaches zero, Eq. (A-54) reduces to

$$(\tau_{yx})_b - (\tau_{zx})_b \frac{\partial y_b}{\partial z} = \frac{\tau_b}{\cos\phi} \tag{A-55}$$

Introduction of Eq. (A-55) into the second term on the right-hand side of Eq. (A-53) gives

$$\int_{z_\ell}^{z_r} \left((\tau_{yx})_b - (\tau_{zx})_b \frac{\partial y_b}{\partial z} \right) dz = \int_{z_\ell}^{z_r} \frac{\tau_b}{\cos\phi} dz = \tau_0 P \tag{A-56}$$

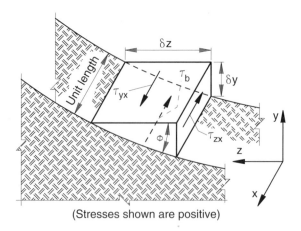

(Stresses shown are positive)

Figure A-5. Shear forces on an element near to the boundary.

$$P = \int_{z_\ell}^{z_r} \frac{dz}{\cos\phi} = \text{length of the wetted perimeter} \qquad (A-57)$$

where

$$\tau_0 = \text{average wall shear stress}$$

In a similar manner it can be shown that the first term on the right-hand side of Eq.(A-53) is the longitudinal shear at the free surface, which is generally considered zero unless the wind forces are significant. Combining Eqs. (A-47), (A-50), (A-51), (A-52), and (A-56), one obtains

$$\rho\frac{\partial Q}{\partial t} + \rho\frac{\partial}{\partial x}\left(\beta V^2 A\right) + \rho q_\ell V_\ell$$

$$= -\rho g\,\cos\theta\,\frac{\partial y_{s\ell}}{\partial x}A - \int_A \frac{\partial C'_s}{\partial x}dA - \int_A \frac{\partial C_I}{\partial x}dA - \rho g A\frac{\partial G_0}{\partial x} - \tau_0 P \quad (A-58)$$

Using Eq. (A-23), the first and second terms on the left-hand side of Eq. (A-58) can be expanded as

$$\rho\frac{\partial Q}{\partial t} + \rho\frac{\partial}{\partial x}\left(\beta V^2 A\right)$$

$$= \rho\left\{V\frac{\partial A}{\partial t} + A\frac{\partial V}{\partial t}\right\} + \rho\left\{\beta V\frac{\partial Q}{\partial x} + \beta VA\frac{\partial V}{\partial x} + V^2 A\frac{\partial \beta}{\partial x}\right\} \qquad (A-59)$$

Substitution for $\partial Q/\partial x$ in Eq. (A-59) from Eq. (A-22) and some manipulation on the last two terms within the second pair of brackets yield

$$\rho \frac{\partial Q}{\partial t} + \rho \frac{\partial}{\partial x}\left(\beta V^2 A\right)$$

$$= \rho\left\{V\frac{\partial A}{\partial t} + A\frac{\partial V}{\partial t}\right\} + \rho\left\{-\beta V\left(\frac{\partial A}{\partial t} + q_\ell\right) + A\frac{\partial\left(\beta V^2/2\right)}{\partial x} + \frac{V^2 A}{2}\frac{\partial\beta}{\partial x}\right\} \qquad \text{(A-60)}$$

Combining Eqs. (A-58) and (A-60), dividing by ρgA, and rearranging the terms, one obtains

$$\frac{I}{g}\frac{\partial V}{\partial t} + \frac{\partial\left[\beta\left(V^2/2g\right)\right]}{\partial x} + \frac{V_\ell - \beta V}{gA}q_\ell + \frac{V^2}{2g}\frac{\partial\beta}{\partial x}$$

$$= -\cos\theta\,\frac{\partial y_{s\ell}}{\partial x} - \int_A \frac{\partial C_s'}{\partial x}dA - \int_A \frac{\partial C_I}{\partial x}dA - \frac{\partial G_0}{\partial x} - \frac{\tau_0}{\rho gR} + \frac{(\beta-1)V}{gA}\frac{\partial A}{\partial t} \qquad \text{(A-61)}$$

where $R = A/P$ =hydraulic radius. Transposing some terms in Eq. (A-61) from one side to the other, one obtains the following one-dimensional momentum equation:

$$\frac{I}{g}\frac{\partial V}{\partial t} + \frac{\partial}{\partial x}\left(\beta\frac{V^2}{2g} + y_{s\ell}\cos\theta + G_0\right) + \frac{V_\ell - \beta V}{gA}q_\ell$$

$$= -\frac{\tau_0}{\gamma R} - \frac{V^2}{2g}\frac{\partial\beta}{\partial x} + \frac{(\beta-1)V}{gA}\frac{\partial A}{\partial t} - \frac{1}{\gamma A}\int_A \frac{\partial}{\partial x}\left(C_s' + C_I\right)dA \qquad \text{(A-62)}$$

Equation (A-62) is an exact and complete one-dimensional momentum equation that is applicable to unsteady, nonuniform, spatially varied flow in a channel of arbitrary shape and alignment. The difference between Eq. (A-62) and Eq. (1-27), which was derived using the simplified approach, is that the former includes an extra term involving the correction terms C_s and C_I. These terms become negligible by making a number of physical assumptions.

For channels with gentle variations in cross section and direction, as shown in Figure A-2, it may be assumed that $v, w \ll u$, and the variation of the stress terms in the longitudinal and the lateral directions is small in comparison to that in the depth directions. With these assumptions the correction terms in Eqs. (A-38) and (A-39) reduce to

$$C_s' = C_s - \sigma_{xx}' = \sigma_{yy}' - \sigma_{yy}'\big|_{y=y_s} - \sigma_{xx}' \qquad \text{(A-63)}$$

$$C_I = -\rho\int_{y_b}^{y_s}\left(u\frac{\partial v}{\partial x}\right)_{z=z_\ell}dy - \rho\int_{z_\ell}^{z_r}\left(u\frac{\partial w}{\partial x}\right)_{y=y_{s\ell}}dz \qquad \text{(A-64)}$$

Experimental measurements show that σ'_{yy} does not vanish at the free surface. However, there exists a force of surface tension that should be included in the analysis. If it is assumed that the surface tension and normal stresses cancel each other at the free surface, then the correction term due to stresses reduces to

$$C'_s = \sigma'_{yy} - \sigma'_{xx} \tag{A-65}$$

This correction term is generally neglected. Morever, it is the variation of C'_s from section to section that comprises its contribution to Eq. (A-62) and it is negligible for above-mentioned channels. The correction term C_l in Eq. (1-45) is important when the streamline curvature in the xy- and the xz-planes is signigicant; in that case the terms $\partial v/\partial x$ and $\partial w/\partial x$ are not small. For negligible streamline curvatures, C_l is zero. These assumptions lead to a horizontal free surface in a cross section, i.e., independent of the z-coordinate $(y_{sc} = y_s)$ and in hydrostatic pressure distribution given by Eq. (A-37) as

$$p = \rho g \, \cos\theta \{ y_s(x,t) - y \}$$

Equation (A-62) with all these assumptions reduces to

$$\frac{1}{g}\frac{\partial V}{\partial t} + \frac{\partial H_\beta}{\partial x} + \frac{V_\ell - \beta V}{gA} q_\ell = -\frac{\tau_0}{\gamma R} + \frac{(\beta-1)V}{gA}\frac{\partial A}{\partial t} - \frac{V^2}{2g}\frac{\partial \beta}{\partial x} \tag{A-66}$$

where

$$H_\beta = \beta\frac{V^2}{2g} + y_s \, \cos\theta + G_0 = \text{total head}$$

Equation (A-66) is identical to Eq. (1-27).

A-3 Differential Mechanical-Energy Equation

The mechanical-energy is obtained by multiplying each of the momentum equations by the corresponding component of velocity and then by adding the three together. Similar to the continuity and momentum equations, the three-dimensional energy equation is derived first, which then is integrated over the cross section to obtain the one-dimensional mechanical-energy equation.

Three-Dimensional Equation

Multiplying Eqs. (A-32), (A-33), and (A-34) by u, v, and w, respectively, adding them, and transporting some terms, one obtains

$$\rho\frac{\partial(U^2/2)}{\partial t} + \rho u\frac{\partial(U^2/2)}{\partial x} + \rho v\frac{\partial(U^2/2)}{\partial y} + \rho w\frac{\partial(U^2/2)}{\partial z}$$

$$= -\left(u\frac{\partial p}{\partial x} + v\frac{\partial p}{\partial y} + w\frac{\partial p}{\partial z} \right) - \rho g\left(u\frac{\partial G}{\partial x} + v\frac{\partial G}{\partial y} + w\frac{\partial G}{\partial z} \right)$$

$$+ \left[u\left(\frac{\partial \sigma'_{xx}}{\partial x} + \frac{\partial \tau_{yx}}{\partial y} + \frac{\partial \tau_{zx}}{\partial z} \right) + v\left(\frac{\partial \tau_{xy}}{\partial x} + \frac{\partial \sigma'_{yy}}{\partial y} + \frac{\partial \tau_{zy}}{\partial z} \right) \right.$$

$$\left. + w\left(\frac{\partial \tau_{xz}}{\partial x} + \frac{\partial \tau_{yz}}{\partial y} + \frac{\partial \sigma'_{zz}}{\partial z} \right) \right] \tag{A-67}$$

where U is the magnitude of the velocity vector and is given by

$$U^2 = u^2 + v^2 + w^2 \tag{A-68}$$

The term inside the brackets on the right-hand side of Eq. (A-67) can be decomposed as

$$\left[u\left(\frac{\partial \sigma'_{xx}}{\partial x} + \frac{\partial \tau_{yx}}{\partial y} + \frac{\partial \tau_{zx}}{\partial z} \right) + v\left(\frac{\partial \tau_{xy}}{\partial x} + \frac{\partial \sigma'_{yy}}{\partial y} + \frac{\partial \tau_{zy}}{\partial z} \right) + w\left(\frac{\partial \tau_{xz}}{\partial x} + \frac{\partial \tau_{yz}}{\partial y} + \frac{\partial \sigma'_{zz}}{\partial z} \right) \right]$$

$$= \left[\frac{\partial}{\partial x}\left(u\sigma'_{xx} + v\tau_{xy} + w\tau_{xz} \right) + \frac{\partial}{\partial y}\left(u\tau_{yx} + v\sigma'_{yy} + w\tau_{yz} \right) \right.$$

$$+ \frac{\partial}{\partial z}\left(u\tau_{zx} + v\tau_{zy} + w\sigma'_{zz} \right) \right] - \left[\sigma'_{xx}\frac{\partial u}{\partial x} + \sigma'_{yy}\frac{\partial v}{\partial y} + \sigma'_{zz}\frac{\partial w}{\partial z} \right.$$

$$\left. + \tau_{xy}\left(\frac{\partial u}{\partial y} + \frac{\partial v}{\partial x} \right) + \tau_{yz}\left(\frac{\partial v}{\partial z} + \frac{\partial w}{\partial y} \right) + \tau_{zx}\left(\frac{\partial w}{\partial x} + \frac{\partial u}{\partial z} \right) \right] \tag{A-69}$$

Introduction of the sum of kinetic and potential energies per unit volume as

$$E = \rho U^2/2 + \rho g\, G \tag{A-70}$$

the addition of a term $\left\{ \rho g\frac{\partial G}{\partial t} + (E + p)\left(\frac{\partial u}{\partial x} + \frac{\partial v}{\partial y} + \frac{\partial w}{\partial z} \right) \right\}$

which is zero by virture of Eqs. (A-25) and (A-3), to Eq. (A-67), and the use of Eq. (A-69) yield

$$\left[\frac{\partial E}{\partial t} + \frac{\partial (uE)}{\partial x} + \frac{\partial (vE)}{\partial y} + \frac{\partial (wE)}{\partial z} \right]$$

$$+\left[\sigma'_{xx}\frac{\partial u}{\partial x}+\sigma'_{yy}\frac{\partial v}{\partial y}+\sigma'_{zz}\frac{\partial w}{\partial z}+\tau_{xy}\left(\frac{\partial u}{\partial y}+\frac{\partial v}{\partial x}\right)+\tau_{yz}\left(\frac{\partial v}{\partial z}+\frac{\partial w}{\partial y}\right)+\tau_{zx}\left(\frac{\partial w}{\partial x}+\frac{\partial u}{\partial z}\right)\right]$$

$$=\left[\frac{\partial}{\partial x}\left(u\sigma'_{xx}+v\tau_{xy}+w\tau_{xz}\right)+\frac{\partial}{\partial y}\left(u\tau_{yx}+v\sigma'_{yy}+w\tau_{yz}\right)+\frac{\partial}{\partial z}\left(u\tau_{zx}+v\tau_{zy}+w\sigma'_{zz}\right)\right]$$

$$+\left[-\frac{\partial(pu)}{\partial x}-\frac{\partial(pv)}{\partial y}-\frac{\partial(pw)}{\partial z}\right] \qquad\qquad\qquad\text{(A-71)}$$

The right-hand side of Eq. (A-71) represents the work done by the surface forces acting on the fluid element and consists of two parts: first, the work done by the viscous forces and second, the work done by the pressure forces. The work that is done by the surface forces on the fluid involves (i) changing the energy of the fluid as represented by the terms inside the first pair of brackets on the left-hand side of Eq. (A-71) and (ii) deforming the fluid as represented by the terms inside the second pair of brackets. The latter terms are customarily called the dissipation function Φ. Equation (A-71) can be written as

$$\frac{\partial E}{\partial t}+\frac{\partial[(E+p)u]}{\partial x}+\frac{\partial[(E+p)v]}{\partial y}+\frac{\partial[(E+p)w]}{\partial z}=\left[\frac{\partial}{\partial x}\left(u\sigma'_{xx}+v\tau_{xy}+w\tau_{xz}\right)\right.$$

$$+\frac{\partial}{\partial y}\left(u\tau_{yx}+v\sigma'_{yy}+w\tau_{yz}\right)+\frac{\partial}{\partial z}\left(u\tau_{zx}+v\tau_{zy}+w\sigma'_{zz}\right)\left.\right]-\Phi \qquad\text{(A-72)}$$

where

$$\Phi=\sigma'_{xx}\frac{\partial u}{\partial x}+\sigma'_{yy}\frac{\partial v}{\partial y}+\sigma'_{zz}\frac{\partial w}{\partial z}+\tau_{xy}\left(\frac{\partial u}{\partial y}+\frac{\partial v}{\partial x}\right)+\tau_{yz}\left(\frac{\partial v}{\partial z}+\frac{\partial w}{\partial y}\right)+\tau_{zx}\left(\frac{\partial w}{\partial x}+\frac{\partial u}{\partial z}\right)$$

Equation (A-72) is the three-dimensional mechanical-energy equation for incompressible flow.

One-Dimensional Equation

The one-dimensional energy equation is derived by integrating term by term Eq. (A-72) over the cross-sectional area of the flow. The integration of the terms in Eq. (A-72) can be written as

$$\int_{z_\ell}^{z_r}dz\int_{y_b}^{y_s}\frac{\partial E}{\partial t}dy=\frac{\partial}{\partial t}\int_{z_\ell}^{z_r}dz\int_{y_b}^{y_s}E\,dy-\int_{z_\ell}^{z_r}\left(E_s\frac{\partial y_s}{\partial t}-E_b\frac{\partial y_b}{\partial t}\right)dz \qquad\text{(A-73)}$$

$$\int_{z_\ell}^{z_r}dz\int_{y_b}^{y_s}\frac{\partial\{u(E+p)\}}{\partial x}dy=\frac{\partial}{\partial x}\int_{z_\ell}^{z_r}dz\int_{y_b}^{y_s}\{u(E+p)\}dy$$

$$-\int_{z_\ell}^{z_r}\left\{u_s\left(E_s+p_s\right)\frac{\partial y_s}{\partial x}-u_b\left(E_b+p_b\right)\frac{\partial y_b}{\partial x}\right\}dz \qquad \text{(A-74)}$$

$$\int_{z_\ell}^{z_r}dz\int_{y_b}^{y_s}\frac{\partial\{v(E+p)\}}{\partial y}dy=\int_{z_\ell}^{z_r}\{v_s\left(E_s+p_s\right)-v_b\left(E_b+p_b\right)\}dz \qquad \text{(A-75)}$$

$$\int_{z_\ell}^{z_r}dz\int_{y_b}^{y_s}\frac{\partial\{w(E+p)\}}{\partial z}dy$$
$$=\int_{z_\ell}^{z_r}-\left\{w_s\left(E_s+p_s\right)\frac{\partial y_s}{\partial z}-w_b\left(E_b+p_b\right)\frac{\partial y_b}{\partial z}\right\}dz \qquad \text{(A-76)}$$

$$\int_A\frac{\partial\left(u\sigma'_{xx}+v\tau_{xy}+w\tau_{xz}\right)}{\partial x}dA=\frac{\partial}{\partial x}\int_A\left(u\sigma'_{xx}+v\tau_{xy}+w\tau_{xz}\right)dA$$
$$-\int_{z_\ell}^{z_r}\left[\left(u\sigma'_{xx}+v\tau_{xy}+w\tau_{xz}\right)_s\frac{\partial y_s}{\partial x}-\left(u\sigma'_{xx}+v\tau_{xy}+w\tau_{xz}\right)_b\frac{\partial y_b}{\partial x}\right]dz \quad \text{(A-77)}$$

$$\int_A\frac{\partial\left(u\tau_{yx}+v\sigma'_{yy}+w\tau_{yz}\right)}{\partial y}dA$$
$$=\int_{z_\ell}^{z_r}\left(u\tau_{yx}+v\sigma'_{yy}+w\tau_{yz}\right)_s-\left(u\tau_{yx}+v\sigma'_{yy}+w\tau_{yz}\right)_b\,dz \qquad \text{(A-78)}$$

$$\int_A\frac{\partial\left(u\tau_{zx}+v\tau_{zy}+w\sigma'_{zz}\right)}{\partial z}dA$$
$$=\int_{z_\ell}^{z_r}-\left[\left(u\tau_{zx}+v\tau_{zy}+w\sigma'_{zz}\right)_s\frac{\partial y_s}{\partial z}-\left(u\tau_{zx}+v\tau_{zy}+w\sigma'_{zz}\right)_b\frac{\partial y_b}{\partial z}\right]dz \quad \text{(A-79)}$$

Using Eqs. (A-73)–(A-79), the integration of Eq. (A-72) can be written as

$$\frac{\partial}{\partial t}\int_{z_\ell}^{z_r}dz\int_{y_b}^{y_s}E\,dy+\frac{\partial}{\partial x}\int_{z_\ell}^{z_r}dz\int_{y_b}^{y_s}\{u(E+p)\}dy$$
$$\leftarrow--\text{I}--\rightarrow\;\;\leftarrow----\text{II}-----\rightarrow$$
$$-\int_{z_\ell}^{z_r}\left(E_s+p_s\right)\left(\frac{\partial y_s}{\partial t}+u_s\frac{\partial y_s}{\partial x}-v_s+w_s\frac{\partial y_s}{\partial z}\right)dz+\int_{z_\ell}^{z_r}p_s\frac{\partial y_s}{\partial t}dz$$
$$\leftarrow---------\text{III}----------\rightarrow\;\leftarrow--\text{IV}--\rightarrow$$

$$+ \int_{z_\ell}^{z_r} (E_b + p_b) \left(\frac{\partial y_b}{\partial t} + u_b \frac{\partial y_b}{\partial x} - v_b + w_b \frac{\partial y_b}{\partial z} \right) dz + \int_{z_\ell}^{z_r} p_b \frac{\partial y_b}{\partial t} dz$$

$$\leftarrow - - - - - - - - \text{V} - - - - - - - - - \rightarrow \quad \leftarrow - - \text{VI} - - \rightarrow$$

$$- \frac{\partial}{\partial x} \int_A \left(u\sigma'_{xx} + v\tau_{xy} + w\tau_{xz} \right) dA$$

$$\leftarrow - - - - - - - \text{VII} - - - - - \rightarrow$$

$$+ \left\{ \int_{z_\ell}^{z_r} \left[\left(u\sigma'_{xx} + v\tau_{xy} + w\tau_{xz} \right)_s \frac{\partial y_s}{\partial x} - \left(u\sigma'_{xx} + v\tau_{xy} + w\tau_{xz} \right)_b \frac{\partial y_b}{\partial x} \right] dz \right.$$

$$\leftarrow -$$

$$- \int_{z_\ell}^{z_r} \left[\left(u\tau_{yx} + v\sigma'_{yy} + w\tau_{yz} \right)_s - \left(u\tau_{yx} + v\sigma'_{yy} + w\tau_{yz} \right)_b \right] dz$$

$$- - - - - - - - - - - - - \text{VIII} - - - - - - - - - - - - -$$

$$\left. - \int_{z_\ell}^{z_r} \left[\left(u\tau_{zx} + v\tau_{zy} + w\sigma'_{zz} \right)_s \frac{\partial y_s}{\partial z} - \left(u\tau_{zx} + v\tau_{zy} + w\sigma'_{zz} \right)_b \frac{\partial y_b}{\partial z} \right] dz \right\}$$

$$- \rightarrow$$

$$= - \int_A \Phi \, dA \tag{A-80}$$

Consider the terms on the left-hand side of Eq. (A-80). The third term by virtue of Eq. (A-19) is zero. The fourth term is zero on the assumption that the pressure at the free surface is zero. The expression inside the large parentheses of the fifth term by virtue of Eq. (A-20) is equal to $W_b R_b$ and the sixth term is zero if one assumes the channel boundaries to be fixed, i.e., $\partial y_b / \partial t = 0$. On the assumption that the rate of change with respect to x of the work done by the internal stresses over cross section A is negligible, the seventh term is zero. Because on the free surface shear stress is usually negligible unless there is strong wind, and on the bottom the velocity components are either zero or small if there is seepage, the eighth term can be neglected. The integration of Eq. (A-72) reduces to

$$\frac{\partial}{\partial t} \int_{z_\ell}^{z_r} dz \int_{y_b}^{y_s} E \, dy + \frac{\partial}{\partial x} \int_{z_\ell}^{z_r} dz \int_{y_b}^{y_s} \{ u(E + p) \} dy + \int_{z_\ell}^{z_r} (E_b + p_b) W_b R_b dz$$

$$= - \int_A \Phi \, dA \tag{A-81}$$

Introducing Eqs. (A-70) and (A-25), the first term on the left-hand side of Eq. (A-81) can be written as

$$\frac{\partial}{\partial t} \int_{z_\ell}^{z_r} dz \int_{y_b}^{y_s} \left\{ \frac{\rho U^2}{2} + \rho g (G_0 + y \cos\theta) \right\} dy$$

$$= \frac{\partial}{\partial t} \left(\bar{\beta} \frac{\rho}{2} V^2 A \right) + \frac{\partial}{\partial t} \left\{ \rho g \left(G_0 + \bar{\bar{y}} \cos\theta \right) A \right\} \tag{A-82}$$

where
$$\bar{\beta} = \frac{\int_A U^2 dA}{V^2 A}; \qquad \bar{\bar{y}} = \frac{\int_A y \, dA}{A} \tag{A-83}$$

Introducing Eqs. (A-70), (A-25), and (A-37), the second and third terms on the left-hand side of Eq. (A-81) can be respectively written as

$$\frac{\partial}{\partial x} \int_A \left[\frac{\rho}{2} U^2 + \rho g (G_0 + y \cos\theta) + \{ \rho g \cos\theta (y_{s\ell} - y) + C_s + C_I \} \right] u \, dA$$

$$= \frac{\partial}{\partial x} \left\{ \frac{\rho}{2} \int_A u U^2 dA + \rho g G_0 Q \right\} + \frac{\partial}{\partial x} \int_A (\rho g \cos\theta \, y_{s\ell} + C_s + C_I) u \, dA$$

$$= \frac{\partial}{\partial x} \left\{ \rho g Q \left(\alpha \frac{V^2}{2g} + G_0 + y_{s\ell} \cos\theta \right) \right\} + \frac{\partial}{\partial x} \int_A (C_s + C_I) u \, dA \tag{A-84}$$

where
$$\alpha = \frac{\int_A U^2 u \, dA}{V^3 A} = \text{energy coefficient} \tag{A-85}$$

and

$$\int_{z_\ell}^{z_r} \left\{ \rho g \left(\frac{U_b^2}{2g} + (G_0 + y_b \cos\theta) \right) + \rho g \cos\theta \{ (y_{s\ell} - y_b) + C_{sb} + C_{Ib} \} \right\} W_b R_b \, dz$$

$$= \rho g \int_{z_\ell}^{z_r} \left(\frac{U_b^2}{2g} + G_0 + y_{s\ell} \cos\theta \right) W_b R_b \, dz + \rho g \int_{z_\ell}^{z_r} (C_{sb} + C_{Ib}) W_b R_b \, dz$$

$$= \rho g \left(\frac{\overline{U_\ell^2}}{2g} + G_0 + y_{s\ell} \cos\theta \right) q_\ell + \rho g \int_{z_\ell}^{z_r} (C_{sb} + C_{Ib}) W_b R_b \, dz \tag{A-86}$$

where
$$\overline{U_\ell^2} = \frac{\int_{z_\ell}^{z_r} U_b^2 W_b R_b \, dz}{q_\ell} \tag{A-87}$$

The term on the right-hand side of Eq. (A-81) denotes the rate of energy dissipation per unit length of channel, $\bar{\varepsilon}$. Combining Eqs. (A-82), (A-84), and (A-86), Eq. (A-81) becomes

$$\frac{\partial}{\partial t}\left(\bar{\beta}\frac{\rho}{2}V^2A\right)+\frac{\partial}{\partial t}\left\{\rho g\left(G_0+\bar{\bar{y}}\cos\theta\right)A\right\}+\frac{\partial}{\partial x}\left\{\rho g Q\left(\alpha\frac{V^2}{2g}+G_0+y_{s\ell}\cos\theta\right)\right\}$$

$$=-\frac{\partial}{\partial x}\int_A\left(C_s+C_I\right)u\,dA-\rho g\left(\frac{\overline{U_\ell^2}}{2g}+G_0+y_{s\ell}\cos\theta\right)q_\ell$$

$$-\rho g\int_{z_\ell}^{z_r}\left(C_{sb}+C_{Ib}\right)W_b R_b\,dz-\bar{\varepsilon} \tag{A-88}$$

The left-hand side (LHS) of Eq. (A-88) can be expanded on using Eq. (A-23) as

$$\text{LHS}=\left\{\rho\bar{\beta}\frac{V^2}{2}\frac{\partial A}{\partial t}+\frac{\rho}{2}VQ\frac{\partial\bar{\beta}}{\partial t}+\rho\bar{\beta}Q\frac{\partial V}{\partial t}\right\}+\left\{\rho g\left(G_0\frac{\partial A}{\partial t}+\cos\theta\frac{\partial\left(\bar{\bar{y}}A\right)}{\partial t}\right)\right\}$$

$$+\rho g\left\{Q\frac{\partial H_\alpha}{\partial x}+\left(\alpha\frac{V^2}{2g}+G_0+y_{s\ell}\,cos\theta\right)\frac{\partial Q}{\partial x}\right\}$$

$$=\rho g\left(\bar{\beta}-\alpha\right)\frac{V^2}{2g}\frac{\partial A}{\partial t}+\rho\alpha\frac{V^2}{2}\left(\frac{\partial A}{\partial t}+\frac{\partial Q}{\partial x}\right)+\rho g G_0\left(\frac{\partial A}{\partial t}+\frac{\partial Q}{\partial x}\right)$$

$$+\rho g y_s\cos\theta\left(\frac{1}{y_s}\frac{\partial\left(\bar{\bar{y}}A\right)}{\partial t}+\frac{\partial Q}{\partial x}\right)$$

$$+\rho g Q\frac{\partial H_\alpha}{\partial x}+\frac{\rho}{2}VQ\frac{\partial\bar{\beta}}{\partial t}+\rho Q\bar{\beta}\frac{\partial V}{\partial t} \tag{A-89}$$

where
$$H_\alpha=\alpha\frac{V^2}{2g}+y_{s\ell}\cos\theta+G_0=\text{total head} \tag{A-90}$$

and a term
$$\rho\alpha\frac{V^2}{2}\frac{\partial A}{\partial t}$$

has been simultaneously added to and subtracted from the equation. Using the continuity equation (A-22) and the equality

$$d\left(\overline{\overline{y}}\,A\right) = y_s\,dA$$

Eq. (A-89) becomes

$$\text{LHS} = \rho g\left(\beta - \alpha\right)\frac{V^2}{2g}\frac{\partial A}{\partial t} - \rho g H_\alpha q_\ell + \rho g Q\frac{\partial H_\alpha}{\partial x} + \rho Q\overline{\beta}\frac{\partial V}{\partial t} + \frac{\rho}{2}VQ\frac{\partial\overline{\beta}}{\partial t} \quad \text{(A-91)}$$

Substitution of Eq. (A-91) into Eq. (A-88) and division of the resulting equation by $\rho g Q$ yields the following one-dimensional energy equation:

$$\frac{\overline{\beta}}{g}\frac{\partial V}{\partial t} + \frac{\partial H_\alpha}{\partial x} + \frac{\overline{U_\ell^2} - \alpha V^2}{2gQ}q_\ell = -\frac{\overline{\varepsilon}}{\gamma Q} + \left(\alpha - \beta\right)\frac{V}{2gA}\frac{\partial A}{\partial t}$$

$$-\frac{V}{2g}\frac{\partial\overline{\beta}}{\partial t} - \frac{1}{\gamma Q}\frac{\partial}{\partial x}\int_A \left(C_s + C_I\right)u\,dA - \frac{1}{\gamma Q}\int_{z_\ell}^{z_r}\left(C_{sb} + C_{Ib}\right)W_b R_b\,dz \quad \text{(A-92)}$$

The difference between Eq. (A-92) and Eq. (1-40) derived by the simplified approach is that the former equation includes some extra terms. If one makes same assumptions that were used in simplifying the momentum equation, including the assumption of hydrostatic pressure distribution, Eq. (A-92) simplifies to

$$\frac{\beta}{g}\frac{\partial V}{\partial t} + \frac{\partial H_\alpha}{\partial x} + \frac{\overline{U_\ell^2} - \alpha V^2}{2gQ}q_\ell = -\frac{\overline{\varepsilon}}{\gamma Q} + \left(\alpha - \beta\right)\frac{V}{2gA}\frac{\partial A}{\partial t} - \frac{V}{2g}\frac{\partial\beta}{\partial t}$$

Equation (A-93) is identical to Eq. (1-40).

CHAPTER 2

STEADY UNIFORM FLOW

It is logical to solve open-channel problems starting with the simplest type of flow. The one-dimensional governing equations derived in the previous chapter reduce to the simplest forms for steady, uniform flow that is discussed in this chapter.

2-1 Governing Equations

Flow in open channel is steady if V and y do not change with time, and it is uniform if V and y do not vary with the longitudinal distance. The governing equations for steady uniform flow are derived in Section 1-8 wherein it is shown that the momentum and mechanical-energy equations for uniform flow are identical. The governing equations for uniform flow are as follows:

Continuity equation: $$Q = VA \quad [\text{Eq.}(1\text{-}44)]$$

Momentum equation: $$S_0 = \frac{\tau_0}{\gamma R} \quad [\text{Eq.}(1\text{-}45)]$$

If one is able to express wall shear stress in Eq. (1-45) in terms of variables of interest, then the resulting equation along with the continuity equation (1-44) can be used to determine the flow condition (V and y) in channels. Theoretical expressions for wall shear stress are available only for laminar flow but are not of much practical interest. Empirical expressions, similar to the Moody diagram for turbulent pipe flow, are used for turbulent open-channel flow.

2-2 Open-Channel Resistance

Experience indicates that wall shear stress should be a function of the following independent variables: hydraulic radius, R; mean flow velocity, V; surface roughness described by a length, k; a nondimensional factor, ζ, describing the shape of the channel cross section; fluid density, ρ; fluid specific weight, γ; and fluid viscosity, μ:

$$\tau_0 = f(R, V, k, \zeta, \rho, \gamma, \mu) \tag{2-1}$$

Additional variables, such as describing the channel profile and plan, can be included in Eq. (2-1) (Rouse, 1965). If R, V, and ρ are chosen as the repeating variables to appear in combination with each of the remaining variables, Eq. (2-1) can be written by dimensional analysis in the following nondimensional form:

$$\frac{\tau_o}{\rho V^2} = \psi\left(\frac{k}{R}, \zeta, \frac{VR\rho}{\mu}, \frac{V}{\sqrt{R\gamma/\rho}}\right) \qquad (2\text{-}2)$$

The function ψ depends on four nondimensional terms within the parentheses; these terms respectively are the relative roughness, the shape factor, the Reynolds number R, and the Froude number F (note that $\gamma/\rho = g$). Elimination of shear stress between Eqs. (1-45) and (2-2) yields

$$V = \sqrt{\frac{g}{\psi}}\sqrt{RS_0} \qquad (2\text{-}3)$$

Replacing $\sqrt{g/\psi}$ by a single variable C, one obtains the classical Chézy formula

$$V = C\sqrt{RS_0} \qquad (2\text{-}4)$$

Note that the Chézy coefficient C, like ψ, is not dimensionless but has the dimension $L^{1/2}\,T^{-1}$. The Chézy equation is similar to the Darcy-Weisbach equation for pipe flow given by

$$h_f = f\,\frac{L}{D}\,\frac{V^2}{2g} \qquad (2\text{-}5)$$

in which h_f is the energy loss due to boundary resistance in a circular pipe of length L and diameter D and f is the Darcy-Weisbach friction factor. Replacing D by $4R$ ($R = A/P = D/4$) and h_f/L by the energy slope S_e, Eq. (2-5) can be written as

$$V = \sqrt{\frac{8g}{f}}\sqrt{RS_e} \qquad (2\text{-}6)$$

Knowing that $S_0 = S_e$ for uniform flow, a comparison of Eqs. (2-4) and (2-6) gives

$$C = \sqrt{\frac{8g}{f}} \qquad (2\text{-}7)$$

The variation of f in circular pipes with the relative roughness k/D and the Reynolds number R is well known and is represented by the Moody diagram. The variation of C with k/D and R can be derived from the Moody diagram according to Eq. (2-7) and is shown schematically in Figure 2-1. The turbulent-flow region ($R \geq 2000$) is divided into three zones (smooth, transition, and rough) depending on the thickness of the laminar sublayer that decreases with increasing Reynolds

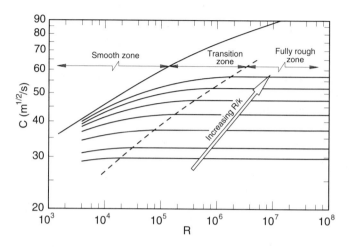

Figure 2-1. Behavior of the Chézy C.

number. In the smooth zone the boundary roughness is buried within the laminar sublayer and therefore does not affect the flow behavior; consequently C does not depend on the relative roughness and is a function of R only. In the rough zone the laminar sublayer is so thin that additional decrease in its thickness with increased Reynolds number does not affect the flow; consequently C is independent of R and is a function of the relative roughness only. Between the smooth and rough zones there exists a transition zone in which C is a function of both. The limits of the transition zone are given as

$$4 < \frac{V_* k}{\nu} < 100 \qquad (2\text{-}8)$$

in which $V_* = \sqrt{\tau_0 / \rho}$ = shear velocity, and $\nu = \mu/\rho$ = fluid kinematic viscosity. For $(V_* k/\nu) < 4$, the flow is in the smooth zone, and the flow is in the rough zone if $(V_* k/\nu) > 100$.

Using hydraulic radius R as the characteristic length as in Eq. (2-2), experimental results on friction factor for turbulent flow through noncircular pipes (Schlichting, 1968) are well represented by the law for circular pipes. In the laminar region, however, friction factor depends on the shape of the conduit. Recent experimental data in smooth rectangular channels (Kazemipour and Aplet, 1982) also show some shape effects on friction factor. For most practical problems in open channels the flow is in the turbulent rough region where the effect on C of the shape factor ζ can be neglected (ASCE, 1963).

There are not sufficient data available to delineate the effect of the Froude number on channel resistance. The Chézy coefficient C is likely to increase with increasing Froude number F due to the formation of surface disturbances, such as roll waves. The effect of roll waves on channel resistance is described by Rouse (1966).

In summary, for most practical problems the Chézy coefficient can be considered a function of relative roughness only. An ASCE (1963) task committee recommended the following equation for C:

$$\frac{C}{\sqrt{8g}} = 2 \log_{10}\left(\frac{12R}{k}\right) \tag{2-9}$$

The values of k for some materials are given in Table 2-1.

Table 2-1. Approximate Values of Roughness Height

Material	k (mm)
Steel	0.05 – 0.2
Cast iron	0.25 – 1.0
Cement	0.30 – 1.2
Concrete	0.50 – 3.0

Manning Equation

There are several empirical relations for C, other than Eq. (2-9), that have been used in practice. The most often used relation is

$$C = \Pi \frac{R^{1/6}}{n} \tag{2-10}$$

which after substitution for C into Eq. (2-4) yields the well-known Manning equation (also known as Strickler's equation in Europe) as

$$V = \frac{\Pi}{n} R^{2/3} S_0^{1/2} \tag{2-11}$$

where n is the Manning roughness coefficient that is a function of surface roughness and Π is a dimensional constant whose magnitude depends on the units of length and time in the equation. The dimensions of Π can be determined from Eq. (2-10) for known dimensions of n. The dimensions of n, as shown later, should be $L^{1/6}$, but it is customary to consider the Manning n as nondimensional, as given in Table 2-2 for various types of channels. From Eq. (2-10) the dimensions of Π are then $L^{1/3} T^{-1}$. The magnitude of Π is unity in the international system (SI) of units. The magnitude of Π in English system of units is, therefore, equal to

Table 2-2. Values of the Manning Roughness Coefficient n

Lined channels	
Cement plaster	0.011
Concrete, trawled	0.012
Concrete, unfinished	0.015
Rubble set in cement	0.017
Unlined nonerodible channels	
Earth, clean	0.016
Earth, gravel	0.022
Natural channels	
Earth, smooth, no weeds	0.020
Earth, some stones and weed	0.025
Gravel beds, straight	0.025
Gravel beds with boulders	0.040
Earth, winding	0.050

$\sqrt[3]{3.281}$, or 1.486, because 1 m = 3.281 ft. Thus, the Manning equation is

$$V = \frac{1}{n} R^{2/3} S_0^{1/2} \quad \text{(SI units)} \tag{2-11a}$$

$$V = \frac{1.486}{n} R^{2/3} S_0^{1/2} \quad \text{(English system of units)} \tag{2-11b}$$

It is evident from Eq. (2-10) that the Manning equation is applicable only in the rough zone where C is a function of surface roughness only. Using $\Pi = 1$ in Eq. (2-10), a comparison of Eqs. (2-9) and (2-10) shows that

$$\frac{R^{1/6}}{n} = 2\sqrt{8g}\, \log_{10}\left(\frac{12R}{k}\right) \tag{2-12}$$

A plot of Eq. (2-12) is shown in Figure 2-2. In the intermediate range of R/k the plotted curve can be approximated by a straight line with a 6:1 slope. According to this approximate relation, $n \propto k^{1/6}$, i.e., Manning n varies as one-sixth power of the roughness height; consequently the dimensions of n should be $L^{1/6}$. A similar relation for gravel-bed rivers, $n = 0.0417k^{1/6}$ (k is measured in meters), was proposed by Strickler who used the median size of the bed material as the roughness height. Using Eq. (2-8) and the Strickler relation, it can be shown (Prob. 2-2) that the Manning equation is applicable if

$$\frac{n^6 \sqrt{gRS_0}}{v} > 5.26 \times 10^{-7}$$

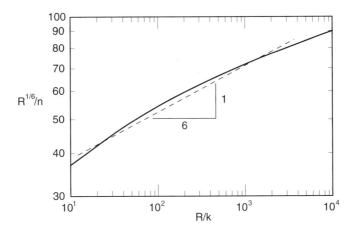

Figure 2-2. Variation of n with k.

In addition to Table 2-2, another excellent source for the selection of n values is a U. S. Geological Survey report (Barnes, 1967) that includes photographs of a number of typical canals and rivers along with brief descriptions of channel conditions and n values. In order to select the proper value of the roughness coefficient, one must know the following various factors that affect its value:

1. *Surface roughness:* It is characterized by the size and shape of the grains forming wetted perimeter.
2. *Vegetation:* It may be regarded as a kind of surface roughness. Its effect depends upon the height, density, distribution, and type of vegetation.
3. *Channel irregularity:* It comprises variation in size and shape of the cross sections along the channel length and irregularity in wetted perimeter. Uniform change in section does not appreciably affect the value of the roughness coefficient, but abrupt changes increase its value.
4. *Channel alignment:* For smooth curvature with large radius the effect on the roughness coefficient is small. Its value increases with sharp curvature.
5. *Silting and scouring:* Silting may change an irregular channel into a comparatively uniform channel and decrease the value of the roughness coefficient. Scouring may do the reverse.
6. *Obstruction:* Presence of, e.g., bridge piers tends to increase the value of the roughness coefficient.
7. *Stage and discharge:* The roughness coefficient decreases with increase in stage and discharge.
8. *Suspended and bed load:* These increase the roughness coefficient.

In summary n increases with increasing surface roughness, vegetation growth in the channel, and channel irregularity, as one would expect.

PROBLEMS

2-1   Assuming that the same values of the Manning n are used on the moon and the earth, derive the Manning equation that would be applicable on the moon (if the moon had water) where the gravity is g'.

2-2   Using Eq. (2-8) and the Srickler relation, $n = 0.0417k^{1/6}$, show that the Manning equation is applicable if

$$\frac{n^6 \sqrt{gRS_0}}{\nu} > 5.26x10^{-7}$$

Note that $V_* = \sqrt{gRS_0}$.

2-3   Is uniform flow possible in (a) a frictionless channel; (b) a horizontal channel? Explain.

2-4   For a given discharge in a channel, and the given bottom slope, roughness, and cross-sectional shape of the channel, would you expect the flow depth in the channel to be higher on the earth or the moon and why?

2-5   A rectangular channel has width 2 m, depth 0.6 m, and bottom slope 0.0005. The channel is lined with concrete with roughness height $k = 3.0$ mm. Estimate (a) the Chézy coefficient C, (b) the Manning coefficient n, and (c) the discharge in the channel.

2-3 Normal Depth

When flow is uniform, the flow depth and velocity in the channel are called normal depth, y_n, and normal velocity, V_n, respectively. For a given discharge in a channel of known cross-sectional shape, bottom slope and Manning roughness coefficient, y_n and V_n, can be determined from the Manning equation (2-11) and the continuity equation (1-44). Substitution for V from Eq. (2-11a) into Eq. (1-44) yields

$$Q = \frac{1}{n} AR^{2/3} S_0^{1/2} \tag{2-13}$$

or
$$AR^{2/3} = \frac{nQ}{S_0^{1/2}} \tag{2-14}$$

In channels, such as circular and trapezoidal channels (rectangular and triangular channels are the particular cases of a trapezoidal channel), for which A and R can be expressed algebraically in terms of y_n (see Figure 2-3), Eq. (2-14) can be solved for y_n. For example, consider a trapezoidal channel of bottom width b and side slopes of $m:1$ (m horizontal and 1 vertical). Then

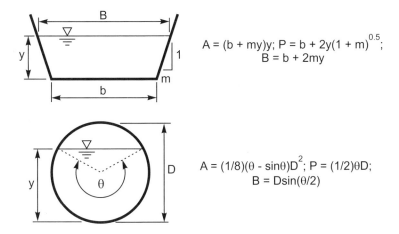

Figure 2-3. Geometric elements of trapezoidal and circular sections.

$$A = (b + my_n)y_n; \quad P = b + 2y_n\sqrt{1+m^2}$$

$$R = \frac{(b + my_n)y_n}{b + 2y_n\sqrt{1+m^2}} \tag{2-15}$$

The resulting equation is a nonlinear algebraic equation that can be solved for y_n by a trial and error method, by using design charts, or by numerical methods. For developing design charts, Eq. (2-14) is written in nondimensional form as

$$\frac{nQ}{L^{8/3}S_0^{1/2}} = \frac{AR^{2/3}}{L^{8/3}} \tag{2-16}$$

where L is a characteristic length and is taken as the diameter D for circular channels and the bottom width b for trapezoidal channels. The right-hand side of Eq. (2-16) is a function of y_n/L; the values of the function for trapezoidal and circular channels are tabulated in Appendix B and can be used to determine y_n. Numerical methods, such as the Newton-Raphson method or the interval-halving method, can be used to solve Eq. (2-14).

In natural channels where it is not convenient to express algebraically A and R in term of y_n, Eq. (2-13) is solved graphically for y_n. Equation (2-13) may be written as

$$Q = K_n\sqrt{S_0} \tag{2-17}$$

in which K_n is the conveyance of the channel at normal depth. The conveyance K is defined as

$$K = \frac{1}{n}AR^{2/3} \tag{2-18}$$

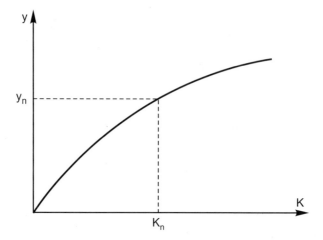

Figure 2-4. The conveyance curve.

For a given channel K is a function of y; the function can be plotted graphically from the known values of A and R for a range of values of y, as shown in Figure 2-4. The normal depth can be obtained from the plot, known as the conveyance curve, for a known value of $K = K_n = Q / \sqrt{S_0}$.

Compound Channels

Natural channels subject to seasonal floods consist of a main channel and two overbank channels, as shown in Figure 2-5. Generally overbank channels are rougher than the main channel, and the average velocities in overbank channels are smaller than that in the main channel. The normal depth in a channel of a compound section is determined from the Manning equation as follows. The channel section is divided into subsections by fictitiously extending the main-channel boundary to the water surface (see Figure 2-5). The shear stress along the

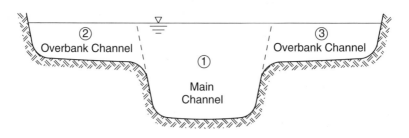

Figure 2-5. A compound channel.

fictitious boundaries is assumed zero; consequently the fictitious boundaries do not contribute to the wetted perimeter of the subsections. The discharge Q_i through the ith subsection is computed from the Manning equation (2-17) and is given by

$$Q_i = K_{ni}\sqrt{S_0} \qquad (2\text{-}19)$$

where K_{ni} is the conveyance of the ith subsection and is given by

$$K_{ni} = \frac{1}{n_i} A_i R_i^{2/3} \qquad (2\text{-}20)$$

where n_i, A_i, and R_i are the Manning roughness coefficient, area, and hydraulic radius of the ith subsection, respectively. For a given value of Manning n and a shape of channel section, the conveyance of an ith subsection K_{ni} is a function of flow depth. Because the total discharge Q in the whole section is equal to the sum of discharges Q_i in the subsections, the conveyance K_n of the whole section can be obtained from

$$K_n = \frac{Q}{\sqrt{S_0}} = \frac{\sum_{i=1}^{N} K_{ni}\sqrt{S_0}}{\sqrt{S_0}} = \sum_{i=1}^{N} K_{ni} \qquad (2\text{-}21)$$

where N is the number of subsections. Conveyance K is a function of y; the function can be determined from the known values of A_i and R_i for a range of y. Normal depth can be obtained from the conveyance function.

Example 2-1. Determine the normal depth for a discharge of 90 m³/s in a trapezoidal channel having a bottom width of 5 m, side slopes of 1 horizontal to 1 vertical (1H:1V), and a bottom slope of 0.0009. The Manning n for the channel is 0.013.

The normal depth will be computed by the trial-and-error method and by the design charts.

(a)   Trial-and-error method: From Eq. (2-15)

$$A = (b + my_n)y_n = (5 + 1 \times y_n)y_n = (5 + y_n)y_n$$

$$P = b + 2y_n\sqrt{(1 + m^2)} = 5 + 2y_n\sqrt{(1 + 1^2)} = 5 + 2\sqrt{2}\,y_n$$

$$R = \frac{A}{P} = \frac{(5 + y_n)y_n}{5 + 2\sqrt{2}\,y_n}$$

Substitution for A and R in Eq. (2-13) gives

$$90 = \frac{1}{0.013}\{(5 + y_n)y_n\}\left\{\frac{(5 + y_n)y_n}{5 + 2\sqrt{2}\,y_n}\right\}^{2/3}\sqrt{0.0009}$$

or
$$\{(5 + y_n)y_n\}^{5/3} - 39\times\left(5 + 2\sqrt{2}\,y_n\right)^{2/3} = 0$$

The solution of this equation by the trial-and-error method is

$$y_n = 3.17 \text{ m}$$

(b) Design charts:

$$\frac{nQ}{b^{8/3}S_o^{1/2}} = \frac{0.013\,x\,90}{5^{8/3}\times\sqrt{0.0009}} = 0.533$$

Upon entering Table B-1 with the above value and $m = 1$, one obtains

$$\frac{y_n}{b} = 0.634$$

whence
$$y_n = 0.634\times5 = 3.17 \text{ m}$$

PROBLEMS

2-6   Determine the discharge for a normal depth of 1.5 m in a trapezoidal channel having a bottom width of 2 m and side slopes of 2H:1V. The bottom slope is 0.0015 and the Manning n is 0.015.

2-7   Determine normal depth to carry 10 m³/s in the channel in Problem 2-6.

2-8   Find the necessary gradient for the channel in Problem 2-6 to deliver 20 m³/s at a normal depth of 1.2 m.

2-9   At a normal depth of 1.5 m 40 m³/s of water flows in the channel in Problem 2-6. Determine the Manning n.

2-10   Determine discharge in the canal, shown in the figure, if the slope is 0.001. The side slopes of the lower part of the section is 1H:1V.

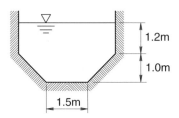

Problems 2-10 and 2-11 ($n=0.013$)

2-11   Plot the conveyance curve for the channel shown in the figure; the side slopes of the lower portion of the section are 1H:1V. Determine the normal depth for a discharge of 10 m³/s and bottom slope of 0.002.

2-12   A rectangular channel 6 m in width carries a discharge of 10 m³/s. The slope of the channel bottom is 0.0006 and the Manning n is 0.02. Compute the normal depth.

2-13   A sewer pipe is laid on a slope of 0.0004 and carries a discharge of 1.5 m³/s. What size of a circular pipe should be used if the pipe flows half full? The Manning n for the pipe is 0.015.

2-14   In a compound channel it is best to compute the discharges through different subsections independently and add them to get the total discharge. The cross section of a compound channel is approximated by two rectangular subsections that are placed side by side to each other; one, 3 m wide and 3 m deep; and the other, 6 m wide and 1 m deep. Assuming a bottom slope of 0.001 and a Manning n of 0.015, determine the difference in total discharges computed for the channel (a) by using the entire cross section as a single unit and (b) by adding the discharges in the two different subsections.

2-15   Using the Manning formula, show that the depth and velocity for a maximum discharge in a circular channel of diameter D are $0.938D$ and $0.81D$, respectively.

2-16   A 4-m-diameter circular culvert ($n = 0.013$) is to carry a discharge of 15 m³/s on a slope of 0.0009. Determine the flow depth in the culvert.

2-17   The following flow conditions exist at the two river sections A and B, 500 m apart:

Section	Flow area m²	Wetted perimeter m	Water-surface level m
A	250	150	43.10
B	275	175	42.72

Calculate the average value of the Manning n for the reach if the discharge is 500 m³/s.

2-18   A natural river section consists of a main channel and two overbank channels, similar to that shown in Figure 2-5. The main and overbank channels can be approximated by rectangular sections. The widths and the Manning roughness coefficients for the overbank channels and the main channels are $b_o = 3$ m; $n_o = 0.018$; $b_m = 1$ m; and $n_m = 0.013$, where the subscripts o and m denote the overbank channels and the main channel, respectively. The main channel is 1 m deeper than the overbank channels. Determine the normal depth in the natural channel if it carries a discharge of 2.5 m³/s at a channel slope of 0.0015.

2-4 Equivalent Roughness

In some channels, as in compound channels described above, different portions of the channel perimeter may have different roughness. Other examples are laboratory flumes with glass walls and a concrete bottom and canals with lined sides and a sand bed. The application of the Manning formula to such channels requires an *equivalent n value* for the entire channel. Based on certain assumptions, different expressions for the equivalent n can be derived. Assuming that the total discharge of the flow is equal to the sum of discharges of the subdivided areas, an expression for the equivalent n can be determined from Eq. (2-21) as

$$n = \frac{PR^{5/3}}{\sum\limits_{i=1}^{N}\left(P_i R_i^{5/3}/n_i\right)} \tag{2-22}$$

For simple channels it may be assumed that $R_1 = R_2 = \cdots = R_N = R$, and from Eq. (2-22) the equivalent n for simple channels is given by

$$n = \frac{P}{\sum\limits_{i=1}^{N}\left(P_i/n_i\right)} \tag{2-23}$$

It can be easily shown (Prob. 2.19) that if one assumes that each subdivided area has the same mean velocity, the equivalent n is given by

$$n = \frac{\left[\sum\limits_{i=1}^{N}\left(P_i n_i^{3/2}\right)\right]^{2/3}}{P^{2/3}} \tag{2-24}$$

PROBLEMS

2-19   Derive Eq. (2-24).

2-20   A rectangular channel is 2 m wide and is laid on a slope of 0.001. The channel bed and walls were made smooth by cement, and the measured flow depth was 1.36 m for a discharge of 6.0 m³/s. The channel bed and walls were then roughened by sand grains and the measured flow depth for a discharge of 3.5 m³/s was 1.30 m. Determine the discharge in the channel for a normal depth of 1.4 m when only the bed was roughened and walls were kept smooth.

2-21   A rectangular channel is 1.0 m wide and is laid on a slope of 0.0001. The bed of the channel is rough with $n = 0.025$, and the sides are smooth with $n = 0.015$. Find the discharge for a flow depth of 0.6 m.

2-5 Best Hydraulic Section

It is obvious from Eq. (2-13) that, for fixed values of A, n, and S_0, the discharge Q in a channel is a maximum when the wetted perimeter P is a minimum. The best hydraulic section is the channel proportions that yield a minimum wetted perimeter for a given flow area. It can be shown that the best hydraulic section has a depth of flow equal to one-half of the channel width (the shape of a half square) for rectangular channels (Prob. 2-22) and has the shape of a half hexagon for trapezoidal channels as shown below. The ideal shape of the best hydraulic section is the half circle.

Trapezoidal Section

For a trapezoidal channel section having a base width b and side slopes $m:1$, the wetted perimeter P and cross-sectional area A from Eq. (2-15) are

$$P = b + 2y\sqrt{1+m^2} \tag{2-25}$$

$$A = (b + my)y \tag{2-26}$$

From Eq. (2-26), $\qquad\qquad\qquad b = \dfrac{A}{y} - my \tag{2-27}$

Substitution for b from Eq. (2-27) into Eq. (2-25) yields

$$P = \frac{A}{y} - my + 2y\sqrt{1+m^2} \tag{2-28}$$

For a given area A, P is a function of y and m. For the wetted perimeter to be a minimum requires that $\partial P/\partial y = 0$ and $\partial P/\partial m = 0$. Differentiation of P with respect to y gives

$$\frac{\partial P}{\partial y} = -\frac{A}{y^2} - m + 2\sqrt{1+m^2} = 0 \tag{2-29}$$

Substitution for A from Eq. (2-26) into Eq. (2-29) and simplification yield

$$b = 2y\left(\sqrt{1+m^2} - m\right) \tag{2-30}$$

Differentiation of P in Eq. (2-28) with respect to m gives

$$\frac{\partial P}{\partial z} = -y + 2y \times \frac{1}{2} \times \left(1+m^2\right)^{-1/2} \times 2m = 0$$

or $\qquad\qquad\qquad\qquad m = \dfrac{1}{\sqrt{3}} \tag{2-31}$

Substitution for m into Eq. (2-30) gives

$$b = \frac{2}{\sqrt{3}} y \quad \text{or} \quad y = \frac{\sqrt{3}}{2} b \qquad (2\text{-}32)$$

Channel section is a half hexagon. A similar approach can be used to determine the best hydraulic sections for other channel shapes (Prob. 2-24).

PROBLEMS

2-22   For the best hydraulic rectangular section, show that the depth of flow should be one half of the width of the channel section.

2-23   Design the best hydraulic trapezoidal section for carrying a discharge of 10 m³/s. The channel has a slope of 0.0016 and a Manning n of 0.025.

2-24   Show that the wetted perimeter for a best hydraulic triangular section is $y/(2\sqrt{2})$.

2-6 Design of Channels

Design of a channel involves determination of the dimensions of the channel section. The channel dimensions can be determined from the continuity equation and the Manning equation. These two equations involve six variables, namely, discharge Q, slope S_0, Manning roughness coefficient n, velocity V, flow depth y, and geometric elements, such as A, P, and R, that depend on the shape of the channel section. When four of the above six variable are known, the remaining two unknowns can be obtained from the above two equations. In most problems the following three variables are known: design discharge Q; channel slope S_0 that is controlled by the topography of the region where the channel is going to be constructed; and the Manning n that depends on the material forming the channel.

If a channel is constructed in erodible material, such as irrigation canals in alluvium, maximum permissible velocity is specified to prevent channel erosion. Normally a trapezoidal section is adopted. The side slopes of the channel section that will be stable under all conditions are specified. The stable slope depends on the type of material. Approximate values of permissible velocities and side slopes for different materials are given in Tables 2-3 and 2-4, respectively. This method

Table 2-3. Maximum Permissible Velocities

Material	V (m/s)
Fine sand	0.4
Sandy loam	0.5
Silt loam	0.6
Firm loam	0.7
Firm clay	1.1
Fine gravel	0.7
Coarse gravel	1.2

Table 2-4. Side Slopes

Material	m
Sandy loam	3
Loose sandy soil	2
Firm soil	1
Firm clay	0.5 – 1
Rock	0.25

of designing a channel is known as the *permissible velocity method*. Once Q, V, n, S_0, and the basic channel shape have been specified, the channel width and depth can be determined from the continuity and Manning equations. The design procedure is illustrated by an example below. It should be mentioned that there are other methods for designing channels, such as the *tractive force method* (Chow, 1959).

Example 2-2. An unlined irrigation canal is to be constructed in a sandy loam soil to carry a discharge of 25 m³/s. The slope of the canal is to be 0.001. Design the canal section.

From Table 2-4, a side slope of 3H:1V is adopted. From Table 2-3 a maximum permissible velocity $V = 0.5$ m/s is selected. The Manning n from Table 2-2 is 0.016. Equation (2-11a) can be written for R as

$$R = \left(\frac{nV}{S_0^{1/2}} \right)^{3/2} = \left(\frac{0.016 \times 0.5}{0.0001} \right)^{3/2} = 0.72 \text{ m} \qquad (1)$$

From the continuity equation, $\quad A = \dfrac{Q}{V} = \dfrac{25}{0.5} = 50 \text{ m}^2 \qquad (2)$

The wetted perimeter P can be obtained from Eqs. (1) and (2) as

$$P = \frac{A}{R} = \frac{50}{0.72} = 69.44 \text{ m}$$

From Eq. (2-15) the area A and wetted perimeter P can be written in terms of b and y as

$$A = (b + my)y = (b + 3y)y = 50$$
$$P = b + 2y\sqrt{1 + m^2} = b + 2y\sqrt{1 + 3^2} = b + 6.32y = 69.44$$

Solution of the above two equations for b and y yields

$$b = 64.70 \text{ m and } y = 0.75 \text{ m}$$

PROBLEMS

2-25　A trapezoidal canal is to be designed for a discharge of 15 m^3/s and it will have slope of 0.002. The canal is to be excavated in firm soil and lined with fine gravel. Choose a canal cross section.

2-26　Determine the cross section of an unlined irrigation canal to carry 5 m^3/s at a slope of 0.0009. The canal is to be excavated in firm clay.

APPENDIX B

DESIGN CHARTS FOR NORMAL DEPTH

Table B-1. Normal Depth in Trapezoidal Channels

y_n/b	$nQ/(b^{8/3} S_0^{1/2})$					y_n/b	$nQ/(b^{8/3} S_0^{1/2})$				
	$m=0$	$m=0.5$	$m=1.0$	$m=1.5$	$m=2.0$		$m=0$	$m=0.5$	$m=1.0$	$m=1.5$	$m=2.0$
0.10	0.019	0.020	0.021	0.022	0.023	0.62	0.263	0.396	0.513	0.617	0.713
0.11	0.022	0.024	0.025	0.026	0.027	0.64	0.274	0.418	0.544	0.657	0.763
0.12	0.025	0.027	0.029	0.030	0.031	0.66	0.285	0.440	0.577	0.699	0.814
0.13	0.029	0.031	0.033	0.035	0.036	0.68	0.297	0.462	0.611	0.743	0.867
0.14	0.032	0.035	0.038	0.039	0.041	0.70	0.308	0.486	0.645	0.788	0.922
0.15	0.036	0.039	0.042	0.045	0.047	0.72	0.319	0.509	0.681	0.835	0.979
0.16	0.039	0.044	0.047	0.050	0.052	0.74	0.330	0.534	0.718	0.884	1.039
0.17	0.043	0.048	0.052	0.055	0.058	0.76	0.342	0.558	0.756	0.933	1.100
0.18	0.047	0.053	0.057	0.061	0.065	0.78	0.353	0.584	0.795	0.985	1.164
0.19	0.051	0.058	0.063	0.067	0.071	0.80	0.365	0.610	0.835	1.038	1.229
0.20	0.055	0.063	0.069	0.074	0.078	0.82	0.376	0.636	0.876	1.093	1.297
0.21	0.059	0.068	0.075	0.080	0.086	0.84	0.388	0.663	0.918	1.150	1.367
0.22	0.063	0.073	0.081	0.087	0.093	0.86	0.399	0.691	0.962	1.208	1.439
0.23	0.067	0.078	0.087	0.095	0.101	0.88	0.411	0.719	1.006	1.268	1.514
0.24	0.071	0.084	0.094	0.102	0.110	0.90	0.422	0.747	1.052	1.329	1.591
0.25	0.076	0.090	0.101	0.110	0.118	0.92	0.434	0.776	1.098	1.393	1.670
0.26	0.080	0.096	0.108	0.118	0.127	0.94	0.446	0.806	1.146	1.458	1.751
0.27	0.085	0.102	0.115	0.126	0.137	0.96	0.457	0.836	1.196	1.524	1.835
0.28	0.089	0.108	0.123	0.135	0.146	0.98	0.469	0.867	1.246	1.593	1.921
0.29	0.094	0.114	0.130	0.144	0.156	1.00	0.481	0.898	1.297	1.664	2.010
0.30	0.098	0.121	0.138	0.153	0.167	1.05	0.510	0.979	1.431	1.848	2.242
0.31	0.103	0.127	0.146	0.163	0.178	1.10	0.540	1.064	1.573	2.044	2.489
0.32	0.108	0.134	0.155	0.173	0.189	1.15	0.569	1.152	1.722	2.251	2.753
0.33	0.112	0.141	0.163	0.183	0.200	1.20	0.599	1.244	1.880	2.471	3.033
0.34	0.117	0.148	0.172	0.193	0.212	1.25	0.629	1.339	2.045	2.704	3.329
0.35	0.122	0.155	0.181	0.204	0.225	1.30	0.659	1.438	2.219	2.949	3.643
0.36	0.127	0.162	0.190	0.215	0.237	1.35	0.689	1.541	2.401	3.207	3.975
0.37	0.132	0.169	0.200	0.226	0.250	1.40	0.719	1.648	2.592	3.479	4.324
0.38	0.137	0.177	0.210	0.238	0.264	1.45	0.750	1.759	2.791	3.764	4.691
0.39	0.142	0.184	0.220	0.250	0.278	1.50	0.780	1.873	2.999	4.063	5.077
0.40	0.147	0.192	0.230	0.262	0.292	1.55	0.810	1.992	3.216	4.375	5.482
0.41	0.152	0.200	0.240	0.275	0.307	1.60	0.841	2.115	3.443	4.702	5.906
0.42	0.157	0.208	0.251	0.288	0.322	1.65	0.871	2.241	3.678	5.044	6.350
0.43	0.162	0.216	0.262	0.301	0.337	1.70	0.902	2.372	3.923	5.400	6.814
0.44	0.167	0.225	0.273	0.314	0.353	1.75	0.932	2.507	4.178	5.772	7.299
0.45	0.172	0.233	0.284	0.328	0.369	1.80	0.963	2.646	4.442	6.158	7.804
0.46	0.177	0.242	0.296	0.342	0.386	1.85	0.994	2.790	4.716	6.561	8.330
0.47	0.183	0.250	0.307	0.357	0.403	1.90	1.024	2.937	5.000	6.978	8.877

(continued)

Table B-1. (Continued)

y_n/b	$nQ/(b^{8/3} S_0^{1/2})$					y_n/b	$nQ/(b^{8/3} S_0^{1/2})$				
	$m=0$	$m=0.5$	$m=1.0$	$m=1.5$	$m=2.0$		$m=0$	$m=0.5$	$m=1.0$	$m=1.5$	$m=2.0$
0.48	0.188	0.259	0.319	0.372	0.421	1.95	1.055	3.089	5.294	7.412	9.446
0.49	0.193	0.268	0.332	0.387	0.439	2.00	1.086	3.246	5.599	7.862	10.037
0.50	0.198	0.277	0.344	0.403	0.457	2.50	1.395	5.061	9.234	13.304	17.234
0.52	0.209	0.296	0.370	0.435	0.495	3.00	1.705	7.364	14.037	20.613	26.985
0.54	0.220	0.315	0.396	0.468	0.535	3.50	2.017	10.197	20.131	30.001	39.592
0.56	0.231	0.334	0.424	0.503	0.577	4.00	2.330	13.601	27.635	41.671	55.340
0.58	0.241	0.354	0.453	0.540	0.621	4.50	2.643	17.617	36.663	55.815	74.501
0.60	0.252	0.375	0.482	0.577	0.666	5.00	2.956	22.282	47.324	72.620	97.338

Table B-2. Normal Depth in Circular Channels

$\dfrac{y_n}{D}$	$\dfrac{nQ}{D^{8/3} S_o^{1/2}}$	$\dfrac{y_n}{D}$	$\dfrac{nQ}{D^{8/3} S_o^{1/}}$	$\dfrac{y_n}{D}$	$\dfrac{nQ}{D^{8/3} S_o^{1/2}}$
0.10	0.007	0.41	0.110	0.72	0.271
0.11	0.008	0.42	0.115	0.73	0.275
0.12	0.010	0.43	0.120	0.74	0.280
0.13	0.011	0.44	0.125	0.75	0.284
0.14	0.013	0.45	0.130	0.76	0.289
0.15	0.015	0.46	0.135	0.77	0.293
0.16	0.017	0.47	0.140	0.78	0.297
0.17	0.020	0.48	0.145	0.79	0.301
0.18	0.022	0.49	0.151	0.80	0.305
0.19	0.025	0.50	0.156	0.81	0.308
0.20	0.027	0.51	0.161	0.82	0.312
0.21	0.030	0.52	0.166	0.83	0.315
0.22	0.033	0.53	0.172	0.84	0.318
0.23	0.036	0.54	0.177	0.85	0.321
0.24	0.039	0.55	0.183	0.86	0.324
0.25	0.043	0.56	0.188	0.87	0.326
0.26	0.046	0.57	0.193	0.88	0.329
0.27	0.050	0.58	0.199	0.89	0.331
0.28	0.053	0.59	0.204	0.90	0.332
0.29	0.057	0.60	0.209	0.91	0.334
0.30	0.061	0.61	0.215	0.92	0.335
0.31	0.065	0.62	0.220	0.93	0.335
0.32	0.069	0.63	0.225	0.94	0.335
0.33	0.073	0.64	0.231	0.95	0.335
0.34	0.078	0.65	0.236	0.96	0.334
0.35	0.082	0.66	0.241	0.97	0.332
0.36	0.086	0.67	0.246	0.98	0.329
0.37	0.091	0.68	0.251	0.99	0.325
0.38	0.096	0.69	0.256	1.00	0.312
0.39	0.100	0.70	0.261		
0.40	0.105	0.71	0.266		

CHAPTER 3

CONTROL SECTIONS

The equations of motion for open-channel flow, except steady uniform flow, are differential equations. Solutions of the differential equations of motion require initial and boundary conditions. The boundary conditions are specified at channel sections, termed control sections. A question arises about the locations of the control sections in the channel reach under consideration. The theories of disturbance propagation and of channel transitions can answer this question and are introduced in this chapter.

3-1 Propagation of Disturbances

A disturbance can be generated in flow by changing flow depth and/or velocity in a section of flow. A disturbance with an abrupt change in depth, such as that can be caused by the sudden closure or opening of a gate in an open channel, is called surge. Consider a surge moving with constant velocity V_w in a wide rectangular channel,[*] as shown in Figure 3-1a. The surge behavior is analyzed by applying the algebraic forms of the continuity and momentum equations to a control volume bounded by sections 1 and 2. The following assumptions are made in the one-dimensional analysis:

(1) Uniform velocity distributions in sections 1 and 2, i.e., $\beta_1 = \beta_2 = 1$.
(2) Hydrostatic pressure distributions in sections 1 and 2.
(3) Negligible effect of channel and friction slopes in a short distance between sections 1 and 2.

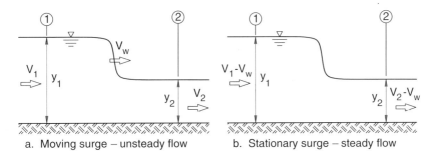

a. Moving surge – unsteady flow   b. Stationary surge – steady flow

Figure 3-1. Surge propagation in a channel.

---

[*] Propagation of a surge in a channel of arbitrary shape of cross section is described in Appendix C.

The unsteady problem can be reduced to a steady-state problem by superimposing on the flow system a constant velocity equal and opposite to surge velocity V_w, as shown in Figure 3-1b. The resulting velocities in sections 1 and 2 become $(V_1 - V_w)$ and $(V_2 - V_w)$, respectively. The surge appears stationary to an observer moving with the surge at velocity V_w. The continuity and momentum equations (1-68) and (1-76) for a unit width of the channel can be written as

$$y_1(V_1 - V_w) = y_2(V_2 - V_w) \tag{3-1}$$

$$\rho y_1(V_1 - V_w)(V_1 - V_2) = \tfrac{1}{2}\gamma(y_2^2 - y_1^2) \tag{3-2}$$

The surge velocity from Eq. (3-1) is

$$V_w = \frac{V_2 y_2 - V_1 y_1}{y_2 - y_1} \tag{3-3}$$

Substituting for V_w from Eq. (3-3) into Eq. (3-2), the latter can be written as

$$y_1\left(V_1 - \frac{V_1 y_1 - V_2 y_2}{y_1 - y_2}\right)(V_2 - V_1) = \frac{g}{2}\left(y_1^2 - y_2^2\right) \tag{3-4}$$

or

$$\frac{y_1 y_2}{y_1 - y_2}(V_2 - V_1)^2 = \frac{g}{2}\left(y_1^2 - y_2^2\right) \tag{3-5}$$

or

$$(V_2 - V_1)^2 = \frac{g}{2}\frac{(y_1 - y_2)^2(y_1 + y_2)}{y_1 y_2} \tag{3-6}$$

Equation (3-6) can be written as

$$V_1 - V_2 = \pm|y_1 - y_2|\sqrt{\frac{g(y_1 + y_2)}{2 y_1 y_2}} \tag{3-7}$$

Substituting for $(V_1 - V_2)$ from Eq. (3-7) into Eq. (3-2), the latter becomes

$$y_1(V_1 - V_w)\left[\pm|y_1 - y_2|\sqrt{\frac{g(y_1 + y_2)}{2 y_1 y_2}}\right] = \frac{g}{2}\left(y_1^2 - y_2^2\right) \tag{3-8}$$

Equation (3-8) can be solved for $(V_1 - V_w)$ as

$$V_1 - V_w = \pm \sqrt{\frac{g}{2} \frac{y_2}{y_1} (y_1 + y_2)} \qquad (3\text{-}9)$$

Addition of Eqs. (3-7) and (3-9) gives

$$V_w - V_2 = \pm \sqrt{\frac{g}{2} \frac{y_1}{y_2} (y_1 + y_2)} \qquad (3\text{-}10)$$

Several types of surges can be identified on an algebraic basis from the above equations (Martin Vide, 1992) and are summarized in Table 3-1. Because both y_1 and y_2 are positive Eq. (3-1) shows that $(V_1 - V_w)$ and $(V_2 - V_w)$ must have the same sign. Now two cases are possible: $y_1 > y_2$ and $y_2 > y_1$.

(a) $y_1 > y_2$: For this case the right-hand side of Eq. (3-3) is positive; therefore $(V_1 - V_w)$ and $(V_2 - V_1)$ must have same sign. Again two cases are considered; (i) both $(V_1 - V_w)$ and $(V_2 - V_1)$ are negative and (ii) both $(V_1 - V_w)$ and $(V_2 - V_1)$ are positive. For the second condition surge velocity V_w can be either positive or negative.

(b) $y_2 > y_1$: For this case the right-hand side of Eq. (3-3) is negative; therefore $(V_1 - V_w)$ and $(V_2 - V_1)$ must have opposite signs. Again two cases are possible; (i) $(V_1 - V_w)$ is negative and $(V_2 - V_1)$ is positive and (ii) $(V_1 - V_w)$ is positive and $(V_2 - V_1)$ is negative. For the second condition surge velocity V_w can be either positive or negative.

Table 3-1. Types of Surge in Open Channels

Type	Sign of $(V_1 - V_w)$	$(V_2 - V_1)$	$(V_2 - V_w)$	Relation among V_1, V_2, and V_w	Sign of V_w
A	−	−	−	$V_w > V_1 > V_2$	+
B	+	+	+	$V_2 > V_1 > V_w$	+
C	+	+	+	$V_2 > V_1 > V_w$	−
D	−	+	−	$V_w > V_2 > V_1$	+
E	+	−	+	$V_1 > V_2 > V_w$	+
F	+	−	+	$V_1 > V_2 > V_w$	−

The six types of surges are shown schematically in Figure 3-2. The examples of the various types of surges are presented in Chapter 7. For a given type of surge, the correct sign in Eqs. (3-7), (3-9), and (3-10) can be selected using the sign of the various terms in Table 3-1 (Example 3-1). It should be noted that in classifying surges Chow (1959) ignored Types B and E surges, and Martin Vide (1992) included Types B and C surges in one category, and so Types D and E surges.

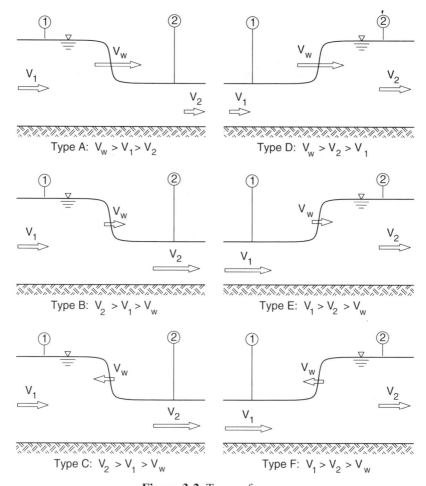

Figure 3-2. Types of surge.

Example 3-1. A river is flowing at a depth of 2.4 m and at a velocity of 1 m/s. It meets a tidal bore (surge produced due to tidal effects) which abruptly increases the depth to 3.6 m. Determine the type of the surge. Find the speed with which the bore moves upstream and the magnitude and direction of the velocity of the water behind the bore.

The surge is moving upstream and $y_1 < y_2$; it is a Type F surge for which $V_1 > V_2 > V_w$. Using the positive sign in Eq. (3-9), the velocity of the Type F surge is given by

$$V_w = V_1 - \sqrt{\frac{g}{2}\frac{y_2}{y_1}(y_1 + y_2)}$$

$$= 1.0 - \sqrt{\frac{9.81}{2} \times \frac{3.6}{2.4}(2.4 + 3.6)} = -5.64 \text{ m/s}$$

For Type F surge $V_2 > V_w$. Using the positive sign in Eq. (3-10), velocity V_2 is given by

$$V_2 = V_w + \sqrt{\frac{g}{2}\frac{y_1}{y_2}(y_1+y_2)} = -5.64 + \sqrt{\frac{9.81}{2} \times \frac{2.4}{3.6}(2.4+3.6)} = -1.21 \text{ m/s}$$

The direction of the flow behind the bore is upstream and its speed is 1.21 m/s.

Celerity of Small Disturbance

If the surge is small, i.e., if $y_1 \cong y_2 = y$, then $V_1 \cong V_2 = V$, and Eqs. (3-9) and (3-10) become identical to

$$V_w - V = \pm\sqrt{gy} \qquad (3-11)$$

where V and y are depth and velocity of the undisturbed flow, respectively. The term on the left-hand side of Eq. (3-11) represents the velocity of a small disturbance relative to the flow velocity. It is called the celerity of a small disturbance and is denoted by c. From Eq. (3-11) the celerity of a small disturbance in a rectangular channel is

$$c = \sqrt{gy} \qquad (3-12)$$

The celerity in channels of arbitrary cross section, as shown in Appendix C, is given by

$$c = \sqrt{gD} \qquad (3-13)$$

in which D is hydraulic depth and equal to A/B and B is the channel width at the free surface.

The six types of surge, shown in Figure 3-2, reduce to three types of small disturbance. Types A and D, B and E, and C and F surges become identical to each other. The three types of small disturbance are shown in Figure 3-3 and are based on the conditions: (i) $V_w > V > 0$; (ii) $V > V_w > 0$; and (iii) $V > 0 > V_w$. These three types are the limiting cases of Types A, B, and C surges (or Types D, E, and F surges).

Three kinds of flow are defined depending on the relative magnitude of flow velocity and the celerity of a small disturbance: (i) subcritical flow if $V < c$; (ii) critical flow if $V = c$; and (iii) supercritical flow if $V > c$. Type A small disturbance travels downstream with a velocity equal to $(V + c)$ and can occur in all three kinds of flow. Type B small disturbance travels downstream with a velocity equal to $(V - c)$ and is possible only in supercritical flow; the limiting positive value equal to zero of $(V - c)$ is possible in critical flow and the disturbance is stationary in that case. Type C small disturbance travels upstream at a speed equal to $(c - V)$ and is possible only in subcritical flow.

The propagation of small disturbance can be visualized by dropping a pebble into water. If the water is stationary, the disturbance pattern on the water surface

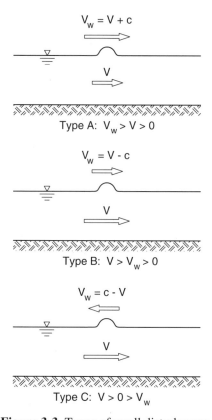

$$V_w = V + c$$

V

Type A: $V_w > V > 0$

$$V_w = V - c$$

V

Type B: $V > V_w > 0$

$$V_w = c - V$$

V

Type C: $V > 0 > V_w$

Figure 3-3. Types of small disturbances.

is concentric circles, as shown in Figure 3-4a. The disturbance travels in all directions with a velocity equal to the celerity c. If the pebble is dropped in flowing water, the disturbance produced by the pebble is displaced in the direction of flow. The disturbance pattern for flow velocity V less than celerity c is shown in Figure 3-4b. The disturbance travels upstream with velocity $(V - c)$ and downstream with velocity $(V + c)$. When flow velocity is equal to celerity, disturbance cannot propagate upstream, as shown in Figure 3-4c; the flow in this case is critical. In supercritical flow the flow velocity is greater than the celerity c, and the disturbance pattern on the water surface is shown in Figure 3-4d. A pair of common tangents envelops the disturbed flow region. These common tangents are called oblique wave fronts or shock fronts.

A nondimensional number, called the Froude number F, is defined as the ratio of velocity of flow to celerity of small disturbance, i.e.,

$$F = \frac{V}{c} = \frac{V}{\sqrt{gD}} \tag{3-14}$$

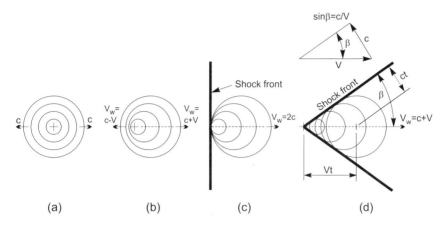

Figure 3-4. Wave patterns created by a small disturbance.

The Froude number is equal to unity for critical flow and is less and greater than unity for subcritical and supercritical flow, respectively. The depth at which the Froude number becomes unity is called the critical depth and is denoted by y_c. The methods for computing critical depth are described in the next section. Note that for rectangular channels hydraulic depth D is equal to depth y, and $F = V/\sqrt{gy}$.

The Froude number is analogous to the Mach number for compressible flow. The Mach number is defined as the ratio of the flow velocity to the velocity of a small pressure wave, i.e., velocity of sound. The three flow regimes in open-channel flow—subcritical, critical, and supercritical—have their counterparts in compressible flow, namely, subsonic, sonic, and supersonic.

Though a small disturbance cannot travel upstream in supercritical flow, a "large" disturbance, as shown later, can move upstream in supercritical flow. However, in so doing, it transforms the supercritical flow to subcritical.

Upstream Propagation of Disturbance

It is shown above that a small disturbance cannot travel upstream in supercritical flow. This observation has an important bearing on channel flows. Consider the steady discharge in a horizontal channel for different tailwater depths, as shown in Figures 3-5 and 3-6 for subcritical and supercritical flow in the channel, respectively. Let the critical depth for the given discharge in the channel be y_c. If the channel is assumed frictionless, the depth in the channel would be uniform, as shown by the solid lines; otherwise depth would change in the upstream direction to overcome the channel friction, as indicated by the dotted curves. The effect of channel friction on flow profile is considered in the next chapter.

For depths in the channel larger than y_c in Figure 3-5, the flow in the channel is subcritical, i.e., $F < 1$; the change in depth Y of the tailwater, as a disturbance, can propagate upstream in the channel. The depth in the channel decreases (or increases) as the tailwater depth decreases (or increases). As Y approaches y_c, the

Figure 3-5. Propagation of disturbance in subcritical flow.

Froude number at the end section becomes unity. Any further decrease in Y, as a disturbance, can no longer propagate across the end section; consequently the flow profile in the channel, shown by the lowest line (or curve), is not affected by tailwater depth. The minimum possible depth at the downstream end of a channel carrying flow in the subcritical regime is the critical depth.

If the flow in the channel is supercritical, only a "large" disturbance can propagate upstream in the channel. According to Figure 3-2, surges that travel upstream are Types C and F. It is shown in Chapter 7 that for Type C surge the flow condition upstream from the surge is subcritical. Therefore, Type F is the only surge that can travel upstream in supercritical flow. The surge velocity from Eq. (3-9) and Table 3-1 is

$$V_w = V_1 - \sqrt{\frac{g}{2}\frac{y_2}{y_1}(y_1 + y_2)} \qquad (3-15)$$

The positive sign in Eq. (3-9) is used because for the Type F surge, $V_1 > V_2 > V_w$. It is seen from Eq. (3-15) that for a given value of y_1, the second term on the right-hand side increases with increasing y_2 and can, therefore, become larger than V_1, leading to a negative V_w, i.e., a surge moving upstream. Hence a disturbance with "large enough" y_2 can propagate upstream in supercritical flow.

The propagation of a disturbance due to increase in tailwater depth in a channel is shown in Figure 3-6. The flow depth in the channel is less than y_c, as the flow in the channel is supercritical, i.e., $F > 1$. The depth is uniform in the frictionless channel, as shown by the solid line; otherwise the depth increases, as shown by the dotted curve. The reason for the increase in depth in supercritical flow will become clear in the next section. If the disturbance is not large enough to propagate upstream, the flow forms a standing "swell" at the downstream end of the channel, as shown by the lower curve in Figure 3-6. As the tailwater depth further increases, the disturbance becomes large enough to propagate upstream. In the frictionless channel the disturbance keeps moving upstream as the velocity V_1 in the channel remains constant. In channels with friction the velocity V_1 increases with the decrease in depth as the disturbance moves upstream, until V_1 becomes

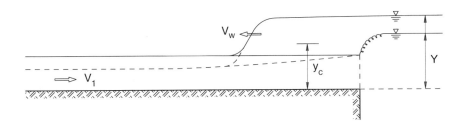

Figure 3-6. Upstream propagation of disturbance in supercritical flow.

equal to the second term on the right-hand side of Eq. (3-15). The disturbance at that section becomes stationary, as shown by the upper curve in Figure 3-6.

Hydraulic Jump

At the location of the stationary surge the flow changes abruptly from the supercritical to the subcritical condition through a feature known as the *hydraulic jump*. Considerable turbulence and energy loss are produced in the hydraulic jump (see Figure 3-7a). A relation between the flow conditions upstream and downstream of the jump is obtained from Eq. (3-15) on setting $V_w = 0$ as

$$\frac{V_1^2}{gy_1} \equiv F_1^2 = \frac{1}{2}\frac{y_2}{y_1}\left(\frac{y_2}{y_1}+1\right) \tag{3-16}$$

Equation (3-16) is a quadratic equation in y_2/y_1; its solution, on discarding the negative root, is

$$\frac{y_2}{y_1} = \frac{1}{2}\left(\sqrt{1+8F_1^2}-1\right) \tag{3-17}$$

The depth downstream of the jump can be determined from Eq. (3-17) if the flow condition upstream of the jump is known. Equation (3-17) can also be written as

$$\frac{y_1}{y_2} = \frac{1}{2}\left(\sqrt{1+8F_2^2}-1\right) \tag{3-18}$$

Equation (3-18) can be used to find the upstream conditions from the known downstream conditions.

Equations (3-16)–(3-18) are applicable only to rectangular channels. Relations for channels of arbitrary cross section can be obtained by applying the continuity and momentum equations to a control volume, shown in Figure 3-7a. The continuity and momentum equations (1-68) and (1-76) can be written as follows:

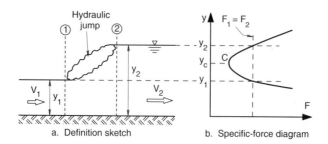

a. Definition sketch   b. Specific-force diagram

Figure 3-7. Application of the specific-force diagram to a hydraulic jump.

Continuity equation: $Q = V_1 A_1 = V_2 A_2$ (3-19)

Momentum equation: $\rho Q(\beta_2 V_2 - \beta_1 V_1) = \gamma A_1 \bar{y}_1 - \gamma A_2 \bar{y}_2$ (3-20)

where $\bar{y}$ is the vertical distance from the free surface of the centroid of the cross section. Substituting for V_1 and V_2 from Eq. (3-19) into Eq. (3-20), dividing the resulting equation by g, and rearranging the terms, one gets

$$A_1 \bar{y}_1 + \frac{\beta_1 Q^2}{gA_1} = A_2 \bar{y}_2 + \frac{\beta_2 Q^2}{gA_2}$$ (3-21)

Define a function F, termed the specific-force function, as

$$F = A\bar{y} + \frac{\beta Q^2}{gA}$$ (3-22)

Using the definition of the specific-force function, Eq. (3-21) can be expressed as $F_1 = F_2$. For a given Q and channel shape, F is a function of y only. A plot of a specific-force curve is presented in Figure 3-7b. The solution of Eq. (3-21) is given by the intersection of a vertical line $F_1 = F_2$ with the specific-force curve.

It is shown in Example 3-2 that the depth represented by point C in Figure 3-7b, where the specific force is a minimum, is the critical depth. The flow along the upper limb AC $(y > y_c)$ and the lower limb BC $(y < y_c)$ is subcritical and supercritical, respectively. For a given value of specific force, the specific-force curve has two possible depths, namely, an initial depth y_1 in the supercritical region and a sequent depth y_2 in the subcritical region. These depths are also called conjugate depths.

The solution of Eq. (3-21) is obtained either numerically by expressing A and $\bar{y}$ in terms of y (Example 3-3) or graphically by plotting the specific-force curve.

For rectangular channels $(\bar{y} = y/2)$, Eq. (3-21) can be written as

$$\frac{y_1^2}{2} + \frac{q^2}{gy_1} = \frac{y_2^2}{2} + \frac{q^2}{gy_2} \qquad (3\text{-}23)$$

where q is the discharge per unit width. It can be easily shown that the above equation is the same as Eq. (3-16), and the energy loss, ΔE_j, in jump in a rectangular channel (Prob. 3-4) is given by

$$\Delta E_j = \frac{(y_2 - y_1)^3}{4 y_1 y_2} \qquad (3\text{-}24)$$

Example 3-2. Prove that the depth at point C on the specific-force curve in Figure 3-7b is the critical depth.

The specific force is a minimum at point C; dF/dy is zero at this point. Differentiation of Eq. (3-22) with respect to y yields

$$\frac{dF}{dy} = \frac{d(A\bar{y})}{dy} - \frac{Q^2}{gA^2}\frac{dA}{dy} = 0$$

It can be shown that $d(A\bar{y}) = A\,dy$ (see Prob. 1-7). Since $dA = B\,dy$, where B is the surface width, the above equation reduces to

$$A - \frac{Q^2 B}{gA^2} = 0$$

or

$$\frac{V}{\sqrt{gD}} = 1 \quad \text{or} \quad F = 1$$

Example 3-3. A trapezoidal channel has a bottom width of 2.0 m and side slopes of 2H:1V and carries a discharge of 15.0 m³/s. Find the depth sequent to the initial depth of 0.6 m.

For a trapezoidal section

$$A\bar{y} = \left[\left(by\frac{y}{2}\right) + my^2\frac{y}{3}\right] = (3b + 2my)\frac{y^2}{6}$$

$$F = A\bar{y} + \frac{Q^2}{gA} = (3b + 2my)\frac{y^2}{6} + \frac{Q^2}{g[(b + my)y]}$$

For $y_1 = 0.6$ m,

$$F_1 = (3\times2 + 2\times2\times0.6)\frac{0.6^2}{6} + \frac{15^2}{9.81\times[(2 + 2\times0.6)\times0.6]} = 12.45$$

$$F_2 = (3\times2 + 2\times2\,y_2)\frac{y_2^2}{6} + \frac{15^2}{9.81\times[(2 + 2\,y_2)y_2]} = F_1 = 12.45$$

The solution of the above equation by the trial-and-error method is

$$y_2 = 2.11 \text{ m}$$

PROBLEMS

3-1   A rectangular channel with sidewalls that are 2 m high carries flow at velocity of 1.5 m/s and depth of 1.8 m. The discharge in the channel is to be reduced by a sudden partial closure of a sluice gate. What is the maximum percentage that the discharge can be reduced without danger of overtopping the channel sidewalls upstream from the gate by the surge developed by the gate closure. Also calculate the velocity of the surge.

3-2   A tidal estuary is 3 m in depth with a current of 0.3 m/s. A bore with wave height of 1.5 m is formed in the estuary. Find the velocity of the bore.

3-3   A tidal bore in an estuary is observed to travel upstream at a velocity of 5.5 m/s. The depth and velocity of flow prior to the arrival of the bore are 2.5 m and 1.5 m/s, respectively. Determine the height of the bore and the flow velocity behind the bore.

3-4   Show that energy loss in hydraulic jump in a rectangular channel is given by Eq. (3-24).

3-5   A rectangular channel is 1.5 m in width and carries discharge of 3.7 m^3/s while flowing at a depth of 0.6 m. Find the conjugate depth if a hydraulic jump forms in the channel. Also calculate the loss of energy in the jump.

3-6   A trapezoidal channel of bottom width 3.0 m and side slopes 1.5 H:1V carries a discharge of 14.5 m^3/s at a depth of 0.25 m. If a hydraulic jump takes place in the channel, determine the sequent depth and the energy loss in the jump.

3-7   A circular jet with velocity V_j and diameter D is directed upstream in a flow in a rectangular channel of unit width to induce a hydraulic jump. Find the relationships between V_2 and y_2 and V_1, y_1, V_j, and D. Neglect the drag force on the jet pipe.

3-8   Downstream from a sluice gate, a hydraulic jump produces a change in depth from 0.6 to 1.8 m in a rectangular channel. Compute the unit discharge.

3-9   Water flows down a spillway at the rate of 3.5 m^3/s per meter width and leaves the apron horizontally at a velocity of 15 m/s. Determine the tailwater depth necessary to form a hydraulic jump.

3-2 Channel Transitions

A large enough disturbance which propagates upstream in supercritical flow can be developed by constricting the flow area. The change in cross section may be due to a change in channel width and/or bottom elevation, such as shown in Figure 3-8. A local change in cross section of flow that produces a variation in flow from one uniform state to another is defined as a channel transition. The flow

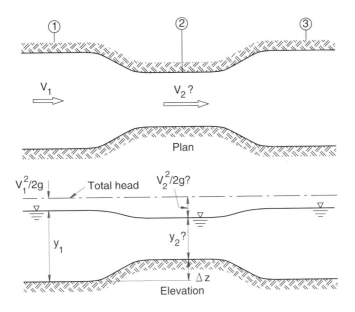

Figure 3-8. A channel transition.

behavior in the channel transition is analyzed by applying the algebraic forms of the continuity and energy equations to the control volume between sections 1 and 2. The energy equation instead of the momentum equation is used, as the latter requires pressure forces on the bottom and sides of the transition, which cannot be correctly estimated due to nonhydrostatic pressure distribution within the transition. The following assumptions are made in the analysis:

(1) Sections 1 and 2 are located sufficiently upstream and downstream from the transition where the pressure distribution is hydrostatic.
(2) The effect on flow of small energy loss in the transition is neglected.

Neglecting the head-loss term, the continuity and energy equations (1-68) and (1-79) can be written as

$$Q = V_1 A_1 = V_2 A_2 \tag{3-25}$$

$$\frac{\alpha_1 V_1^2}{2g} + y_1 \cos\theta_1 + z_1 = \frac{\alpha_2 V_2^2}{2g} + y_2 \cos\theta_2 + z_2 \tag{3-26}$$

in which z = elevation of the channel bottom above a datum (= G_0). Substituting for V_1 and V_2 from Eq. (3-25) into Eq. (3-26), the latter becomes

$$\frac{\alpha_1 Q^2}{2gA_1^2} + y_1 \cos\theta_1 = \frac{\alpha_2 Q^2}{2gA_2^2} + y_2 \cos\theta_2 + \Delta z \tag{3-27}$$

where $\Delta z = z_2 - z_1$. Equation (3-27) is a cubic equation in y_2 (or y_1) and has one negative and two positive roots. The criterion for selecting the correct positive root is developed from the concept of specific energy that is dealt with next.

Specific Energy

The specific energy, E, is defined as the energy with respect to channel bottom, i.e.,

$$E = y \cos\theta + \frac{\alpha V^2}{2g} \tag{3-28}$$

Substituting for V as Q/A, Eq. (3-28) becomes

$$E = y \cos\theta + \frac{\alpha Q^2}{2gA^2} \tag{3-29}$$

Though E should be termed specific head instead of specific energy, the latter term is more common in practice and, therefore, is used herein. For a given Q and the geometry of a cross section, E is a function of y. A plot of the function for section 1 is shown by curve 1 in Figure 3-9. It is seen from Eq. (3-29) that $E \to y \cos\theta$ as $y \to \infty$ and $E \to \infty$ as $y \to 0$. Therefore the function has two asymptotes: $E = y$ for $\cos\theta = 1$ and $E = 0$. The function has two branches: one,

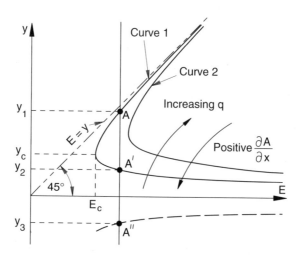

Figure 3-9. The specific-energy curve.

shown by the solid line between the two asymptotes and the second shown by the dotted line. The following observations are made from the specific-energy curve:

(1) A minimum specific energy, denoted by E_c, is required to maintain a given flow in a given shape of cross section. The depth at the minimum specific energy, represented by point C on the E–y curve, is critical, as shown later.

(2) For a given $E > E_c$, there exist three depths, y_1, y_2, and y_3, that are represented by points A, A', and A", respectively. The depth y_3 is negative and therefore has no physical meaning. The other two positive depths are called alternate depths.

(3) The solid curve has two limbs: the upper, where $y > y_c$, and the lower, where $y < y_c$. The flow regimes defined by the upper and lower limbs are subcritical- and supercritical-flow regimes, respectively.

(4) The specific energy increases with increasing depth in the subcritical-flow regime and decreases with increasing depth in the supercritical-flow regime.

The channel transition in Figure 3-8 is nonprismatic, i.e., flow area A is a function of both x and y. The term $\partial A/\partial x$ is negative between sections 1 and 2, i.e., for a given y, area A decreases in the flow direction. Curve 1 in Figure 3-9 is applicable only to section 1 and section 3. The curves for other sections can be drawn, such as curve 2 for section 2. Because, for a given value of y, E increases with decreasing A [Eq. (3-29)], the curve for section 2 is drawn to the right of that for section 1. For rectangular channels, the unit discharge q is constant along these curves. The directions of positive $\partial A/\partial x$ and increasing q are shown in Figure 3-9.

Critical Depth

The specific energy E is a minimum at point C. The flow condition at point C is obtained by differentiating Eq. (3-29) and equating the resulting equation to zero; i.e.,

$$\frac{dE}{dy} = \cos\theta + \frac{\alpha Q^2}{2g}\left(-\frac{2}{A^3}\right)\left(\frac{dA}{dy}\right) = 0$$

Using $dA = B\,dy$, the above equation can be written as

$$\alpha Q^2 B / gA^3 \cos\theta = 1 \qquad (3\text{-}30)$$

or
$$F = 1$$

where the Froude number, F, is defined as

$$F = \sqrt{\frac{\alpha(Q^2/A^2)}{g(A/B)\cos\theta}} = \frac{V}{\sqrt{gD\cos\theta/\alpha}} \tag{3-31}$$

Assuming $\alpha \cong 1$ and $\cos\theta = 1$, Eq. (3-31) reduces to

$$F = \frac{V}{\sqrt{gD}}$$

This expression for the Froude number is the same as in Eq. (3-14). Therefore, the depth in a section where the specific energy is the minimum is critical. Note that hydraulic depth D for rectangular channels is equal to y.

For a given discharge and cross-sectional shape, y_c can be determined from Eq. (3-30). For channels, such as circular and trapezoidal channels, in which A and B can be expressed algebraically in terms of y, Eq. (3-30) is a nonlinear algebraic equation that can be solved for y_c by the trial-and-error method, by using design charts, or by numerical methods. For developing design charts Eq. (3-30) is written in nondimensional form as

$$\frac{\alpha Q^2 m^3}{g\cos\theta\, b^5} = \frac{y_c'^3 (y_c' + 1)^3}{2y_c' + 1} \quad \text{(trapezoidal channels)} \tag{3-32}$$

$$\frac{\alpha Q^2}{g\cos\theta\, D^5} = f\left(\frac{y_c}{D}\right) \quad \text{(circular channels)} \tag{3-33}$$

where $y_c' = m\, y_c / b$. A plot of Eqs. (3-32) and (3-33) is presented in Figure 3-10 which can be used to determine y_c in trapezoidal and circular channels. Numerical methods, such as the Newton-Raphson and interval-halving methods, can be used to solve Eqs. (3-32) and (3-33); the details of these methods can be found in books on numerical methods

In natural channels where it is not convenient to express algebraically A and B in terms of y, Eq. (3-30) is solved graphically for y_c. Define a function Z, termed the section factor for critical flow, as

$$Z^2 = \frac{A^3}{B} \tag{3-34}$$

where Z is a function of y that can be plotted for the known values of A and B for a range of y. Equation (3-30) can be written in terms of Z as

$$\frac{\alpha Q^2}{g\cos\theta} = \frac{A_c^3}{B_c} = Z_c^2 \tag{3-35}$$

in which Z_c is the section factor for critical flow for discharge Q at the critical

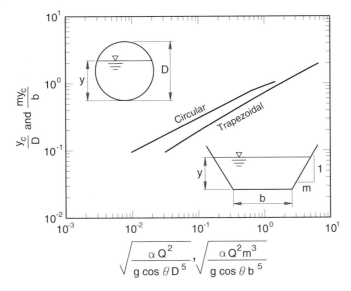

Figure 3-10. Dimensionless critical-depth curves.

depth y_c. The critical depth can be obtained from the plot of the function Z (Figure 3-11) for a value of Z_c given by Eq. (3-35).

The critical depth for a given E_c can be determined as follows (Prob. 3-16). Elimination of Q between Eqs. (3-29) and (3-30) yields

$$E = \left(y + \frac{A}{2B} \right) \cos \theta \tag{3-36}$$

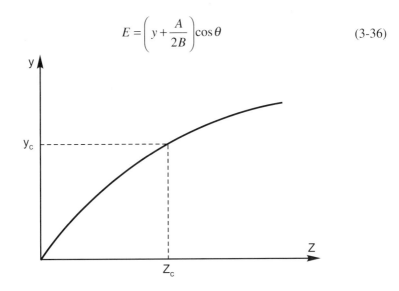

Figure 3-11. Curve y versus Z.

The above equation is plotted as a function of y and the critical depth is obtained from the plot.

The critical depth can also be defined in another way. For a given specific energy, the depth at which discharge is the maximum is the critical depth. Equation (3-29) can be written as

$$Q = A\sqrt{2g(E - y\cos\theta)/\alpha}$$

For a given E, Q is a function of y, which is plotted in Figure 3-12. The discharge is zero at $y = E/\cos\theta$ and $y = 0$. The discharge is the maximum at an intermediate depth where dQ/dy is zero. Differentiating the above equation with respect to y and setting dQ/dy equal to zero, one obtains (Prob. 3-24)

$$\frac{\alpha Q^2 B}{gA^3 \cos\theta} = 1$$

The condition given in the above equation is identical to that given in Eq. (3-30) for the critical depth. In other words, for a given specific energy discharge is the maximum at the critical depth.

Example 3-4. Determine the critical depth in the channel of Example 2-1. Assuming $\alpha = 1$ and $\cos\theta = 1$,

$$\sqrt{\frac{\alpha Q^2 m^3}{g\cos\theta\, b^5}} = \sqrt{\frac{1 \times 90^2 \times 1^3}{9.81 \times 1 \times 5^5}} = 0.514$$

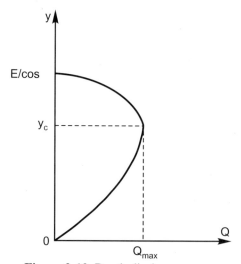

Figure 3-12. Depth-discharge curve.

Upon entering the above value in Figure 3-10, one obtains

$$\frac{my_c}{b} = 0.53$$

whence
$$y_c = \frac{0.53 \times 5}{1} = 2.63 \text{ m}$$

Rectangular Channels

The expressions for critical depth and minimum specific energy in rectangular channels are relatively simple. Assuming $\alpha \cong 1$ and $\cos \theta \cong 1$, and using $A = by$ for rectangular channels, Eq. (3-30) can be written as

$$\frac{Q^2}{g} = \frac{(by_c)^3}{b}$$

or
$$y_c = \sqrt[3]{\frac{q^2}{g}} \tag{3-37}$$

where $q = Q/b$ = unit discharge and b = channel width. Equation (3-37) shows that y_c increases with increasing q. Minimum specific energy from Eq. (3-29) is

$$E_c = y_c + \frac{Q^2}{2g(by_c)^2}$$

Using $Q = bq$ and Eq. (3-37), the above equation becomes

$$E_c = \tfrac{3}{2} y_c \tag{3-38}$$

For a given Q and E, the alternate depths in rectangular channels can be determined from Eq. (3-29), which can be written in nondimensional form by dividing it by y_c as

$$\frac{E}{y_c} = \frac{y}{y_c} + \frac{Q^2}{2gy_c(b^2 y^2)}$$

Using Eq. (3-37), the above equation reduces to

$$\frac{E}{y_c} = \frac{y}{y_c} + \frac{1}{2(y/y_c)^2} \tag{3-39}$$

The values of E/y_c for a range of values of y/y_c are included in Table D-1 in Appendix D. This table can be used to compute the alternate depths in rectangular channels, as illustrated in Examples 3-5 and 3-6.

Compound Channels

Critical flow in compound channels is possible at more than one depth, as can be seen from Figure 3-13, which includes the specific-energy curves for three discharges (100, 220, and 500 m³/s) in an idealized cross section, shown as an inset in the figure. The specific energy E is computed by Eq. (3-29). The energy coefficient α in Eq. (3-29) is obtained from Eq. (1-43) as

$$\alpha = \frac{\int_A u^3\,dA}{V^3 A} = \frac{\sum_{i=1}^{N} V_i^3 A_i}{V^3 A} = \frac{\sum_{i=1}^{N}\left(Q_i^3/A_i^2\right)}{\left(\sum_{i=1}^{N} Q_i\right)^3 \Big/ \left(\sum_{i=1}^{N} A_i\right)^2} \tag{3-40}$$

Substitution for Q_i from Eq. (2-19) into Eq. (3-40) yields

$$\alpha = \frac{\left(\sum_{i=1}^{N} A_i\right)^2}{\left(\sum_{i=1}^{N} K_i\right)^3} \sum_{i=1}^{N}\left(\frac{K_i^3}{A_i^2}\right) \tag{3-41}$$

While the specific-energy curves for Q values of 100 and 500 m³/s show only one

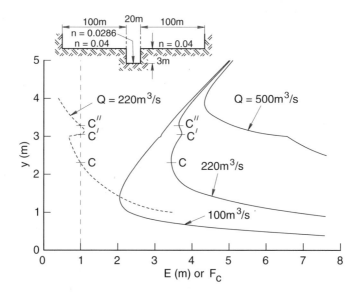

Figure 3-13. Specific-energy and Froude number curves

point of minimum specific energy, the specific-energy curve for $Q = 220$ m^3/s reveals two points of minimum specific energy indicated by C and C" and one point of local maximum specific energy indicated by C'. More than one point of minimum specific energy are possible only if the energy coefficient is considered as a function of y. The points of local minimum and maximum specific energy are obtained by differentiating Eq. (3-29) with respect to y and setting the derivative to zero, as given by

$$\frac{dE}{dy} = \cos\theta - \frac{\alpha Q^2}{gA^3}\frac{dA}{dy} + \frac{Q^2}{2gA^2}\frac{d\alpha}{dy} = 0 \qquad (3\text{-}42)$$

Noting that $dA/dy = B$ and rearranging the terms in Eq. (3-42), one gets

$$\frac{Q^2(\alpha B - A\alpha'/2)}{gA^3\cos\theta} = 1 \qquad (3\text{-}43)$$

where $\alpha' = d\alpha/dy$. Differentiating Eq. (3-40) with respect to y, an expression for α' (Blalock and Sturm, 1981) can be derived as (Prob. 3-23)

$$\frac{d\alpha}{dy} = \frac{A^2\sigma_1}{K^3} + \sigma_2\left(\frac{2AB}{K^3} - \frac{A^2\sigma_3}{K^4}\right) \qquad (3\text{-}44)$$

where σ_1, σ_2, and σ_3 are defined in Problem 3-23. Because the Froude number at the point of minimum specific energy should be unity, Eq. (3-43) can be used to define a compound-channel Froude number F_c as

$$F_c = \left\{\frac{Q^2(\alpha B - A\alpha'/2)}{gA^3\cos\theta}\right\}^{1/2} \qquad (3\text{-}45)$$

It should be noted that for $\alpha = $ constant Eq. (3-45) reduces to Eq. (3-31). Equation (3-45) also is plotted as a dotted curve in Figure 3-13. As expected, Eq. (3-45) for F_c correctly locates points C, C', and C", and should be used in identifying the flow regime in compound channels. The flow at point C" also is critical because F_c is unity, although the specific energy is not a minimum. The concept that the specific energy is a minimum for critical flow is valid only in simple channels. The flow depth at point C' is slightly greater than the floodplain level. The flow is subcritical for $y > y_{c'}$ and $y_{c'} > y > y_c$ and is supercritical for $y_{c''} > y > y_{c'}$ and $y_c > y$. Chaudhry and Bhallamudi (1988) used the momentum approach and derived another expression for F_c, which is given by

$$F_c = \left\{\frac{Q^2(\beta B - A\beta')}{gA^3\cos\theta}\right\}^{1/2} \qquad (3\text{-}46)$$

where $\beta' = d\beta/dy$ and β is the momentum coefficient defined by Eq. (1-42). The similarity between Eqs (3-45) and (3-46) is obvious.

Change in Bottom Elevation

Referring back to the discussion of channel transition at the beginning of this section, consider a channel transition consisting of a "hump" in channel floor, as shown in Figure 3-14. The hump is long enough for parallel flow to occur in the middle of the hump. The channel transition is prismatic; a single specific-energy curve is applicable to all sections. The problem under consideration is to determine the flow conditions in section 2 (i.e., y_2 and V_2) for known flow condition in section 1 (i.e., y_1 and V_1) and the rise Δz in the channel bottom. The solution is obtained by solving Eq. (3-27), which can be written as

$$E_2 = E_1 - \Delta z \qquad (3-47)$$

The solution of Eq. (3-47) will be obtained for three different values of Δz. Let $E_1 - E_c = \Delta z_c$.

(i) Case A: $\Delta z < \Delta z_c$ so that $E_2 > E_c$. The graphical solution is shown in Figure 3-14. The solution lies at the intersection of a vertical line $E = E_2$ and the E–y curve, i.e., point B or B', which represent the two positive roots of Eq. (3-47). To pick the correct root, consider that flow in section 1 is subcritical (i.e., $y_1 > y_c$) and is represented by point A. As flow moves over the hump, E decreases and the flowpoint moves on the E–y curve from point A to point B (or B'). The flowpoint can reach point B' only by moving around the E–y curve, which requires that E should decrease to E_c up to point C and then increase to E_2 beyond point C. This condition cannot be met for the given geometry of the hump. Hence the correct solution is represented by point B, where the flow is subcritical, i.e., $y_2 > y_c$. The specific energy in section 3 is the same as in section 1, i.e., $E_3 = E_1$. Now as flow moves down the hump from section 2 to section 3, E increases and the flowpoint

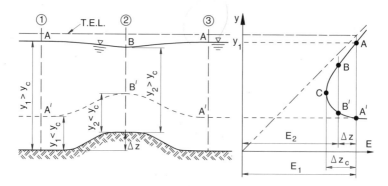

Figure 3-14. Case A: Flow past a hump in channel floor for $\Delta z < \Delta z_c$.

moves on the E-y curve from point B to point A; the depth in section 3 is equal to y_1.

A similar argument is applicable if the flow condition in section 1 is supercritical (i.e., $y_1 < y_c$) and represented by point A' in Figure 3-14. The correct solution for section 2 in this case is represented by point B' where the flow is supercritical, i.e., $y_1 < y_c$, and the depth in section 3 is equal to $y_1 < y_c$. In summary, for the case where $E_2 > E_c$, $y_2 > y_c$ if $y_1 > y_c$ and $y_2 < y_c$ if $y_1 < y_c$. In other words, a subcritical flow remains subcritical and a supercritical flow remains supercritical.

(ii) Case B: $\Delta z = \Delta z_c$ so that $E_2 = E_c$. The solution is presented graphically in Figure 3-15. The vertical line $E = E_2$ intersects the E-y curve only at one point, point C, which represents the flow condition at section 2, i.e., $y_2 = y_c$. As flow moves over the hump, the flowpoint moves from point A (or A') to point C on the E-y curve. Now as flow moves down the hump, the flowpoint on the E–y curve is free either to move back to point A (or A') or to go around to point A' (or A). As explained in Case C, which is discussed next, the flow condition represented by point A' is more stable.

(iii) Case C: $\Delta z > \Delta z_c$ so that $E_2 < E_c$. The vertical line $E = E_2$ in Figure 3-16 does not intersect the E–y curve. It means that the hump (barrier) is too large and it "chokes" the flow; there is not enough energy to maintain the given flow over the barrier in section 2. To maintain the flow, the specific energy E_1 must increase by ΔE_1 given by

$$\Delta E_1 = E_c - E_2 = \Delta z - \Delta z_c \tag{3-48}$$

From Eqs. (3-48), the minimum total head, H_{min}, required to maintain the given flow is

$$H_{min} = E_1 + \Delta E_1 = E_c + \Delta z \tag{3-49}$$

The channel bottom in section 1 is assumed to be the datum in defining the total

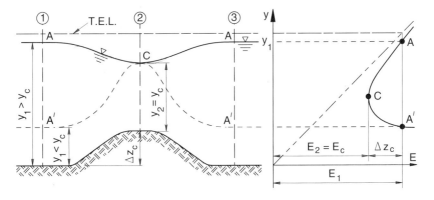

Figure 3-15. Case B: Flow past a hump in channel floor for $\Delta z = \Delta z_c$.

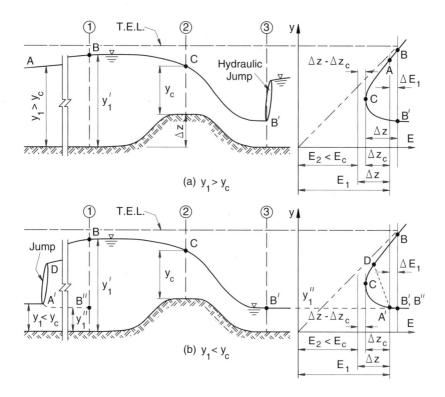

Figure 3-16. Case C: Flow past a hump in channel floor for $\Delta z > \Delta z_c$.

head in Eq. (3-49). In the subcritical regime, E increases with increasing y; therefore if $y_1 > y_c$, the depth in section 1 increases to y_1' as represented by point B in Figure 3-16a; the depth in section 2 becomes y_c (point C) corresponding to the requirement of the minimum specific energy. Upstream from section 1 a gradually varied flow (GVF) profile AB develops. The depth varies from y_1' (point B) in section 1 to y_1 (point A) in an upstream location that depends on ΔE_1. The specific energy ΔE_1 is gained in the GVF profile, as explained in the next chapter.

In the supercritical flow regime, E increases with decreasing y; energy gain ΔE_1 requires that y_1 must decrease to y_1'' (Figure 3-16b), which would require a GVF profile A'B" shown by the dotted line. Such a profile physically is not possible; the reason will become clearer in the next chapter. The only other way to gain ΔE_1 is to increase the depth in section 1 to y_1'. This condition requires that the incoming supercritical flow should become subcritical. The disturbance developed by the hump (barrier) is large enough to travel upstream in supercritical flow. However, in so doing, the flow changes to subcritical. This change in the flow regime happens through a hydraulic jump (represented by A'D) which occurs upstream from section 1. A GVF profile DB exists between the hydraulic jump and section 1, as shown in Figure 3-16b. The type of the GVF profile and the length of the profile are such that the difference between the gain of the specific energy ΔE_{DB} in the GVF profile and the loss of energy ΔE_j in the hydraulic jump is equal to ΔE_1. Once again the depth in section 2 is y_c.

For a hump where $\Delta z \geq \Delta z_c$, the depth in section 2 is critical, and section 2 controls the flow condition in the channel. Any small disturbance from downstream cannot propagate past this section. Such a control is termed critical-depth control (CDC).

The flow regime upstream from a CDC is subcritical. The convergence of the upstream flow toward the hump tends to carry on into a further convergence downstream from the hump; the flow regime downstream from a CDC is supercritical. The flow condition in section 3 is represented by point B'. The energy gain ΔE_1 is lost downstream in a GVF and in a hydraulic jump if it occurs.

Example 3-5. Water flows with a velocity of 1.5 m/s and at a depth of 2.5 m in a rectangular channel. Determine (a) the maximum size of rise in the channel bottom without affecting flow condition upstream of the rise in the bottom and (b) the depth over the rise when the height of the rise is one-half of the height in (a).

Let the sections upstream from the rise and over the rise be sections 1 and 2, respectively.

(a)
$$E_1 = 2.5 + \frac{1.5^2}{2g} = 2.61 \, \text{m}$$

$$y_c = \sqrt[3]{\frac{q^2}{g}} = \sqrt[3]{\frac{(1.5 \times 2.5)^2}{9.81}} = 1.13 \, \text{m}$$

and
$$E_2 = E_c = \tfrac{3}{2} \, y_c = 1.69 \, \text{m}$$
$$\text{Maximum size of rise} = E_1 - E_c = 2.61 - 1.69 = 0.92 \, \text{m}$$

(b)
$$\Delta z = 0.92/2 = 0.46 \, \text{m}$$
$$E_2 = E_1 - \Delta z = 2.61 - 0.46 = 2.15 \, \text{m}$$
$$\frac{E_2}{y_c} = \frac{2.15}{1.13} = 1.903$$

Upon entering the above value of E_2/y_c in Table D-1,

$$\frac{y_2}{y_c} = 1.738; \quad \text{or} \quad y_2 = 1.738 \times 1.13 = 1.96 \, \text{m}$$

Change in Channel Width

Consider the channel transition shown in Figure 3-17. Within the transition the channel bottom is horizontal (no hump) and the channel width is a variable. The channel in the transition is nonprismatic; the E-y curves differ from section to section. The problem under consideration is to determine flow conditions in section 2 (i.e., y_2 and V_2) for known flow conditions in section 1 (i.e., y_1 and V_1)

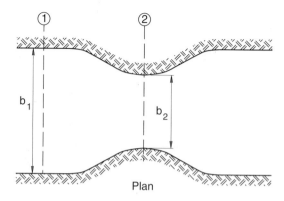

Figure 3-17. A channel transition with a change in width.

and the change in the channel width from b_1 to b_2. The solution is obtained again by solving Eq. (3-27), which can be written as

$$E_2 = E_1 \qquad (3\text{-}50)$$

because $\Delta z = 0$. The solution of Eq. (3-50) will be obtained for three different values of b_2. Let $b_2 = b_c$ when $E_2 = E_{c2}$, where E_{c2} is the minimum specific energy in section 2. Note that E_{c2} increases with decreasing b_2.

(i) Case A: $b_2 > b_c$ so that $E_1 = E_2 > E_{c2}$. The solution is presented in Figure 3-18. Two E–y curves, curves 1 and 2 for sections 1 and 2 respectively, are drawn. The solution lies at the intersection of the vertical line $E = E_1 = E_2$ and curve 2, i.e., point B or B', which represent the two positive roots of Eq. (3-50). To pick the correct root, consider that flow in section 1 is subcritical (i.e., $y_1 > y_c$) and is represented by point A. The quantity $\partial A/\partial x$ is negative within the contraction.

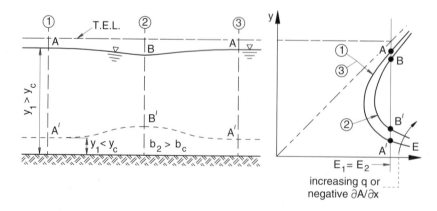

Figure 3-18. Case A: Flow past a constriction in width for $b_2 > b_c$.

As flow moves along the contraction, point A moves in the direction of negative $\partial A/\partial x$ to point B (or B'). The flowpoint can reach point B' only by moving along the vertical line, which requires that width be less than b_2. This condition cannot be met for the given geometry of the transition. Hence the correct solution is represented by point B, where the flow is subcritical, i.e., $y_2 > y_c$. Now as the flow moves along the expansion, the flowpoint moves from point B to point A as $E_3 = E_1$.

A similar argument is applicable if the flow condition in section 1 is supercritical (i.e., $y_1 < y_c$) and represented by point A' in Figure 3-18. The correct solution in this case is represented by point B' where the flow is supercritical, i.e., $y_2 < y_c$. As the flow moves along the expansion, point B' moves to point A'. In summary, for the case where $b_2 > b_c$, $y_2 > y_c$ if $y_1 > y_c$ and $y_2 < y_c$ if $y_1 < y_c$. In other words a subcritical flow remains subcritical and a supercritical flow remains supercritical.

(ii) Case B: $b_2 = b_c$ so that $E_1 = E_2 = E_{c2}$. The solution is presented graphically in Figure 3-19. The vertical line $E = E_1 = E_2 = E_{c2}$ intersects curve 2 only at one point, point C, which represents the flow condition in section 2, i.e., $y_2 = y_c$. As flow moves along the contraction, the flowpoint moves from point A (or A') to point C on the E-y curve. Now as flow moves along the expansion, the flow-point on the E-y curve is free either to move back to point A (or A') or to go around to point A' (or A). The flow condition represented by point A' is more stable, as explained in the next case. The width that creates critical flow in section 2 in a rectangular channel can be determined as follows:

$$E_1 = E_c = \frac{3}{2} y_c = \frac{3}{2} \sqrt[3]{(Q/b_c)^2/g}$$

whence,

$$b_c = \left(\frac{3}{2}\right)^{3/2} \frac{Q}{\sqrt{gE_1^3}} \qquad (3\text{-}51)$$

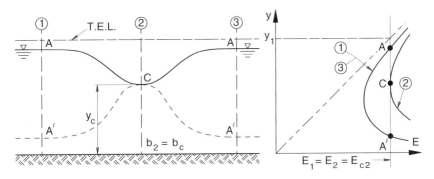

Figure 3-19. Case B: Flow past a constriction in width for $b_2 = b_c$.

(iii) Case C: $b_2 < b_c$. The vertical line $E = E_1 = E_2 < E_{c2}$ in Figure 3-20 does not intersect curve 2. This means that the constriction (barrier) is too large so that there is not enough energy to maintain the given flow in section 2. The energy E_1 must increase by ΔE_1 given by

$$\Delta E_1 = E_{c2} - E_1 \tag{3-52}$$

The minimum total head required to maintain given flow is

$$H_{min} = E_1 + \Delta E_1 = E_c \tag{3-53}$$

The mechanisms for gaining the specific energy by ΔE_1 and for its subsequent loss, and the flow profiles are the same as described above in Figure 3-16 for the hump in channel bottom. There exists a critical-depth control at a section of the minimum width. Again the flow upstream and downstream from the critical-depth control is subcritical and supercritical, respectively.

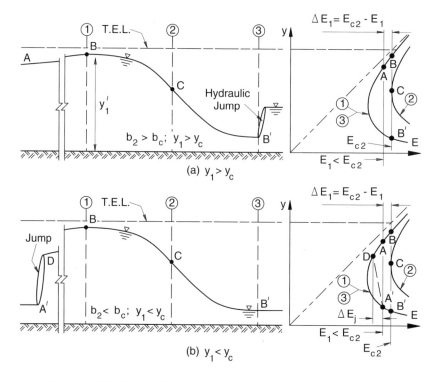

Figure 3-20. Case C: Flow past a constriction in width for $b_2 < b_c$.

Control Structures

There is a class of channel transition that includes control structures, such as a weir, spillway, or sluice gate. These structures in most cases produce a large enough disturbance to choke the flow, similar to that developed in Case C of the above-described transitions. Consequently the flow upstream and downstream from such structures, as shown in Figure 3-21, is subcritical and supercritical, respectively. A unique relation between the elevation of the water surface at a particular section, termed head, and the discharge, similar to that between discharge and critical depth, exists for these structures. This relation serves as a boundary condition in profile computations. In this respect these structures act as a channel control and are referred to as artificial channel control (ACC). The head-discharge relations for such control structures are discussed in Chapter 9.

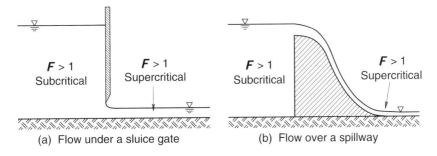

Figure 3-21. Artificial channel control.

PROBLEMS

3-10   A rectangular channel 2 m wide carries a discharge of 5 m³/s. Determine (a) critical depth; (b) minimum specific energy; and (c) any other depth at which the same specific energy is sustained as at a depth of 1.2 m.

3-11   Determine the critical depth for a discharge of 50 m³/s in a trapezoidal channel with a side slope of 2H:1V and a bottom width of 10 m.

3-12   The width of a rectangular channel is reduced from 3 to 2 m and the floor is raised 0.3 m in elevation at a section. What discharge is indicated by a drop of 0.3 m in surface elevation at the contracted section if the depth of the approach flow is 1.8 m?

3-13   Determine the unit discharge that produces the alternate depths of 0.5 and 1.5 m in a rectangular channel.

3-14   Water flows in a rectangular channel with a smoothly rounded 15-cm step. The flow conditions upstream from the step are velocity = 4 m/s and depth = 0.5 m. Determine the change in depth for the case of an upward step and a downward step. Also find the maximum allowable upward step for the specified upstream flow to be possible.

3-15   Water is flowing at velocity of 3 m/s and depth of 3 m in a 3-m-wide rectangular channel. Determine change in depth produced by (a) a smooth contraction to a width of 2.7 m and (b) a smooth expansion to a width of 3.3 m. Also find the maximum allowable contraction in width for the upstream flow to be possible as specified.

3-16   Water discharges from a lake into a steep channel so that the depth at the lake outlet is critical. The channel section is as follows:

Height above channel bottom, cm	0	20	40	60	80	100	120	140
Channel width, m	1.0	2.0	2.8	3.5	4.3	4.8	5.2	5.5

Determine the discharge in the channel for the water surface in the lake 1.6 m above the channel bottom at the outlet.

3-17   Find the critical depth for a flow of 6 m^3/s in a triangular channel; one side of the channel is vertical and the other side makes an angle of 30° with the horizontal.

3-18   A rectangular channel that is 6 m wide enters through a smooth transition into a contracted section wherein the width is 3 m. Find the maximum discharge that the channel can carry without having the depth of flow exceed the height of the sidewalls that are 1.8 m high. Assume that the approach flow is subcritical, the bed level remains constant, and there are no energy losses.

3-19   Determine the flow regime (subcritical or supercritical) in the channel in Problem 2.10.

3-20   A long channel is carrying the water at a depth of 3 m. A barrier is installed across the channel raising the flow depth to 3.3 m at the barrier. Sketch the flow profile if the velocity of the undisturbed flow is (a) 0.3 m/s and (b) 6 m/s.

3-21   For purpose of discharge measurements, the width of a rectangular channel is reduced from 3 to 2.1 m and the floor is raised 0.3 m in elevation at a given section. What flow rate would be indicated by 0.3-m drop in surface elevation of the contracted section if the depth of the approaching flow is 2 m?

3-22   If the channel of Problem 3-21 carries a flow of 12 m^3/s, which is the minimum possible depth for subcritical flow upstream of the given channel contraction?

3-23   Show that the rate of change of the energy coefficient with respect to depth in a compound channel can be expressed as

$$\frac{d\alpha}{dy} = \frac{A^2\sigma_1}{K^3} + \sigma_2\left(\frac{2AB}{K^3} - \frac{A^2\sigma_3}{K^4}\right)$$

where

$$\sigma_1 = \sum_{i=1}^{N} \left[\left(\frac{K_i}{A_i} \right)^3 \left(3B_i - 2R_i \frac{dP_i}{dy} \right) \right]$$

$$\sigma_2 = \sum_{i=1}^{N} \left(\frac{K_i^3}{A_i^2} \right)$$

$$\sigma_3 = \sum_{i=1}^{N} \left[\left(\frac{K_i}{A_i} \right) \left(5B_i - 2R_i \frac{dP_i}{dy} \right) \right]$$

and P_i and B_i are the wetted perimeter and the surface width of the ith subsection, respectively.

3-24   Show that for a given specific energy discharge is the maximum at the critical depth.

3-25   If the channel of Problem 3-12 carries a flow of 15 m³/s, which is the minimum possible depth for subcritical flow upstream of the given channel contraction.

3-3 Locations and Types of Control Sections

The above-stated fact that a small disturbance can move upstream only in subcritical flow has an important practical significance. This fact implies that supercritical flow cannot be affected by a downstream feature and can therefore be controlled only from upstream. If the flow in the channel reach under consideration is in the supercritical region, the control section, where the boundary conditions are generally specified, must exist at the upstream end of the channel reach. On the other hand, a small disturbance in subcritical flow travels both upstream and downstream. Subcritical flow is affected by both downstream and upstream features and therefore can be controlled both from upstream and downstream. In the case of a subcritical flow in a channel reach, the boundary conditions are generally specified at both ends of the channel reach, i.e., the control sections must be located at both upstream and downstream ends of the channel reach. These remarks will become clearer in Chapter 7, in which the method of characteristics to solve the partial differential equations for unsteady flow is described.

The governing equation for steady flow, as shown in Chapter 1, is an ordinary differential equation that needs only one boundary condition. If the ordinary differential equation can be solved analytically in closed form, the constant of integration in the analytical solution can be determined from the boundary condition; consequently the location of the control section, where the boundary condition is prescribed, is of no concern. However, the differential equation for most practical cases is integrated numerically. McBeans and Perkins (1975) showed that the growth behavior of numerical errors is a function of the direction of calculation. Jain (1976) computed flow profiles for stratified flows and found that the numerical solution of the governing equation is very sensitive to the direction of integration. Although it may be theoretically possible to calculate flow profiles in either the upstream or downstream direction, the direction of integration should be chosen carefully to avoid large numerical errors. As a rule, the appropriate direction of integration is upstream for subcritical flow and downstream for supercritical flow. In other words, the control section, where the boundary condition is assigned, should be located upstream from the channel reach for supercritical flow and downstream from it for subcritical flow.

The boundary condition is determined by the control section where a relation between Q and y is known. Three types of control, namely, normal-depth control (NDC), critical-depth control (CDC), and artificial channel control (ACC) are identified. Such relations for the normal-depth and critical-depth controls have already been discussed, and these are the Manning equation (2-13) for the NDC and the critical-depth relation (3-30) for the CDC. The relations for the artificial channel controls are described in Chapter 9. The probable location of a critical-depth control is at the choke of the channel transition, as shown in Figures 3-16 and 3-20, or at the downstream end of a subcritical-flow regime, as shown in Figure 3-5. The location of an artificial channel control is in the vicinity of the control structure and is known empirically. The location of the section of the normal-depth control is determined using the following explanation.

It is possible for a flow to occur at a normal depth in a uniform channel reach if there exists no other control (ACC or CDC) in the channel reach. If there are other controls that produce a depth other than the normal depth, the most probable location where the flow can occur at the normal depth should be furthest from these ACC and CDC. The flow upstream and downstream from these controls is in the subcritical and supercritical regions, respectively. Therefore, the most probable location of an NDC is at the upstream end of the channel reach in subcritical flow and at the downstream end of the channel reach in supercritical flow.

The flow is subcritical upstream from a CDC or an ACC, which implies that a CDC or an ACC exists only at the downstream end of a subcritical-flow regime. Similarly, the flow is supercritical downstream from a CDC or an ACC; consequently a CDC or an ACC occurs only at the upstream end of a supercritical-flow regime. These observations imply that subcritical flow is subject to a downstream control and supercritical flow is governed by an upstream control.

3-4 Flow Profiles without Channel Resistance

The topic of flow profiles without the channel resistance is included herein to illustrate the procedure for identifying the existence of a critical-depth control. The concepts presented in this section will be helpful in analyzing flow profiles with channel resistance, which are described in the next chapter. Out of the three possible controls, only critical-depth and artificial channel controls (CDC and ACC) are possible in channels without resistance, as normal depth has no physical meaning in such channels. The governing equations for flow under consideration are as follows:

Continuity equation: $\qquad\qquad Q = VA$ $\qquad\qquad$ (3-54)

Energy equation: $\dfrac{\alpha V^2}{2g} + y\cos\theta + z = \text{constant}$ (3-55)

The constant in Eq. (3-55) is determined from the boundary condition, i.e., from the known flow conditions in the control section. The first step, which is the most crucial, in solving Eq. (3-55) is the determination of the location and flow condition in the control section. The method of locating the control section is illustrated through the following examples.

Consider a flow controlled by a sluice gate in a channel that terminates in a free overfall, as shown in Figure 3-22. For simplicity, the channel is assumed to be rectangular in cross section, and α and $\cos\theta$ in Eq. (3-55) are supposed to be unity. Four channel sections are selected; sections 1 and 2 are immediately upstream and downstream from the gate, respectively; section 3 is a little downstream from the gate; and section 4 is at the free overfall. Let us analyze the effect on the flow profiles of a rise in the channel bottom between sections 3 and 4. The specific-energy curve, shown in Figure 3-22, will be used in analyzing the flow profiles. Because the channel is prismatic, one E–y curve is applicable to all sections. The flow conditions are determined for a known fixed gate opening w and unit discharge q. The flow depth in section 2 is determined from $y_2 = C_c w$, where C_c is the coefficient of contraction and is known for a given boundary geometry of the gate. The velocity V_2 is obtained from the continuity equation as $V_2 = q/y_2$.

The very first step in analyzing flow profiles is the determination of the locations of the control sections. As explained earlier, only ACCs and CDCs are possible. It is easy to locate the ACCs. There is only one ACC in the present illustration, which is the sluice gate. The only possible location of a CDC is at the free overfall (see Figure 3-5). Only one boundary condition, i.e., one control section, is needed to evaluate the constant in Eq. (3-55). If there are more than one control section, the section that requires the largest total head to maintain the given flow acts as a control section. Assuming the channel bottom at section 2 as a datum, the total head required to maintain a given flow at section 2 is $H_2 = E_2 = y_2 + V_2^2/2g$. The flow condition in section 2 is represented by point B. The minimum total head required to maintain the given flow at the free overfall (section 4) is $(H_4)_{\min} = E_c + \Delta z$, where Δz is the rise in the channel bottom at section 4.

For $\Delta z = 0$, $(H_4)_{\min} = E_c$; the flow condition is represented by point C. It can be easily seen that for $\Delta z = 0$, $H_2 > (H_4)_{\min}$; consequently section 2 is the control section. Assuming negligible loss of energy across the gate, the specific energy at all sections is the same and equal to E_2. The flow upstream from the sluice gate is subcritical; the depth in section 1 is the alternate depth of y_2 and is given by the intersection of the vertical line $E = E_2$ with the upper limb of the E–y curve. The flow condition in section 1 is represented by point A. The flow downstream from the gate is supercritical; the depth in sections 3 and 4 is equal to y_2. In sections 3

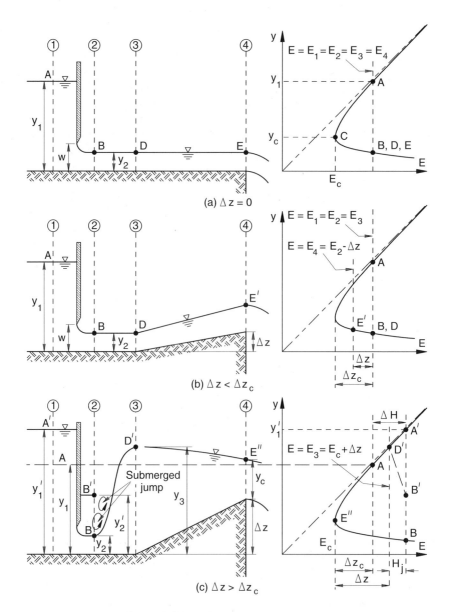

Figure 3-22. Shift in the control section.

and 4 in Figure 3-22a the flow conditions are represented by points D and E, respectively.

Now let $E_2 - E_c = \Delta z_c$. For $\Delta z < \Delta z_c$, $H_2 > (H_4)_{min}$; section 2 is the control section. The specific energy in sections 1 and 3 does not change; the flow condition is still represented by points A and D, respectively. Specific energy in section 4 is equal to $E_4 = E_2 - \Delta z$. Flow in section 4 is supercritical; the depth in section 4 is determined by the intersection of the vertical line $E = E_2 - \Delta z$ and the lower limb of the E–y curve and is represented by point E' in Figure 3-22b. The depth between sections 3 and 4 increases from point D to point E'. In the limiting case as $\Delta z \to \Delta z_c$, $E_4 \to E_c$ and $y_4 \to y_c$.

For $\Delta z > \Delta z_c$, $H_2 < (H_4)_{min}$; section 4 becomes the CDC section. The depth in section 4 is the critical depth; and $(H_4)_{min} = E_c + \Delta z$. The flow upstream from the CDC section is subcritical. The specific energy in section 3 is $E_3 = H_4 = E_c + \Delta z$; the depth in section 3 is given by the intersection of the vertical line $E = E_3$ and the upper limb of the E–y curve and is represented by point D' in Figure 3-22c. The flow leaves the gate in the supercritical regime and the flow in section 3 is in the subcritical regime. The change in the flow regime from supercritical to subcritical occurs through a feature described earlier as a hydraulic jump. However, the jump in the present situation is a submerged hydraulic jump, because the gate outlet, as shown in Figure 3-22c, must be submerged (why?). The total head in section 2 is not enough to maintain the given flow; it should increase by $\Delta H = H_4 - H_2 + H_j$, where H_j is the head loss in the jump. The gain in the head is achieved by increasing the depth in section 1 from y_1 to y_1'. The flow conditions in sections 1 and 2 are obtained from the energy equation between sections 1 and 2 and the momentum equation between sections 2 and 3. On the assumption of a negligible energy loss between sections 1 and 2, the energy equation can be written as

$$y_1' + \frac{q^2}{2gy_1'^2} = y_2' + \frac{q^2}{2gy_2^2} \tag{3-56}$$

The momentum equation (3-20) between sections 2 and 3 becomes

$$\rho q\left(\frac{q}{y_3} - \frac{q}{y_2}\right) = \frac{1}{2}\gamma\left(y_2'^2 - y_3^2\right) \tag{3-57}$$

Note that in Eqs. (3-56) and (3-57) while the velocity in section 2 is based on the flow depth y_2, the piezometric head is equal to the total depth y_2'. The two unknowns, namely y_1' and y_2', can be determined from Eqs. (3-56) and (3-57). The flow condition in sections 1 and 2 is represented by points A' and B', respectively. The submerged hydraulic jump between sections 2 and 3 is shown schematically by the dashed line B'D' in the E–y diagram. The E–y diagram can be a helpful tool in analyzing flow profiles, as illustrated in this example.

The flow in a channel with a contraction in width, instead of a rise in the channel floor, between sections 3 and 4 in Figure 3-22 can be analyzed following the above-described procedure. The control shifts from the gate to the overfall when the width at the overfall becomes less than a critical width that yields a critical depth at the overfall (Prob. 3-28). If a transition between sections 3 and 4

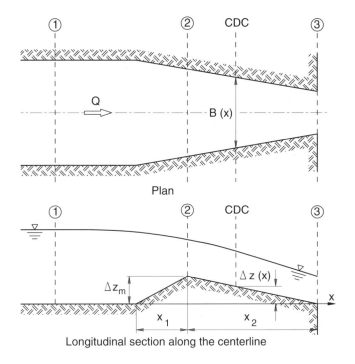

Figure 3-23. Location of the critical-depth control.

includes changes in both the channel bottom elevation and the channel width, the location of the CDC need not be at the freefall; but can be in a section within the transition, as illustrated in the following example.

Consider the flow in a channel shown in Figure 3-23. The flow profile in this channel is controlled by a CDC, as there is no ACC. The depth in the CDC section is critical. On using Eqs. (3-38) and (3-37), the total head in the CDC section can be expressed as

$$H = \frac{3}{2} y_c + \Delta z = \frac{3}{2} \sqrt[3]{\frac{(Q/B)^2}{g}} + \Delta z \qquad (3\text{-}58)$$

in which B and Δz can be expressed in terms of x for a given geometry of the transition. The section where the total head given by Eq. (3-58) is a maximum is the CDC section (Prob. 3-31).

Example 3-6. Locate the control section in Figure 3-22 for the following data: $E_1 = 1.2$ m; $y_2 = 0.25$ m; and $\Delta z = 0.60$ m. Determine the depths in sections 1–4.

Assume that the control is at the gate. The energy equation between sections 1 and 2 can be written as

$$E_1 = E_2, \quad \text{or} \quad 1.2 = y_2 + \frac{q^2}{2gy_2^2}$$

whence, $\quad q = \sqrt{2gy_2^2(E_1 - y_2)} = \sqrt{2 \times 9.81 \times 0.25^2(1.2 - 0.25)} = 1.08 \, \text{m}^3/\text{s/m}$

$$(H_4)_{min} = E_c + \Delta z = \frac{3}{2}\sqrt[3]{\frac{q^2}{g}} + \Delta z = \frac{3}{2}\sqrt[3]{\frac{1.08^2}{9.81}} + 0.60 = 1.34 \, \text{m} > E_1$$

The control is at the freefall. A submerged hydraulic jump, as shown in Figure 3-22c, will form. The unknown quantities are: depth y_1' in section 1; depth y_2' in section 2; depths y_3 and y_4 in sections 3 and 4; and discharge q^*. The energy equation between sections 1 and 2 can be written as

$$E_1 = y_2' + \frac{q^{*2}}{2gy_2^2} \quad \text{or} \quad 1.2 = y_2' + \frac{q^{*2}}{2 \times 9.81 \times 0.25^2} \tag{1}$$

The momentum equation between sections 2 and 3 is given by Eq. (3-44), which can be written as

$$q^*\left(\frac{q^*}{y_3} - \frac{q^*}{0.25}\right) = \frac{9.81}{2}\left(y_2'^2 - y_3^2\right) \tag{2}$$

The energy equation between sections 3 and 4 yields

$$y_3 + \frac{q^{*2}}{2gy_3^2} = E_c + \Delta z \quad \text{or} \quad y_3 + \frac{q^{*2}}{2 \times 9.81 y_3^2} = \frac{3}{2}\sqrt[3]{\frac{q^{*2}}{9.81}} + 0.6 \tag{3}$$

The solution of Eqs. (1), (2), and (3) gives

$$q^* = 0.54 \, \text{m}^3/\text{s/m}; \quad y_2' = 0.96 \, \text{m} \text{ and } y_3 = 1.05 \, \text{m}$$

The depth in section 4 is the critical depth:

$$y_4 = y_c = \sqrt[3]{\frac{q^{*2}}{g}} = \sqrt[3]{\frac{0.54^2}{9.81}} = 0.31 \, \text{m}$$

The specific energy in section 1 is 1.2 m;

$$\frac{E_1}{y_c} = \frac{1.2}{0.31} = 3.871$$

Upon entering the above value of E_1/y_c in Table D-1

$$\frac{y_1'}{y_c} = 3.837 \quad \text{or} \quad y_1' = 3.837 \times 0.31 = 1.19 \, \text{m}$$

PROBLEMS

3-26   Determine depths in sections 1 – 4 in Figure 3-22 for the following data:
        $q = 1$ m³/s/m ; $y_2 = 25$ cm ; and $\Delta z = 0.5$ m.

3-27   Locate the control section in Figure 3-22 for the following data: $E_1 = 1.2$ m;
        $y_2 = 25$ cm; and $\Delta z = 0.6$ m.

3-28   A horizontal rectangular channel of 3 m in width is contracted to a width of 1.8 m
        before it terminates in a sudden freefall. The flow in the channel is controlled by a
        sluice gate located in the wider section of the channel. The channel carries a
        discharge of 9 m³/s. Neglecting all losses, determine the maximum gate opening at
        which the flow downstream from the gate is supercritical.   Assume that the
        coefficient of contraction C_c for the gate is 0.8.

3-29   A 4-m-wide horizontal rectangular channel carrying a discharge of 12 m³/s is
        contracted to a width of 2.5 m before it terminates in a freefall. The flow in the
        channel is controlled by a sluice gate located in the wider reach of the channel.
        Neglecting all losses, determine the flow depth upstream from the gate for a gate
        opening of 0.8 m. The coefficient of contraction for the gate is 0.7.

3-30   A rectangular channel carries water at a rate of 1.5 m³/s per meter width, has an
        adverse bottom slope of 0.003, and terminates in a freefall. Determine the maximum
        gate opening for the flow in the channel downstream from the gate to remain
        supercritical. What would happen if the gate opening is made larger than the
        maximum gate opening calculated earlier?

3-31   A 6-m-wide rectangular channel, carrying 30 m³/s, is contracted by a straight-wall
        contraction to 3 m width. The channel terminates in a freefall immediately
        downstream from the transition. The length of the transition is 30 m. The channel
        bottom within the transition is raised and the bottom profile is given by the equation
        $z = (x/15)(2 - x/15)$, where z = change in elevation and x = distance from the
        upstream end of the transition. Determine the location of the control section and
        sketch the flow profile.

3-32   Water flows through a rectangular channel of constant width at the rate of 1 m³/s per
        meter width. The channel in a section has a "hump" of 30 cm in height. What is the
        largest flow depth downstream from the hump that would not affect the flow profile
        upstream from the hump?

3-33   A flow with Froude number 2.0 flows up a wide channel with an adverse slope of
        0.001. The channel terminates in a freefall. Find the maximum possible distance
        between the freefall and the section where the depth is 30 cm. Neglect friction
        losses.

3-34   Water flows through a rectangular channel of 3 m width into a reservoir at the rate
        of 35 m³/s. A bridge pier of 1 m width is located in the center of the channel. What
        is the highest water level in the reservoir above the channel outlet that would not
        affect the flow profile upstream from the pier?

3-35   The depth of water upstream from a sluice gate in a rectangular channel is 3.6 m.
        The gate opening is 0.6 m. Some distance downstream from the gate the depth of

water is 3 m. Assume that the contraction coefficient for the gate is 0.62. What is the discharge per unit width?

3-36 A long rectangular channel terminates in a freefall. Determine the depth at the freefall if (a) the unit discharge in the channel is 0.8 $m^3/s/m$ at a normal depth of 1.0 m and (b) the unit discharge is 1.9 $m^3/s/m$ at a normal depth of 0.5 m.

3-37 Applying the continuity and momentum equations to the analysis of a submerged jump which occurs downstream of a sluice gate in a rectangular channel (Figure 3-22c), prove that

$$\frac{y_2'}{y_3} = \sqrt{1 + 2F_3^2\left(1 - \frac{y_3}{y_2}\right)}$$

where

$$F_3^2 = q^2/\left(gy_3^3\right)$$

APPENDIX C

SURGE PROPAGATION

The movement of a surge in rectangular channels was analyzed in Section 3-1. In this appendix the equations of motion of a surge in channels of arbitrary cross-sectional shape are presented. The equations are based on the same assumptions given in Section 3-1 for rectangular channels. The continuity and momentum equations (1-68) and (1-76) for the control volume between sections 1 and 2 in Figure 3-1b can be written as

$$A_1 (V_1 - V_w) = A_2 (V_2 - V_w) \tag{C-1}$$

$$\rho A_1 (V_1 - V_w)(V_2 - V_1) = \gamma (A_1 \bar{y}_1 - A_2 \bar{y}_2) \tag{C-2}$$

The surge velocity from Eq. (C-1) is

$$V_w = \frac{V_1 A_1 - V_2 A_2}{A_1 - A_2} \tag{C-3}$$

Substituting for V_w from Eq. (C-3) into Eq. (C-2), the latter can be written as

$$V_2 - V_1 = \pm \sqrt{\frac{g(A_1 \bar{y}_1 - A_2 \bar{y}_2)}{A_1 A_2}(A_1 - A_2)} \tag{C-4}$$

Substituting for $(V_2 - V_1)$ from Eq. (C-4) into Eq. (C-2), the latter becomes

$$V_1 - V_w = \pm \sqrt{\frac{gA_2}{A_1} \frac{(A_1 \bar{y}_1 - A_2 \bar{y}_2)}{(A_1 - A_2)}} \tag{C-5}$$

Addition of Eq. (C-4) to Eq. (C-5) gives

$$V_2 - V_w = \pm \sqrt{\frac{gA_1}{A_2} \frac{(A_1 \bar{y}_1 - A_2 \bar{y}_2)}{(A_1 - A_2)}} \tag{C-6}$$

The above equations involve five variables, namely, y_1, y_2, V_1, V_2, and V_w. For given values of any three variables, the remaining two variables can be determined from the above equation. For a small surge these equations can be simplified as follows. For a small surge let $y_1 = y$, $y_2 = y + dy$, $A_1 = A$, and $A_2 = A + dA$; then $A_2 = A + B\, dy$ and

$$A_1 \bar{y}_1 = A\bar{y}; \quad A_2 \bar{y}_2 = A(\bar{y}+dy)+Bdy\frac{dy}{2} = A(\bar{y}+dy); \quad (A_1 \bar{y}_1 - A_2 \bar{y}_2)= Ady$$

Using the above relations for small surge, Eqs. (C-5) and (C-6) become identical to

$$V_w - V = \sqrt{g\frac{A}{B}} = \sqrt{gD} \qquad\qquad\qquad (C\text{-}7)$$

As defined earlier, the celerity c of a disturbance is the velocity of the disturbance relative to the undisturbed flow velocity, i.e., $c = V_w - V$. Thus celerity c from Eq. (C-7) is given by

$$c = \sqrt{gD} \qquad\qquad\qquad (C\text{-}8)$$

APPENDIX D

COORDINATES OF THE DIMENSIONLESS
SPECIFIC-ENERGY CURVE

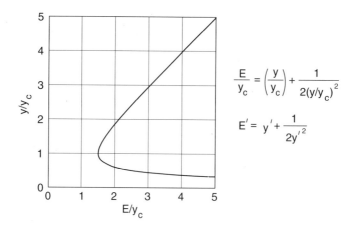

$$\frac{E}{y_c} = \left(\frac{y}{y_c}\right) + \frac{1}{2(y/y_c)^2}$$

$$E' = y' + \frac{1}{2y'^2}$$

Figure D-1. Dimensionless specific-energy curve.

Table D-1. $E' = E/y_c$ as a function of $y' = y/y_c$

y'	0.00	0.01	0.02	0.03	0.04	0.05	0.06	0.07	0.08	0.09
						E'				
0.0		5000	1250	555.6	312.5	200.1	138.9	102.1	78.21	61.82
0.1	50.10	41.43	34.84	29.71	25.65	22.37	19.69	17.47	15.61	14.04
0.2	12.70	11.55	10.55	9.682	8.920	8.250	7.656	7.129	6.658	6.235
0.3	5.856	5.513	5.203	4.921	4.665	4.432	4.218	4.022	3.843	3.677
0.4	3.525	3.384	3.254	3.134	3.023	2.919	2.823	2.734	2.650	2.572
0.5	2.500	2.432	2.369	2.310	2.255	2.203	2.154	2.109	2.066	2.026
0.6	1.989	1.954	1.921	1.890	1.861	1.833	1.808	1.784	1.761	1.740
0.7	1.720	1.702	1.684	1.668	1.653	1.639	1.626	1.613	1.602	1.591
0.8	1.581	1.572	1.564	1.556	1.549	1.542	1.536	1.531	1.526	1.521
0.9	1.517	1.514	1.511	1.508	1.506	1.504	1.502	1.501	1.501	1.500
1.0	1.500	1.500	1.501	1.501	1.502	1.504	1.505	1.507	1.509	1.511
1.1	1.513	1.516	1.519	1.522	1.525	1.528	1.532	1.535	1.539	1.543
1.2	1.547	1.552	1.556	1.560	1.565	1.570	1.575	1.580	1.585	1.590
1.3	1.596	1.601	1.607	1.613	1.618	1.624	1.630	1.636	1.642	1.649
1.4	1.655	1.662	1.668	1.674	1.681	1.688	1.695	1.701	1.708	1.715
1.5	1.722	1.729	1.736	1.744	1.751	1.758	1.766	1.773	1.780	1.788

(continued)

Table D-1. (Continued)

y'	0.00	0.01	0.02	0.03	0.04	0.05	0.06	0.07	0.08	0.09
						E'				
1.6	1.795	1.803	1.810	1.818	1.826	1.834	1.841	1.849	1.857	1.865
1.7	1.873	1.881	1.889	1.897	1.905	1.913	1.921	1.930	1.938	1.946
1.8	1.954	1.963	1.971	1.979	1.988	1.996	2.004	2.013	2.022	2.030
1.9	2.038	2.047	2.056	2.064	2.073	2.082	2.090	2.099	2.108	2.116
2.0	2.125	2.134	2.142	2.151	2.160	2.169	2.178	2.187	2.196	2.204
2.1	2.213	2.222	2.231	2.241	2.249	2.258	2.267	2.276	2.285	2.294
2.2	2.303	2.312	2.322	2.330	2.340	2.349	2.358	2.367	2.376	2.385
2.3	2.394	2.404	2.413	2.422	2.431	2.440	2.450	2.459	2.468	2.478
2.4	2.487	2.496	2.505	2.515	2.524	2.533	2.543	2.552	2.561	2.571
2.5	2.580	2.589	2.599	2.608	2.618	2.627	2.636	2.646	2.655	2.665
2.6	2.674	2.683	2.693	2.702	2.712	2.721	2.731	2.740	2.750	2.759
2.7	2.769	2.778	2.788	2.797	2.807	2.816	2.826	2.835	2.845	2.854
2.8	2.864	2.873	2.883	2.892	2.902	2.912	2.921	2.931	2.940	2.950
2.9	2.960	2.969	2.979	2.988	2.998	3.008	3.017	3.027	3.036	3.046
3.0	3.056	3.064	3.075	3.084	3.094	3.104	3.113	3.123	3.133	3.142
3.1	3.152	3.162	3.171	3.181	3.191	3.200	3.210	3.220	3.229	3.239
3.2	3.249	3.258	3.268	3.278	3.288	3.297	3.307	3.317	3.326	3.336
3.3	3.346	3.356	3.365	3.375	3.385	3.395	3.404	3.414	3.424	3.434
3.4	3.443	3.453	3.463	3.472	3.482	3.492	3.502	3.512	3.521	3.531
3.5	3.541	3.551	3.560	3.570	3.580	3.590	3.600	3.609	3.619	3.629
3.6	3.639	3.648	3.658	3.668	3.678	3.688	3.697	3.707	3.717	3.727
3.7	3.737	3.746	3.756	3.766	3.776	3.786	3.795	3.805	3.815	3.825
3.8	3.835	3.844	3.854	3.864	3.874	3.884	3.894	3.903	3.913	3.923
3.9	3.933	3.943	3.953	3.962	3.972	3.982	3.992	4.002	4.012	4.021
4.0	4.031	4.041	4.051	4.061	4.071	4.080	4.090	4.100	4.110	4.120
4.1	4.130	4.140	4.150	4.159	4.169	4.179	4.189	4.199	4.209	4.218
4.2	4.228	4.238	4.248	4.258	4.268	4.278	4.288	4.297	4.307	4.317
4.3	4.327	4.337	4.347	4.357	4.366	4.376	4.386	4.396	4.406	4.416
4.4	4.426	4.436	4.446	4.456	4.465	4.475	4.485	4.495	4.505	4.515
4.5	4.525	4.535	4.544	4.554	4.564	4.574	4.584	4.594	4.604	4.614
4.6	4.624	4.634	4.643	4.653	4.663	4.673	4.683	4.693	4.703	4.713
4.7	4.723	4.732	4.742	4.752	4.762	4.772	4.782	4.792	4.802	4.812
4.8	4.822	4.832	4.842	4.851	4.861	4.871	4.881	4.891	4.901	4.911
4.9	4.921	4.931	4.941	4.951	4.960	4.970	4.980	4.990	5.000	5.010

CHAPTER 4

GRADUALLY VARIED FLOW

Flow in open channels is generally nonuniform. In gradually varied flow the rate of change of depth with distance is small; therefore the streamline curvature is negligible and the pressure distribution is hydrostatic. The qualitative sketching of gradually varied flow profiles can be learned by simply analyzing the nature of the governing equations, while the quantitative sketching of gradually varied flow profiles requires integration of the governing equations. Gradually varied flow can occur with or without lateral flow. The former flow is termed as spatially varied flow and is dealt with in Chapter 6. The qualitative sketching of gradually varied flow profiles without lateral flow is discussed in this chapter, while the integration techniques of the gradually varied flow equations are presented in Chapter 5.

4-1 Governing Equations

The governing equations for gradually varied flow without lateral flow ($q_\ell = 0$) are derived in Section 1-8, and these equations are as follows:

Continuity equation: $\qquad Q = VA = \text{constant} \quad$ [Eq. (1-44)]

Energy equation: $\qquad \dfrac{dH}{dx} = -S_f \qquad$ [Eq.(1-55)]

where $\qquad H = \alpha \dfrac{V^2}{2g} + y\cos\theta + G_0$

Using the continuity equation (1-44), the energy equation (1-55) can be written as

$$\alpha \frac{Q^2}{2g} \frac{d}{dx}\left(\frac{1}{A^2}\right) + \cos\theta \frac{dy}{dx} + \frac{dG_0}{dx} = -S_f \qquad (4\text{-}1)$$

The energy coefficient α is assumed to be not a function of x. In the most general case the cross sectional area A is a function of both x and y, i.e., $A = A(x,y)$. Then

$$\frac{dA}{dx} = \left(\frac{\partial A}{\partial x}\right)_{y=\text{const}} + \frac{dy}{dx}\left(\frac{\partial A}{\partial y}\right)_{x=\text{const}}$$

An expression for $(d/dx)(1/A^2)$ in Eq. (4-1) can be derived as follows:

$$\frac{d}{dx}\left(\frac{1}{A^2}\right) = \frac{d}{dA}\left(\frac{1}{A^2}\right)\frac{dA}{dx} = -\frac{2}{A^3}\left[\left(\frac{\partial A}{\partial x}\right)_{y=const} + \frac{dy}{dx}\left(\frac{\partial A}{\partial y}\right)_{x=const}\right] \quad (4\text{-}2)$$

A small change δA in area A due to a small change δy in depth y can be written as

$$\delta A = B\,\delta y \quad (4\text{-}3)$$

where B is the water-surface width. In the limit as $\delta y \to 0$, Eq. (4-3) yields

$$\left(\frac{dA}{dy}\right)_{x=const} = B \quad (4\text{-}4)$$

Then Eq. (4-2) becomes

$$\frac{d}{dx}\left(\frac{1}{A^2}\right) = -\frac{2}{A^3}\left[\left(\frac{\partial A}{\partial x}\right) + B\frac{dy}{dx}\right] \quad (4\text{-}5)$$

Substituting Eq. (4-5) into Eq. (4-1), noting that $dG_0/dx = -S_0$ and rearranging the terms in the resulting equation, one obtains

$$\frac{dy}{dx} = \frac{S_0 - S_f + (\alpha Q^2/gA^3)(\partial A/\partial x)}{\cos\theta\{1 - (\alpha BQ^2)/(gA^3\cos\theta)\}} \quad (4\text{-}6)$$

For prismatic channels $\partial A/\partial x$ is zero. If one defines the Froude number, F, according to Eq. (3-31) as

$$F^2 = \frac{\alpha BQ^2}{gA^3\cos\theta} = \frac{V^2}{gD\cos\theta/\alpha} \quad [\text{Eq.}(3\text{-}31)]$$

where $D = A/B$ = hydraulic depth, then Eq. (4-6) can be written as

$$\frac{dy}{dx} = \frac{S_0 - S_f}{\cos\theta\left(1 - F^2\right)} \quad (4\text{-}7)$$

For most practical problems, $\cos\theta \approx 1$ as the bottom slopes are small. With this approximation Eq. (4-7) reduces to

$$\frac{dy}{dx} = \frac{S_0 - S_f}{1 - F^2} \quad (4\text{-}8)$$

Equation (4-8), referred to hereafter as the gradually varied flow equation, is a first-order nonlinear ordinary differential equation whose solution requires one boundary condition in terms of y at the control section.

A solution of Eq. (4-8) requires that the friction slope S_f should be expressed in terms of flow variables. Common practice is to approximate S_f by the uniform-flow formula [Chézy equation,Eq. (2-4); Manning equation (2-11a) or (2-15)], i.e.,

$$S_f = \frac{V^2}{C^2 R} = \frac{n^2 V^2}{R^{4/3}}$$ (4-9)

or

$$S_f = \left(\frac{Q}{K}\right)^2$$ (4-10)

4-2 Classification of Flow Profiles

For purposes of discussion, bottom slopes are categorized into *sustaining* and *nonsustaining* slopes. A positive bottom slope, i.e., $S_0 > 0$, is termed a sustaining slope, and a bottom slope equal to or less than zero, i.e., $S_0 \leq 0$, is called a nonsustaining slope. The sustaining slopes ($S_0 > 0$) are further classified depending on the relative magnitudes of the normal and critical depths: *mild slope* if $y_n > y_c$; *critical slope* if $y_n = y_c$; and *steep slope* if $y_n < y_c$. The nonsustaining slopes ($S_0 \leq 0$) are *horizontal slopes* ($S_0 = 0$) and *adverse slopes* ($S_0 < 0$).

The normal-depth line (referred to as NDL) and the critical-depth line (referred to as CDL) divide the flow region in three zones, as shown in Figure 4-1: Zone 1 above the two lines; Zone 2 between the two lines; and Zone 3 below the two lines. However, there are only two zones in channels with a nonsustaining slope or a critical slope. There is no Zone 1 in channels with nonsustaining slopes as the normal depth is infinite in horizontal channels and is nonexistent in channels with adverse slope. There is no Zone 2 in channels with critical slope as $y_n = y_c$.

A letter with a numeral subscript is used to designate flow profiles. The first letter of the names of the slopes is used to indicate the type of bottom slope. The numeral subscript refers to the zone. According to this notation, as an example, an M_3 profile is a water-surface profile in a mild channel in Zone 3.

On the basis of the above slope and zone types, 12 types of flow profiles are possible. There are 3 in mild channels, namely, M_1, M_2, and M_3; 3 in steep

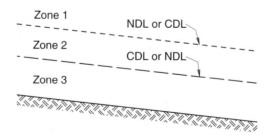

Figure 4-1. Zones for classification of surface profiles.

channels, namely, S_1, S_2, and S_3; 2 in critical-slope channels, namely, C_1 and C_3; 2 in horizontal channels, namely, H_2 and H_3; and 2 in adverse-slope channels, namely, A_2 and A_3. Some persons include the uniform-flow profile in critical-slope channels as the C_2 profile.

Example 4-1. Flow depth in a section of the nonuniform-flow reach of the channel in Example 2-1 is 2.9 m. Determine the type of flow profile in the channel
From Examples 2-1 and 3-4, $y_n = 3.17$ m and $y_c = 2.63$ m. The slope is mild, because $y_n > y_c$. For $y = 2.9$ m, $y_n > y > y_c$; the profile lies in Zone 2. The type of profile is, therefore, M_2.

Backwater and Drawdown Curves

Water-surface profiles can be further classified as *backwater curves* if dy/dx is positive and as *drawdown curves* if dy/dx is negative. The sign of the term dy/dx can be determined from Eq. (4-8) using the following inequalities:

$$S_f \gtrless S_0 \text{ if } y \lessgtr y_n \qquad (4\text{-}11)$$

$$F \gtrless 1 \text{ if } y \lessgtr y_c \qquad (4\text{-}12)$$

The above inequalities are based on the facts that both S_f [see Eq. (4-10)] and F [see Eq. (3-31)] decrease with increasing depth y, $S_f = S_0$ when $y = y_n$ and $F = 1$ when $y = y_c$. It can be easily shown (Example 4-2) from Eq. (4-8) that the profiles in Zones 1 and 3 are backwater curves and the profiles in Zone 2 are drawdown curves.

Example 4-2. Is the profile in Example 4-1 a backwater or a drawdown curve?
Here, $S_f > S_0$ because $y < y_c$, and $F < 1$ because $y > y_c$. The numerator and the denominator in Eq. (4-8) are negative and positive, respectively; consequently, the sign of dy/dx is negative. The flow profile is a drawdown curve

PROBLEMS

4-1   A 2-m-wide rectangular channel ($n = 0.013$) carries a discharge of 4 m^3/s at a slope of 0.008. What type of flow profile exists in the channel if the actual depth in a section is 1.5 m?

4-2   A trapezoidal channel with a base width of 5 m and side slopes of 1:1 carries a discharge of 90 m^3/s. Determine the critical slope for $n = 0.013$.

4-3   Show that the critical slope, S_{cn}, at a given normal depth y_n may be expressed by

$$S_{cn} = \frac{g n^2 D_n}{R_n^{4/3}}$$

and that this slope for a wide channel is

$$S_{cn} = \frac{gn^2}{y_n^{1/3}}$$

4-4   A trapezoidal channel with a base width of 2 m, side slopes of 2H:1V, and Manning n of 0.015 carries water at a normal depth of 3.5 m when the channel slope is critical. Determine the slope.

4-5   Show that the gradually varied flow equation for flow in a nonprismatic channel may be expressed as

$$\frac{dy}{dx} = \frac{S_0 - S_f + (F^2/B)(\partial A/\partial x)}{1 - F^2}$$

4-6   Assuming that the Chézy C is constant, show that Eq. (4-8) for wide rectangular channel reduces to

$$\frac{dy}{dx} = S_0 \frac{1 - (y_n/y)^3}{1 - (y_c/y)^3}$$

What are the shapes of the flow profiles in channels with critical slope?

4-3 Characteristics of Flow Profiles

Having established the sign of dy/dx in each zone from Eqs. (4-11) and (4-12), the shape of a profile can be established by determining from Eq. (4-8) the water-surface slope at the boundaries of each zone

Water-Surface Slope at Zonal Boundaries

Both $S_f \rightarrow 0$ and $F \rightarrow 0$ as $y \rightarrow \infty$; Eq. (4-8) shows that $dy/dx \rightarrow S_0$, i.e., the flow surface is asymptotic to a horizontal line. Here, $S_f \rightarrow S_0$ as $y \rightarrow y_n$; dy/dx from Eq. (4-8) is zero, i.e., the flow surface is asymptotic to the normal-depth line. The Froude number $F \rightarrow 1$ as $y \rightarrow y_c$; Eq. (4-8) gives that $dy/dx \rightarrow \infty$, i.e., the profile intersects the critical-depth line at a right angle. Both $S_f \rightarrow \infty$ and $F \rightarrow \infty$ as $y \rightarrow 0$; Eq. (4-8) shows that $dy/dx \rightarrow \infty/\infty$, an indeterminate form. It can be shown that dy/dx has a positive finite limit (Chow, 1959).

Shapes of Flow Profiles

With the foregoing introduction, the characteristics of flow profiles are summarized in Table 4-1, and the shapes of the various profiles are sketched in Figure 4-2. For example, consider a profile in which $y_n > y > y_c$. Because $y_n > y_c$, the profile belongs to a mild channel. The profile lies between the NDL and the CDL, i.e., in Zone 2. The type of the flow profile is, therefore, M_2. It is a drawdown curve, because $S_f > S_0$ and $F < 1$, as explained in Example 4-2. The profile approaches the NDL tangentially, because $dy/dx \to 0$ as $y \to y_n$. It approaches the CDL at a right angle, since $dy/dx \to \infty$ as $y \to y_c$. In a similar fashion the shapes of the other profiles in Figure 4-2 can be sketched. The flow in the vicinity of the critical depth becomes so highly curvilinear as to make Eq. (4-8) inapplicable. The computations based on Eq. (4-8) are not accurate in the vicinity of the critical depth and the profiles there are shown by dotted curves.

The profiles in Zone 1, namely M_1 and S_1, normally occur upstream from a control structure such as weirs, dams, etc. The profiles in Zone 2, except the S_2 curve, form upstream from free overfalls. The S_2 curve occurs at the entrance region of a steep channel leading from a reservoir. The profiles in Zone 3 normally occur downstream from control structures such as gates. Several examples illustrating the locations of different profiles are presented in the ensuing sections.

Mechanism of Specific Energy Gain

Though it is a common knowledge that hydroelectric energy can be generated by creating a reservoir behind a dam, the knowledge about the source of the energy is not that common. To understand the mechanism responsible for the energy, let us go back to the mechanical-energy equation   (1-55), which on substituting $H = E + G_0$ can be written as

$$\frac{dE}{dx} = S_0 - S_f \qquad (4\text{-}13)$$

Table 4-1. Characteristics of the Flow Profiles

Channel Slope	Profile Type	Relation of y to y_n and y_c	S_f	F dy/dx	Sign
Mild	M_1	$y>y_n>y_c$	$<S_0$	<1	$+$
	M_2	$y_n>y>y_c$	$>S_0$	<1	$-$
	M_3	$y_n>y_c>y$	$>S_0$	>1	$+$
Steep	S_1	$y>y_c>y_n$	$<S_0$	<1	$+$
	S_2	$y_c>y>y_n$	$<S_0$	>1	$-$
	S_3	$y_c>y_n>y$	$>S_0$	>1	$+$
Critical	C_1	$y>y_n=y_c$	$<S_0$	<1	$+$
	C_3	$y_n=y_c>y$	$>S_0$	>1	$+$
Horizontal	H_2	$y_n>y>y_c$	$>S_0$	<1	$-$
	H_3	$y_n>y_c>y$	$>S_0$	>1	$+$
Adverse	A_2	$y>y_c$	$>S_0$	<1	$-$
	A_3	$y_c>y$	$>S_0$	>1	$+$

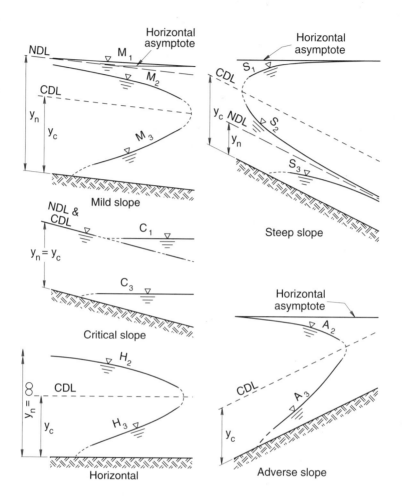

Figure 4-2. Water-surface profiles in gradually varied flow.

Assuming uniform flow in the channel prior to the construction of the dam, the energy slope is equal to the channel slope. The energy slope S_f represents the energy loss in friction per unit length of the channel, and it decreases with increasing depth. The energy loss in friction in the channel reach upstream of the dam is significantly lower after the construction of the dam. The energy saved by reducing the energy loss in friction is the source of the energy in question. It can be easily shown that the flow profiles for which dE/dx is positive are M_1, S_1, S_2, and C_1 (Prob. 4-8). For the remaining profiles dE/dx is negative. This information is used later in analyzing the flow profiles.

PROBLEMS

4-7   A 2-m-wide rectangular channel has a slope of 0.009, discharge of 4.2 m³/s, and Manning n of 0.015. What type of flow profile exists for the actual depth of flow of (i) 1.5 m, (ii) 0.6 m, and (iii) 0.2 m. Where might these profiles occur in engineering practice?

4-8   Show that dE/dx is positive for M_1, S_1, S_2, and C_1 curves and negative for the remaining curves.

4-4 Sketching Flow Profiles

Before starting to integrate Eq. (4-8) for computing flow profiles, one should sketch the possible flow profiles in a given channel. The general procedure for sketching the possible profiles is as follows:

1.   Plot the longitudinal profile of the channel bottom using an exaggerated vertical scale.
2.   For each channel reach compute y_n using the Manning equation (2-13), and plot the NDL.
3.   For each channel reach compute y_c using the critical-depth relation (3-30), and plot the CDL.
4.   Locate all possible controls. Three types of controls, namely, the artificial channel control (ACC), the normal-depth control (NDC) and the critical-depth control (CDC), and their locations are described in Section 3-3. The locations of the different controls are reiterated below.
       a. An ACC is located in the vicinity of a control structure.
       b. An NDC is located at (i) the upstream end of a mild reach and (ii) the downstream end of a steep reach.
       c. A CDC is located at (i) the choke of a channel transition and (ii) the downstream end of a subcritical-flow regime. The latter is possible only in channels with a mild, horizontal, or adverse slope, because water-surface profiles that pass through the critical depth at the downstream end and occur in a subcritical-flow regime are only M_2, H_2, and A_2 (see Figure 4-2). As a limiting case of the curve A_2, the downstream end of a subcritical-flow regime can be an outlet of a reservoir discharging into a channel. In other words there exists a CDC at the outlet of a reservoir.
5.   Starting from the control depth in each control section, trace a possible profile in each reach. If more controls than one are possible in any section, the control that requires the maximum total head to maintain the given flow acts as an active control. Because a CDC requires the minimum total head to maintain the flow, the CDC is never active if another control exists in the section in addition to the CDC. In such cases it is not necessary to show the CDC in the figure.
6.   If the flow is supercritical in an upstream reach and subcritical in a downstream reach, a hydraulic jump usually develops, and the flow in the hydraulic jump changes from supercritical to subcritical. The jump can

form either in the upstream steep reach or downstream mild reach. The procedure to identify the reach in which the jump forms is discussed later.

Prismatic Channels with Change in Slope and Roughness

If there is a change in slope and/or roughness in a prismatic channel, the normal depths in the two reaches of the channel are different. The type of flow profile in each reach depends upon the flow condition in the reaches. Flow profiles for the various flow conditions in the two reaches are shown in Figure 4-3 and also summarized in Table 4-2, which is self-explanatory. In Table 4-2 reaches 1 and 2 refer to upstream and downstream reaches, respectively, and the quantities with subscripts 1 and 2 belong to reaches 1 and 2, respectively. The two reaches are considered very long so that the controls at the two ends of either reach do not interact with each other, and the flow in the reaches away from the controls occurs at normal depth as it is controlled by the channel friction. The effect on flow profile of only those controls that exist in the section with the break in the slope is being examined. The controls at the upstream end of the upstream reach and the downstream end of the downstream reach are, therefore, not shown in the sketch.

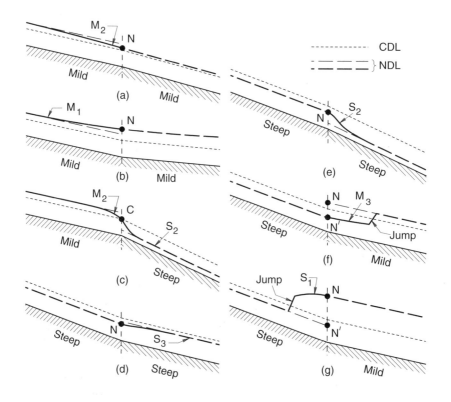

Figure 4-3. Flow profiles in channels with change in slope and roughness.

Table 4-2. Types of Flow Profile in Prismatic Channels with a Change of Slope and Roughness

Reach 1 Slope	Reach 2 Slope	Possible Control(s)	Figure No.	Relative Magnitudes of Normal Depths and Specific Energies	Active Control	Flow Profiles Reach 1	Reach 2
Mild	Mild	NDC at N	4-3a	$y_{n1} > y_{n2}; E_{n1} > E_{n2}$	NDC at N	M_2	UF
		CDC at C$^{\#}$	4-3b	$y_{n2} > y_{n1}; E_{n2} > E_{n1}$	NDC at N	M_1	UF
Mild	Steep	CDC at C	4-3c	$y_{n1} > y_{n2}$ $E_{n1} > E_{n2}$ * or $E_{n1} < E_{n2}$ **	CDC at C	M_2	S_2
Steep	Steep	NDC at N	4-3d	$y_{n2} > y_{n1}; E_{n1} > E_{n2}$	NDC at N	UF	S_3
			4-3e	$y_{n1} > y_{n2}; E_{n2} > E_{n1}$	NDC at N	UF	S_2
Steep	Mild	NDC at N	4-3f	$y_{n2} > y_{n1}; (E_{n1} - \Delta E_{j1}) > E_{n2}$	NDC at N´	UF	M_3 & J
		NDC at N´	4-3g	$y_{n2} > y_{n1}; E_{n2} > (E_{n1} - \Delta E_{j1})$	NDC at N	J & S_1	UF

Legend: J = Jump; UF = Uniform flow; ΔE_{j1} = Energy loss in the jump in reach 1.

*The loss of the specific energy within the M_2 curve is more than the gain of the specific energy within the S_2 curve.

**The loss of the specific energy within the M_2 curve is less than the gain of the specific energy within the S_2 curve.

CDC is not shown in Figures 4-3a and 4-3b.

Furthermore, if there is an ACC or an NDC in addition to a CDC in a section, then the CDC, because it is not active, is not shown in the sketch.

As an example consider the flow conditions given in row 1 of Table 4-2 and Figure 4-3a. Reaches 1 and 2 are mild; therefore the possible controls in the two reaches are the NDCs at the upstream end of the mild channels and the CDCs at the downstream end of the mild channels. Only an NDC at N at the upstream end of reach 2 is shown in Figure 4-3a and Table 4-2; the other controls are not shown due to the reasons mentioned above. There is only one active control at N in the section at the break in the channel slope. The only possible flow profile in reach 1 that passes through N and approaches the NDL is an M_2 curve. Flow is uniform in reach 2, unless it is affected by a downstream control other than the channel friction. The profile is shown in Figure 4-3a. The length of the M_2 profile is such that the specific-energy loss in the channel reach covered by the M_2 profile is equal to $(E_{n1} - E_{n2})$, where E_{n1} and E_{n2} are the specific energies at the normal depth in reach 1 and 2, respectively.

In the case where reach 1 is mild and reach 2 is steep, the only control in the section with the break in the channel slope is a CDC, as shown in Figure 4-3c and row 3 in Table 4-2. The only possible flow profiles that pass through C and approach the normal-depth lines are an M_2 curve in reach 1 and an S_2 curve in reach 2. There is a loss of the specific energy in the M_2 curve and a gain of the specific energy in the S_2 curve. If $E_{n1} > E_{n2}$, the loss of energy in the M_2 curve is larger than the gain of energy in the S_2 curve, and if $E_{n1} < E_{n2}$, the loss of energy in the M_2 curve is smaller than the gain of energy in the S_2 curve.

In cases where reach 1 is steep and reach 2 is mild, there can be two possible normal-depth controls in the same section, as shown in Figures 4-3f and 4-3g and rows 6 and 7 in Table 4-2. One possible NDC is at point N and the other NDC is at point N', but only one NDC can be active at a time. If the NDC at point N is active, an S_1 curve preceded by a hydraulic jump forms in reach 1, as shown in Figure 4-3g. If the NDC at point N' is active, an M_3 curve followed by a hydraulic jump forms in reach 2, as shown in Figure 4-3f. The active control between points N and N' is determined as follows. Let the conjugate depth of y_{n1} be y' and the specific energy at depth y' be $E_{y'}$. Then $E_{y'} = (E_{n1} - \Delta E_{j1})$, where ΔE_{j1} is the energy loss in the jump in reach 1. If $E_{n2} > (E_{n1} - \Delta E_{j1})$, an S_1 curve in which the specific energy is gained along the flow direction should exist downstream from the jump. The control is, therefore, at point N. If $E_{n2} < (E_{n1} - \Delta E_{j1})$, an M_3 curve in which the specific energy is lost, in addition to the energy loss in the jump, should exist; hence the control is at point N'. The active control can also be determined by comparing y' to y_{n2} (Example 4-4). If $y' < y_{n2}$, the control is at point N and if $y' > y_{n2}$, the control is at point N' (why?). The depth y' can be obtained from Eq. (3-17) for rectangular and from Eq. (3-21) for nonrectangular channels.

Example 4-3. A rectangular channel 6 m wide conveys 100 m³/s of water. The channel slope is 0.003 for the first reach and then there is a sudden change in the slope to 0.01 in the second reach. The Manning n for the channel is 0.015. Sketch the water-surface profile in the channel. The longitudinal profile of the channel bottom is plotted first, as shown in the figure below.

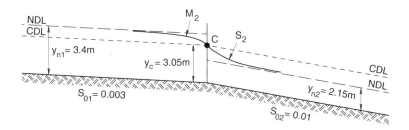

Example 4-3

Normal depths: Let y_{n1} and y_{n2} be the normal depths in the first and second reaches, respectively. The Manning equation for the first reach can be written as

$$100 = \frac{1}{0.015} \times 6 y_{n1} \times \left(\frac{6 y_{n1}}{6 + 2 y_{n1}} \right)^{2/3} \times \sqrt{0.003}$$

or

$$\frac{y_{n1}^{5/3}}{\left(6 + 2 y_{n1}\right)^{2/3}} = \frac{100 \times 0.015}{6^{5/3} \times \sqrt{0.003}} = 1.38$$

By the trial-and-error method $y_{n1} = 3.40$ m. Similarly for the second reach

$$\frac{y_{n2}^{5/3}}{\left(6 + 2 y_{n2}\right)^{2/3}} = \frac{100 \times 0.015}{6^{5/3} \times \sqrt{0.01}} = 0.76$$

By the trial-and-error method $y_{n2} = 2.15$ m.

Critical depth: $\qquad q = 100/6 = 16.67 \text{ m}^3/\text{s/m}$

$$y_c = \sqrt[3]{\frac{q^2}{g}} = \sqrt[3]{\frac{16.67^2}{9.81}} = 3.05 \text{ m}$$

The NDL and the CDL for each reach are sketched in the figure. Because $y_{n1} > y_c$ and $y_{n2} < y_c$, the first reach is mild and the second reach is steep.
Control sections: There is no ACC. The NDCs (not shown in the figure) are at the upstream end of the mild reach and at the downstream end of the steep reach. The CDC is at point C at the downstream end of the mild reach.
Flow profile: The only possible flow profile that passes through C is an M_2 curve in the mild reach and an S_2 curve in the steep reach:

$$E_{n1} = y_{n1} + \frac{q^2}{2 g y_{n1}^2} = 2.15 + \frac{16.67^2}{2 \times 9.81 \times 2.15^2} = 5.21 \text{ m}$$

$$E_{n2} = y_{n2} + \frac{q^2}{2 g y_{n2}^2} = 3.40 + \frac{16.67^2}{2 \times 9.81 \times 3.40^2} = 4.63 \text{ m}$$

$$E_c = \tfrac{3}{2}y_c = \tfrac{3}{2}\times 3.05 = 4.58 \text{ m}$$

There is loss of the specific energy of $(4.63 - 4.58) = 0.05$ m in the channel reach covered by the M_2 curve, and there is gain of the specific energy of $(5.21 - 4.58) = 0.63$ m in the channel reach covered by the S_2 curve. The loss of the specific energy within the M_2 curve is less than the gain of the specific energy within the S_2 curve, because $E_{n2} > E_{n1}$.

Example 4-4. Sketch the flow profile if the slopes in the first and second reaches of the channel in Example 4-3 are interchanged.

The channel bottom, the NDL, and the CDL for both reaches are drawn in the figure below. Reaches 1 and 2 are steep and mild, respectively.

Control sections: There is no ACC; the NDCs are at points N and N'; and there is a CDC at the downstream end of reach 2 (not shown).

Flow profile: The flow changes from supercritical in reach 1 to subcritical in reach 2; a hydraulic jump will form. Out of the two controls in the same section, points N and N', only one can be active. The active control is determined as follows. Let the conjugate depth of y_{n1} be y'. If $y' < y_{n2}$, the control is at point N, and if $y' > y_{n2}$, the control is at point N'. The depth y' can be obtained from Eq. (3-17):

$$F_1^2 = \frac{q^2}{gy_{n1}^3} = \frac{16.67^2}{9.81\times 2.15^3} = 2.85$$

From Eq. (3-17), $\dfrac{y'}{y_{n1}} = \dfrac{1}{2}\left(\sqrt{1+8F_1^2} - 1\right) = \dfrac{1}{2}\left(\sqrt{1+8\times 2.85} - 1\right) = 1.94$

or $y' = 1.94\times y_{n1} = 1.94\times 2.15 = 4.17 \text{ m}$

Since $y' > y_{n2}$, the control is at point N'; an M_3 curve followed by a hydraulic jump forms in reach 2.

The active control can also be determined by the energy approach. Let E_{n1} and E_{n2} be the specific energy at the normal depth in reaches 1 and 2, respectively, and let E_y be the specific energy at depth y'. Then $E_{y'} = (E_{n1} - \Delta E_{j1})$, where ΔE_{j1} is the energy loss in the jump in reach 1. If $E_{n2} > (E_{n1} - \Delta E_{j1})$, an S_1 curve in which the specific energy is gained

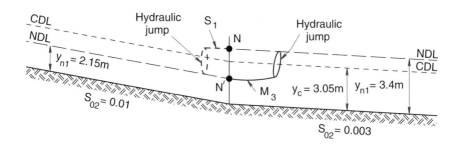

Example 4-4.

should exist downstream of the jump and the control should be at point N. If $E_{n2} < (E_{n1} - \Delta E_{j1})$, an M_3 curve in which the specific energy is lost, in addition to the energy loss in the jump, should exist, and the control should be at point N′:

$$E_{n2} = y_{n2} + \frac{q^2}{2gy_{n2}^2} = 3.40 + \frac{16.67^2}{2 \times 9.81 \times 3.40^2} = 4.63 \text{ m}$$

$$E_{y'} = y' + \frac{q^2}{2gy'^2} = 4.17 + \frac{16.67^2}{2 \times 9.81 \times 4.17^2} = 4.98 \text{ m}$$

Because $E_{y'} > E_{n2}$, the control is at point N′. It can be shown that the energy loss ΔE_{j1} in the jump in reach 1 is not enough to decrease the specific energy from E_{n1} to E_{n2}:

$$E_{n1} = y_{n1} + \frac{q^2}{2gy_{n1}^2} = 2.15 + \frac{16.67^2}{2 \times 9.81 \times 2.15^2} = 5.21 \text{ m}$$

$$E_{n1} - E_{n2} = 5.21 - 4.63 = 0.58 \text{ m}$$

$$\Delta E_{j1} = E_{n1} - E_{y'} = 5.21 - 4.98 = 0.23 \text{ m}$$

Therefore $$\Delta E_{j1} < (E_{n1} - E_{n2})$$

Interaction of Controls

Generally there is more than one control in channels. Depending upon the distance between the controls and the flow conditions in the channel, the controls can interact among themselves. Referring to the discussion of flow profiles without resistance in Section 3-4, consider the flow controlled by a sluice gate in a channel that terminates in a free overfall, as shown in Figure 4-4. Similar to Figure 3-20, four channel sections are selected in Figure 4-4. Sections 1 and 2 are immediately upstream and downstream from the gate, respectively; section 3 is downstream from section 2; and section 4 is at the free overfall. The effect on the flow profiles of an adverse slope between sections 3 and 4, including the channel resistance, is to be analyzed.

The very first step in analyzing flow profiles is the plotting of the NDL and the CDL in each channel reach. The slopes of the channel reaches are horizontal and adverse, and the normal depth is infinite in the horizontal channel and imaginary in the adverse channel. Therefore, the NDL is not shown in Figure 4-4, and there are no NDCs. Only ACCs and CDCs are possible. The sluice gate is an artificial control structure. ACCs are at points A and A′ in sections 1 and 2, respectively. Only the location of a possible CDC is at the free overfall at point C in section 4. The CDC is active only if the flow in the adverse channel is in the subcritical-flow regime.

Let the adverse slope between sections 3 and 4 initially be zero, as shown in Figure 4-4a. The flow profiles can be sketched starting from the controls. An H_2

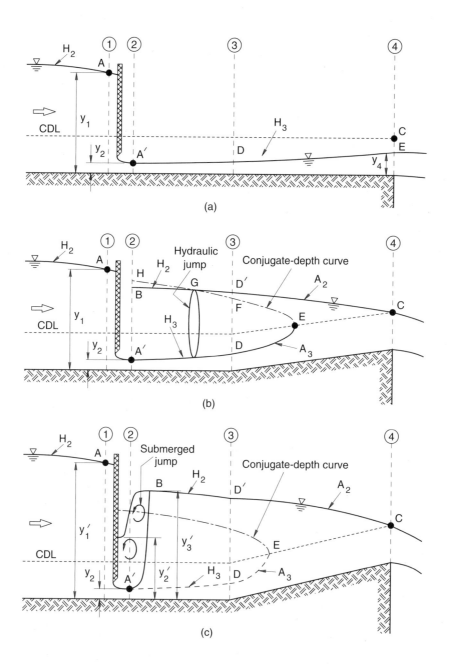

Figure 4-4. The drowning of one control by another control.

curve from point A and an H_3 curve from point A′ are drawn. The flow depth within the H_3 curve increases along the flow direction until it reaches the critical depth. Note that the flow depth downstream from the gate in Figure 3-20a was a constant as the channel was frictionless. The flow depth at the downstream end of the H_3 curve depends on the distance between sections 2 and 4. In Figure 4-4a the distance between sections 2 and 4 is assumed to be small so that the flow depth y_4 at the downstream end of the H_2 curve is less than the critical depth. The CDC in the present case is not active, because the flow upstream of it is supercritical and the energy available to pass the flow at depth y_4 is more than that required to pass the flow at the critical depth. If the distance between sections 2 and 4 is long enough for the flow depth at the downstream end of the H_3 curve to reach the critical depth, then the CDC will become active and an H_2 curve will exist upstream from section 4. The flow will change from supercritical to subcritical, developing a hydraulic jump somewhere between the gate and the free overfall.

The flow profile for an adverse slope between sections 3 and 4 is shown in Figure 4-4b. The H_2 curve upstream from point A and the H_3 curve downstream from point A′ up to section 3 are the same as in Figure 4-4a. The profile DE in the adverse channel is an A_3 curve. For the same Manning n for the reaches with horizontal and adverse slopes, the rate of increase in depth is higher in the A_3 curve than the H_3 curve. The depth reaches the critical depth in a shorter distance; the point E is upstream from point C. The loss of energy in the H_3 and A_3 curves is such that the total head left in section 4 is less than the critical head and therefore the CDC becomes active. Note that the CDC in Figure 3-20b is not active, because there is no energy loss in curve A′DE as the channel therein is frictionless. The profiles CD′ and D′B are an A_2 curve and an H_2 curve, respectively. In Figure 4-4b a curve is drawn in a dash-dot line above the H_3 and A_3 curves. This curve is a plot of the depth sequent to A′DE and is termed as conjugate-depth curve. A hydraulic jump forms where the H_2 or A_2 curve intersects the conjugate-depth curve, as shown in Figure 4-4b. The conjugate-depth curve serves a useful device for locating a hydraulic jump. It should be pointed out that the energy loss is less in the H_2 curve than the H_3 curve and it is less in the A_2 curve than the A_3 curve.

The jump moves upstream with the increasing adverse slope, until it reaches the gate. Any further increase in adverse slope drowns the control at A′ by forming a submerged hydraulic jump, as shown in Figure 4-4c. The depth in H_2 curve BD′ and in A_2 curve CD′ is more than that in the corresponding curves in Figure 4-4b. Consequently, neither of the H_2 and A_2 curves intersects the conjugate-depth curve. The procedure for analyzing the submerged hydraulic jump is the same as given in Section 3-4. The depth in section 1 increases; consequently additional energy is gained to overcome the losses between the gate and the free overfall.

Example 4-5. The NDL and CDL in the reaches of a channel are shown in the figure below. The last reach includes a sluice gate. Sketch the possible flow profiles.

Control sections. The ACCs are at points A and A′ upstream and downstream of the sluice gate, respectively; the NDCs are at points N and N″ at the upstream end of the mild reaches and at point N′ at the downstream end of the steep reach; and the CDCs are at the reservoir outlet (not shown) and at points C, and C′ at the downstream end of the

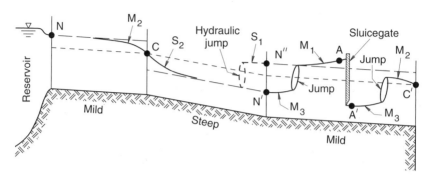

Example 4-5.

mild reaches.

Flow profiles: As the flow enters the channel from the reservoir, the water surface drops by an amount equal to the velocity head plus the entrance loss. The curves from the control point C are M_2 in the mild reach and S_2 in the steep reach. The curve M_1 traced from the control point A shows that the channel is long enough so that the control point N" is not "drowned out." There are two controls (points N' and N") at the junction of the steep and mild reaches. If point N" is active, the S_1 curve preceded by a hydraulic jump forms in the steep reach. If point N' is active, the M_3 curve that is followed by a hydraulic jump forms in the mild reach. The active control can be determined by following the procedure outlined in Example 4-4. The curve from the control point A' is M_3. If the channel downstream from the sluice gate is long enough for the M_3 curve to reach the critical depth, then point C' is a control; the curve upstream from point C' is M_2. The location of the hydraulic jump from the M_3 curve to the M_2 curve requires computations of these profiles; the computation procedures are described in the next chapter.

PROBLEMS

4-9   Determine the type of curve that will form in a wide rectangular channel ($n = 0.013$) with the following change in the bottom slopes of reaches 1 and 2: (i) $y_{n1} = 2$ m, $S_{01} = 0.003$, and $S_{02} = 0.006$; (ii) $y_{n1} = 2$ m, $S_{01} = 0.006$; and $S_{02} = 0.003$, and (iii) $y_c = 1.5$ m, $S_{01} = 0.0009$, and $S_{02} = 0.0005$.

4-10   A wide rectangular channel is laid with a change in slope from 0.01 to 0.002. The Manning n for the channel is 0.025. The depth of uniform flow in the mild channel is 2 m. Sketch the flow profile in the channel.

4-11   Water flows under a sluice gate in a rectangular channel ($n = 0.02$) of large width. The sluice gate is regulated to discharge 6 m³/s per meter width of the channel. Sketch the flow profile in the channel for a gate opening of 0.5 m and a channel slope of 0.001. Assume that the contraction coefficient for the gate is 0.62.

4-12   A wide rectangular channel with a bottom slope of 0.0004 consists of two long reaches and carries a unit discharge of 3 m³/s/m. The upstream reach is rougher than the downstream reach. The values of the Manning n for the upstream and downstream reaches are 0.024 and 0.012, respectively. Identify the control section and sketch the flow profile in the channel.

4-13   A rectangular channel 6 m in width conveys 100 m³/s of water. The channel slope is 0.003 for the first reach and then there is a sudden break in the slope. The slope becomes 0.01. The Manning n for the channel is 0.015. Sketch the flow profile in the channel.

4-14   A wide rectangular channel has a change in bottom slope from 0.005 to 0.03. Sketch the flow profile for $q = 1.5$ m³/s/m and Manning $n = 0.014$.

4-15   Water flows under a sluice gate in a rectangular channel. The channel has a width of 6 m, Manning n of 0.013, and bottom slope of 0.0005. The gate opening is 1 m and the contraction coefficient for the gate is 0.7. Determine the discharge if the flow upstream from the gate is uniform.

4-16   A long reach of channel having a steep slope is followed by a fairly short reach of horizontal channel, which in turn is followed by a long reach of mild channel terminating in an overfall at the downstream end. The shape of the channel cross section remains constant. Sketch all possible flow profiles and label them.

Profiles in Channels with Transitions

Though changes of cross section and/or bottom elevation within a short reach are rare in natural streams and uncommon in artificial channels, it is instructive to analyze flow profiles in channels with such transitions as well as changes of grades. Depending on the size of the contraction of channel width and/or the rise of channel bottom, the downstream end of such transitions can "choke" the flow and act as a critical-depth control (CDC). Similarly, depending on the expansion of channel width and/or drop of channel bottom, the downstream end of the reach upstream from such transitions can act as a CDC.

The procedure for sketching flow profiles outlined above can be used for channels with transitions, but with one additional computational step. After plotting the NDL and CDL, compute the specific energy E_n at y_n for each reach and the minimum specific energy E_c at the CDC. It should be noted that the classification of the flow profiles in Figure 4-2 does not apply to the profiles in transitions; the profiles in transitions are termed backwater or drawdown curves. Within transitions the shape of the water-surface curve may be approximated neglecting the energy loss. The method to be followed in sketching the flow profile is illustrated through the following example.

Consider flow in a rectangular channel with a drop Δz in the channel bottom, as shown in Figure 4-5a. Both the upstream reach (referred to as reach 1) and the downstream reach (referred to as reach 2) are mild. The NDL and CDL are sketched in Figure 4-5a. Note that y_c is the same in both the reaches. The next step is to locate the controls that are an NDC at point N and a CDC at point C. Only one control can be active, unless there exists a hydraulic jump between sections 1 and 2, as shown later. The control that requires a larger head to pass the flow is the active control. Neglecting losses within the transition, it can be easily seen from the energy equation that the CDC becomes active only if $E_c > (E_{n2} - \Delta z)$. The profiles will be drawn for a fixed y_{n1} and for a decreasing y_{n2}. Because y_{n1} is fixed, E_{n1} is constant. The depth y_{n2}, and hence E_{n2}, in reach 2 is varied by changing either the slope, the roughness, or both of reach 2.

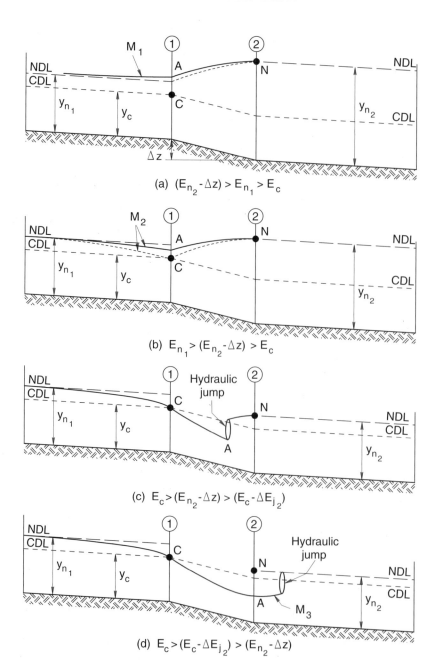

(a) $(E_{n_2} - \Delta z) > E_{n_1} > E_c$

(b) $E_{n_1} > (E_{n_2} - \Delta z) > E_c$

(c) $E_c > (E_{n_2} - \Delta z) > (E_c - \Delta E_{j_2})$

(d) $E_c > (E_c - \Delta E_{j_2}) > (E_{n_2} - \Delta z)$

Figure 4-5. Flow past a drop in the channel bottom.

Let us assume that the depth y_{n2} is high such that $(E_{n2} - \Delta z) > E_{n1}$. This inequality implies $(E_{n2} - \Delta z) > E_c$, because $E_{n1} > E_c$. Hence point C is not a control. Starting from point N, draw the flow profile. The flow downstream from point N is uniform, unless it is affected by a downstream control other than the channel friction. The depth in the transition upstream from point N decreases, as can be seen from a specific-energy diagram. The depth at the upstream end of the transition (section 1), and hence the flow profile in reach 1, depends upon the relative magnitudes of E_{n1} and E_{n2}. Because $(E_{n2} - \Delta z) > E_{n1}$, the flow profile in reach 1 is such that there is a gain of energy in the profile. Such a curve is M_1. The depth in section 1 is higher than y_{n1} and is represented by point A in Figure 4-5a. If $(E_{n2} - \Delta z) = E_{n1}$, then the depth in section 1 is equal to y_{n1} and flow is uniform in reach 1.

Let the slope and/or the roughness of reach 2 be changed to decrease y_{n2}, and hence E_{n2}, such that $E_{n1} > (E_{n2} - \Delta z) > E_c$; the CDC still is not active. Now the flow profile in reach 1 is such that there is a loss of energy in the profile. Such a curve is M_2. The depth in section 1 is lower than y_{n1} and is represented by point A in Figure 4-5b. If $(E_{n2} - \Delta z) = E_c$, the depth in section 1 is y_c and the energy loss in the M_2 curve is at a maximum.

Let the slope and/or the roughness of reach 2 be further changed such that $E_c > (E_{n2} - \Delta z)$. The CDC at point C becomes active. Now the flow profile is drawn starting from point C, as shown in Figure 4-5c. The flow profile in reach 1 is an M_2 curve. The flow depth within the transition decreases as the flowpoint moves downstream and the flow becomes supercritical. Because the flow in reach 2 is subcritical, a hydraulic jump forms. As the difference between E_c and $(E_{n2} - \Delta z)$ increases, the jump moves downstream. For $(E_c - \Delta E_{j2}) = (E_{n2} - \Delta z)$, the jump forms in section 2, where ΔE_{j2} is the head loss in the jump in section 2. For $(E_c - \Delta E_{j2}) < (E_{n2} - \Delta z)$, the jump forms in the transition, as shown in Figure 4-5c, and both the CDC and the NDC become active. For $(E_c - \Delta E_{j2}) > (E_{n2} - \Delta z)$, the jump forms in reach 2 and is preceded by an M_3 curve (see Figure 4-5d). The depth in section 2 is less than y_c and is represented by point A. Length ℓ of the M_3 curve is determined from

$$\int_\ell \left(\frac{dE}{dx} \right) dx = E_{n2} + \Delta E_j - \left(E_c + \Delta z \right) \tag{4-14}$$

where ΔE_j is the head loss in the jump. The smaller the difference is between $(E_c - \Delta E_{j2})$ and $(E_{n2} - \Delta z)$, the shorter the length of the M_3 curve will be.

A summary of possible flow profiles for the various flow conditions in a channel with a transition is presented in Tables 4-3 and 4-4. The use of these tables is illustrated through the above example. The various flow profiles and flow conditions in this example are summarized in the first four rows of Table 4-4. The slopes of reaches 1 and 2 are mild and given in columns 1 and 2. The possible controls are a CDC at C in section 1 and an NDC at N in section 2, as mentioned in

Table 4-3. Types of Flow Profiles in Prismatic Channels with a Contraction and/or a Rise

Reach 1 Slope	Reach 2 Slope	Possible Control(s)	Relative Magnitudes Of Specific Energies	Active Control(s)	Flow Profiles Reach 1	Transition	Reach 2
Mild	Mild	N in section 2 C in section 2	$E_{n1} > E_{n2}+\Delta z > E_{c2}+\Delta z$	N in section 2	M_2	DD	UF
			$E_{n2}+\Delta z > E_{n1} > E_{c2}+\Delta z$	N in section 2	M_1	DD	UF
			$E_{n2}+\Delta z > E_{c2}+\Delta z > E_{n1}$	N in section 2	M_1	DD	UF
Mild	Steep	C in section 2	$E_{n1} > E_{n2}+\Delta z > E_{c2}+\Delta z$	C in section 2	M_2	DD	S_2
			$E_{n2}+\Delta z > E_{n1} > E_{c2}+\Delta z$	C in section 2	M_2	DD	S_2
			$E_{n2}+\Delta z > E_{c2}+\Delta z > E_{n1}$	C in section 2	M_1	DD	S_2
Steep	Steep	N in section 2 C in section 2	$E_{n1} > E_{n2}+\Delta z > E_{c2}+\Delta z$	N in section 1	UF	BW	S_3
			$E_{n2}+\Delta z > E_{n1} > E_{c2}+\Delta z$	N in section 1	UF	BW	S_2
			$E_{c2}+\Delta z > E_{n1}-\Delta E_{j1}$	C in section 2	J & S_1	DD	S_2
			$E_{n1} > E_{n2}+\Delta z > E_{c2}+\Delta z$ $> E_{n1}-\Delta E_{j1}$	C in section 2 or* N in section 1	J & S_1 UF	DD BW	S_2 S_3
			$E_{n2}+\Delta z > E_{n1} > E_{c2}+\Delta z$ $> E_{n1}-\Delta E_{j1}$	C in section 2 or* N in section 1	J & S_1 UF	DD BW	S_2 S_2
Steep	Mild	N in section 1 N in section 2 C in section 2	$E_{n1}-\Delta E_{j1} > E_{n2}+\Delta z$	N in section 1	UF	BW	M_3 & J
			$E_{n2}+\Delta z > E_{n1}-\Delta E_{j2}$	N in section 2	J & S_1	DD	UF
			$E_{n1}-\Delta E_{j2} > E_{n2}+\Delta z$ $> E_{n1}-\Delta E_{j1}$	N in section 2 or* N in section 1	J & S_1 UF	DD BW	UF M_3 & J

Legend: C = CDC; N = NDC; DD = Drawdown curve; BW = Backwater curve; J = Jump; UF = Uniform flow; ΔE_{j1} = Energy loss in jump in section 1; ΔE_{j2} = Energy loss in jump in section 2; *see Section 4-5.

column 3. The relative magnitudes of the specific energy for the case shown in Figure 4-5a are presented in row 1 and column 4. The active control for this case is the NDC at N in section 2 as given in column 5. The flow profiles are included in columns 6 – 8, and are an M_1 curve in reach 1, a backwater curve in the transition, and a uniform flow in reach 2.

Table 4-4. Types of flow profiles in prismatic channels with an expansion and/or a drop.

Reach 1 Slope	Reach 2 Slope	Possible Control(s)	Relative Magnitudes Of Specific Energies	Active Control(s)	Flow Profiles Reach 1	Transition	Reach 2
Mild	Mild	C in section 1 N in section 2	$E_{n2}-\Delta z > E_{n1} > E_{c1}$	N in section 2	M_1	BW	UF
			$E_{n1} > E_{n2}-\Delta z > E_{c1}$	N in section 2	M_2	BW	UF
			$E_{c1} > E_{n2}-\Delta z$ $> E_{c1}-\Delta E_{j2}$	C in section 1 & N in section 2	M_2	DD, J, & BW	UF
			$E_{c1} > E_{c1}-\Delta E_{j2}$ $> E_{n2}-\Delta z$	C in section 1	M_2	DD	M_3 & J
Mild	Steep	C in section 1	$E_{n1} > E_{n2}-\Delta z$ $> E_{c2}+\Delta z$	C in section 1	M_2	DD	S_3
			$E_{n2}-\Delta z > E_{n1}$	C in section 1	M_2	DD	S_2
Steep	Steep	N in section 1 C in section 1	$E_{n1} > E_{n2}-\Delta z$	N in section 1	UF	DD	S_3
			$E_{n2}-\Delta z > E_{n1}$	N in section 1	UF	DD	S_2
Steep	Mild	N in section 1 C in section 1 N in section 2	$E_{n2}-\Delta z > E_{n1}-\Delta E_{j1}$	N in section 2	J & S_1	BW	UF
			$E_{n1}-\Delta E_{j2} > E_{n2}-\Delta z$	N in section 1	UF	DD	M_3 & J
			$E_{n1}-\Delta E_{j1} > E_{n2}-\Delta z$ $> E_{n1}-\Delta E_{j2}$	N in section 1 & N in section 2	UF	DD, J, & BW	UF

Legend: C = CDC; N = NDC; DD = Drawdown curve; BW = Backwater curve; J = Jump; UF = Uniform flow; ΔE_{j1} = Energy loss in jump in section 1; ΔE_{j2} = Energy loss in jump in section 2.

Example 4-6. Two long, rectangular channels of different widths are connected by a relatively short channel contraction. Sketch the flow profile for the following flow conditions: $q_1 = 1.0$ m³/s/m; $y_{n1} = 0.8$ m; $q_2 = 1.5$ m³/s/m; and $y_{n2} = 0.5$ m. The subscripts 1 and 2 denote the conditions in the upstream and downstream reaches, respectively. Neglect the energy loss in the transition.

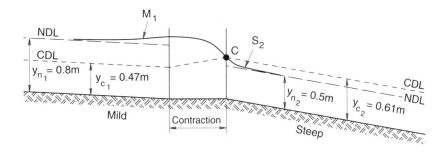

Example 4-6

$$y_{c1} = \sqrt[3]{\frac{1^2}{9.81}} = 0.47 \text{ m} < y_{n1}; \text{Reach 1 is mild.}$$

$$y_{c2} = \sqrt[3]{\frac{1.5^2}{9.81}} = 0.61 \text{ m} > y_{n2}; \text{Reach 2 is steep.}$$

Control sections: No ACC; the NDCs (not shown) are at the upstream end of the mild reach and at the downstream end of the steep reach; and a CDC at the "choke" at point C.

Flow profile: Trace the flow profile starting from point C. The only possible curve in reach 2 is S_2. The depth in the transition increases as the flowpoint moves upstream from point C, because the flow is subcritical. If $E_{c2} > E_{n1}$, the depth in section 1 will be higher than y_{n1} and vice versa. Compute E_{c2} and E_{n1}:

$$E_{c2} = \tfrac{3}{2} y_{c2} = \tfrac{3}{2} \times 0.61 = 0.92$$

$$V_{n1} = \frac{1}{0.8} = 1.25 \text{ m/s}; \quad \frac{V_{n1}^2}{2g} = \frac{1.25^2}{2 \times 9.81} = 0.08 \text{ m}; \quad E_{n1} = 0.8 + 0.08 = 0.88 \text{ m}$$

Because $E_{c2} > E_{n1}$, the depth in section 1 is higher than y_{n1}; the curve in reach 1 is M_1.

PROBLEMS

4-17   A rectangular channel consists of two long reaches that are connected by a short transition. Sketch the water-surface profile in the channel for the following conditions. The subscripts 1 and 2 denote the conditions in the upstream and downstream reaches, respectively. Neglect the energy loss in the transition.

(a) $y_{n1} > y_{c1}; y_{n2} < y_{c2}; b_1 > b_2; E_{n2} > E_{n1}$.

(b) $y_{c1} > y_{n1}; y_{n2} > y_{c2}; b_1 > b_2; E_{n1} > E_{n2}$.

(c) $y_{c1} > y_{n1}; y_{n2} > y_{c2}; b_2 > b_1; E_{n1} > E_{n2}$.

(d) $y_{n1} > y_{c1}; y_{n2} < y_{c2}; b_2 > b_1; E_{n2} > E_{n1}$.

4-18   Two long rectangular channels of different widths are connected by a short channel transition. Sketch the water-surface profile for the following flow conditions. The effects in the transition of the bed slope and the bottom friction can be neglected.

(a) $q_1 = 1.5 \text{ m}^3/\text{s/m}; y_{n1} = 0.5 \text{ m}; q_2 = 1.0 \text{ m}^3/\text{s/m}; \text{ and } y_{n2} = 0.8 \text{ m}.$
(b) $q_1 = 1.0 \text{ m}^3/\text{s/m}; y_{n1} = 0.8 \text{ m}; q_2 = 1.5 \text{ m}^3/\text{s/m}; \text{ and } y_{n2} = 0.8 \text{ m}.$
(c) $q_1 = 1.0 \text{ m}^3/\text{s/m}; y_{n1} = 1.0 \text{ m}; q_2 = 1.4 \text{ m}^3/\text{s/m}; \text{ and } y_{n2} = 0.4 \text{ m}.$

4-19   A trapezoidal channel has a base width of 6 m, side slopes of 1H:1V, Manning n of 0.015, and bottom slope of 0.0006 and carries a discharge of 20 m^3/s. There exists a transition in a channel in which the shape of the channel cross section changes from trapezoidal to rectangular. The rectangular-channel reach is steep. Determine the width of the rectangular section so that the flow in the entire trapezoidal reach remains uniform. Neglect the energy losses in the transition.

4-20   Two long rectangular channels of different widths are connected by a relatively short transition. Neglecting the effects of the bed slope and bottom friction in the transition, sketch the flow profile if

$$y_{c1} = 1.5 y_{c2} = 1.5 y_{n1} = 3 y_{n2}$$

4-21   Two long rectangular channels are connected by a short transition. The effect of the bed slope and friction in the transition can be neglected. For the following data determine the width of the downstream channel so that the flow in the entire upstream channel remains uniform. $y_{n1} = 1.2$ m; $q_1 = 1.0$ m^3/s/m; $b_1 = 6$ m; $n_2 = 0.013$; $S_{02} = 0.0005$

4-22   Two long and steep reaches of a rectangular channel are connected by a short contraction. Neglecting the effect of bed slope and bottom friction in the transition, sketch the flow profile in the channel and determine the relations between E_{n1}, E_{n2}, E_{c1}, and E_{c2} if the control is (a) a CDC and (b) an NDC.

4-23   Consider a flow in a rectangular channel with a rise Δz in its bottom. The channel reaches upstream and downstream from the rise are mild and steep, respectively. Locate the possible controls in the channel. Sketch the possible water-surface profiles in the channel. For each profile determine the flow conditions in terms of specific energies at the normal and critical depths in the two reaches and Δz.

4-5 Nonunique Water-Surface Profiles

It is possible that for certain flow conditions the flow profile in a channel with a transition consists of a hydraulic jump and a curve. Because there is a loss of energy in a jump and a gain or loss of energy in a curve, it is possible that different combinations of a jump and a curve in a channel yield the same change in energy. Hence more than one solution is possible. Jain (1993) found that more than one flow profile is possible in a channel that includes either a contraction in its width or a rise in its bottom and the channel reach upstream from the transition is steep. These flow profiles and the conditions leading to nonunique solutions are presented in this section. The channel reach upstream of the transition is referred to as reach 1. The downstream reach, referred to as reach 2, can be mild or steep. The energy loss within the transition reach is assumed to be negligible.

Mild Downstream Reach

The flow profiles are first analyzed in a channel consisting of a contraction in width between sections 1 and 2 and a mild downstream reach, as shown in Figure 4-6. The profiles are presented for a fixed normal depth y_{n1} in reach 1 and for a range of normal depths, $y_{n2} = y_{Ni}$ ($i = 1, 2, \ldots$) in reach 2. A hydraulic jump will form in the channel as the flow changes from supercritical in reach 1 to subcritical in reach 2. The possible controls are (Figure 4-6) the NDCs at points N_i and N' and a CDC at the "choke" at point C. Because the specific energy at depth y_{c2} (point C) is always less than that at depth y_{n2} (points N_i), the CDC is never active.

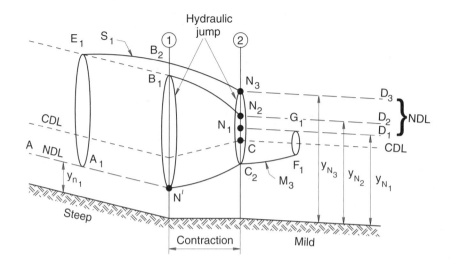

Figure 4-6. Limiting flow profiles for a mild downstream reach.

If the control at points N_i, for instance N_3, is active, an S_1 curve preceded by a hydraulic jump, as represented by $AA_1E_1B_2N_3D_3$, will form. As the depth in reach 2 decreases from y_{n3} to y_{n2}, the hydraulic jump moves downstream and forms in section 1. The control is at point N_2. The flow profile is represented by $AN'B_1N_2D_2$. The specific energy at point N_2 is equal to $(E_{n1} - \Delta E_{j1})$, where ΔE_{j1} is the energy loss in the jump in section 1. The control remains at N_i as long as

$$y_{Ni} > y_{N2} \qquad (4\text{-}15)$$

If the control at point N' is active, an M_3 curve followed by a hydraulic jump, as shown by $AN'C_2F_1G_1D_1$, will form. The depth in reach 2 is y_{N1}. As the depth in reach 2 increases from y_{N1} to y_{N3}, the jump moves upstream and forms in section 2; the control still remains at point N'. The specific energy at point N_3 is equal to $(E_{n1} - \Delta E_{j2})$, where ΔE_{j2} is the energy loss in the jump in section 2. The control remains at point N' as long as

$$y_{Ni} < y_{N3} \qquad (4\text{-}16)$$

It can be seen from Eqs. (4-15) and (4-16) that the two statements, that the control is at point N_i for y_{Ni} higher than y_{N2} and that the control is at point N for y_{Ni} lower than y_{N3}, are contradictory for the range $y_{N3} > y_{Ni} > y_{N2}$. In other words, both controls, and consequently two profiles, are possible for the depth y_{Ni} in the range

$$y_{N3} > y_{Ni} > y_{N2} \qquad (4\text{-}17)$$

Equation (4-17) can be written in terms of specific energy as

$$(E_{n1} - \Delta E_{j2}) > E_{n2} > (E_{n1} - \Delta E_{j1}) \qquad (4\text{-}18)$$

It is tacitly assumed in Figure 4-6 that $y_{N3} > y_{N2}$. This assumption is correct if $\Delta E_{j1} > \Delta E_{j2}$. It can be shown that the energy loss in a jump formed within the transition decreases as the jump moves downstream. The rate of change along the flow direction of the energy loss in a jump can be written as

$$\frac{d(\Delta E_j)}{dx} = \frac{d(\Delta E_j)}{dy_u} \frac{dy_u}{dx} \qquad (4\text{-}19)$$

where y_u is the flow depth immediately upstream from the jump. The first term on Eq. (4-19) is negative, as can be seen from a specific-force diagram (see Figure 3-7b) and Eq. (3-24) for the energy loss in a jump. As y_u increases, its conjugate depth, and consequently the height of the jump and the energy loss in the jump, decreases. The second term for a transition with a contraction is positive, as can be seen from a specific-energy diagram. Therefore, the term on the left-hand side of Eq. (4-19) is negative, and $\Delta E_{j1} > \Delta E_{j2}$.

The two possible flow profiles for the depth y_{Ni} given by Eq. (4-17) are shown in Figure 4-7, which also includes a conjugate-depth curve PQRS. This curve is a plot of the depth sequent to AN'C$_2$S. The flow profile AN'C$_2$F$_1$G$_1$D$_i$ is for the control at point N' and is established while increasing depth in reach 2. The flow profile AA$_1$E$_1$B$_2$N$_i$D$_i$ is for the control at point N$_i$ and is established while decreasing depth in reach 2. The curve C$_2$F$_1$ is an M$_3$ curve and the curve E$_1$B$_2$ is an S$_1$ curve. The energy loss in the jump A$_1$E$_1$ plus the energy gain in S$_1$ is equal to the sum of the energy loss in M$_3$ and the jump F$_1$G$_1$.

Now examine whether a jump can also form within the transition for the flow condition given by Eq. (4-18). A jump, such as A$_2$E$_2$ in Figure 4-7, within the transition satisfies Eq. (4-18) because the loss of energy in such a jump is between

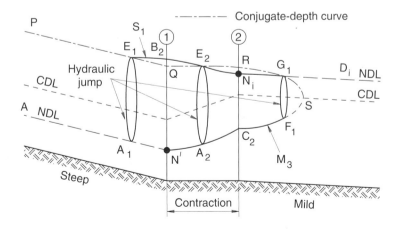

Figure 4-7. Two possible flow profiles for a mild downstream reach.

ΔE_{j1} and ΔE_{j2}. However, this profile is not stable. A negative or a positive perturbation in depth in reach 2 forces the jump in reach 2 or reach 1, respectively (why?).

The same flow profiles as presented in Figure 4-7 occur in a channel with a rise in its bottom. For this transition Eq. (4-18) modifies to

$$(E_{n1} - \Delta E_{j2}) > (E_{n2} + \Delta z) > (E_{n1} - \Delta E_{j1}) \qquad (4\text{-}20)$$

in which Δz is the rise in the channel bottom in the transition.

Steep Downstream Reach

The flow profiles in a channel with a contraction in width and a downstream reach that is steep are presented in Figure 4-8. Two possible controls are an NDC at point N and a CDC at point C. If the CDC is active, the flow upstream from C is subcritical. The flow in reach 1 changes from supercritical to subcritical through a hydraulic jump that is followed by an S_1 curve. The flow downstream from C is supercritical and an S_2 curve forms. The energy is lost in the jump and it is gained in both the S_1 and S_2 curves. If $E_{n2} > E_{n1}$, the gains of energy in S_1 and S_2 are larger than the loss of energy in the jump and vice versa. If the jump forms in section 1, $E_{c2} = (E_{n1} - \Delta E_{j1})$. For $E_{c2} > (E_{n1} - \Delta E_{j1})$, there is not enough energy in the transition to pass the flow and point C is a control.   The NDC is active if $E_{n1} > E_{c2}$. The profile is traced starting at point N. The depth increases in the transition moving downstream from point N. The profile in reach 2 is S_2 if $E_{n2} > E_{n1}$ and is S_3 if $E_{n2} < E_{n1}$. Both controls are, therefore, possible for the following range of E_{c2}:

$$E_{n1} > E_{c2} > E_{n1} - \Delta E_{j1} \qquad (4\text{-}21)$$

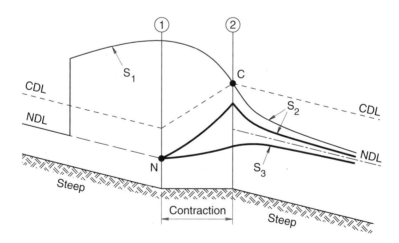

Figure 4-8. Two possible flow profiles for a steep downstream reach.

For $E_{n1} > E_{n2} > E_{c2} > (E_{n1} - \Delta E_{j1})$ the profile in reach 2 is an S_3 curve, and for $E_{n2} > E_{n1} > E_{c2} > (E_{n1} - \Delta E_{j1})$ the profile in reach 2 is an S_2 curve. In normal operating conditions the NDC at point N is active. The CDC at point C can be made active by initially creating subcritical flow in reach 2 by means of a downstream control that is later removed.

The same flow profiles as shown in Figure 4-8 occur in a channel with a rise in the channel bottom for the condition

$$E_{n1} > (E_{c2} + \Delta z) > (E_{n1} - \Delta E_{j1})$$

(4-22)

in which Δz is the rise in the channel bottom in the transition.

Example 4-7. Two reaches of a rectangular channel are connected by a short contraction. The flow conditions in the upstream reach are $q_1 = 1.0$ m³/s/m and $y_{n1} = 20$ cm. The unit discharge and the normal depth for the downstream reach are $q_2 = 1.5$ m³/s/m and $y_{n2} = 85$ cm. Show that two flow profiles are possible in the channel, and sketch both profiles. Neglect the energy loss in the transition.

$$y_{c1} = \sqrt[3]{\frac{1^2}{9.81}} - 0.47 \text{ m}; y_{c1} > y_{n1}; \text{ the upstream reach is steep}$$

$$y_{c2} = \sqrt[3]{\frac{1.5^2}{9.81}} = 0.61 \text{ m}; y_{n2} > y_{c2}; \text{ the downstream reach is mild}$$

Control sections: the NDCs at the downstream end of reach 1 and the upstream end of reach 2; a CDC at the upstream end of reach 2; and no ACC. The CDC is never active because $E_N > E_{c2}$.

$$V_{n1} = \frac{1}{0.20} = 5.0 \text{ m/s}; \frac{V_{n1}^2}{2g} = \frac{5.0^2}{2 \times 9.81} = 1.27 \text{ m}; E_{n1} = 0.20 + 1.27 = 1.47 \text{ m}$$

Let y' be the conjugate depth of y_{n1}:

$$y' = \frac{0.20}{2} \left(\sqrt{1 + \frac{8 \times 1^2}{9.81 \times 0.20^3}} - 1 \right) = 0.91 \text{ m}; E_{y'} = 0.919 + \frac{1^2}{2 \times 9.81 \times 0.91^2} = 0.97 \text{ m}$$

$$E_{y'} = E_{n1} - \Delta E_{j1} = 0.97 \text{ m}$$

Let y'' be the conjugate depth of y_{n2}:

$$y'' = \frac{0.85}{2} \left(\sqrt{1 + \frac{8 \times 1.5^2}{9.81 \times 0.85^3}} - 1 \right) = 0.42 \text{ m}; \quad E_{y''} = 0.42 + \frac{1.5^2}{2 \times 9.81 \times 0.42^2} = 1.07 \text{ m}$$

$$E_{n2} = 0.85 + \frac{1.5^2}{2 \times 9.81 \times 0.85^2} = 1.01 \text{ m}; \Delta E_{j2} = E_{y''} - E_{n2} = 1.07 - 1.01 = 0.06 \text{ m}$$

$$E_{n1} - \Delta E_{j2} = 1.47 - 0.06 = 1.41 \, \text{m}$$

Two solutions are possible, because $(E_{n1} - \Delta E_{j2}) > E_{n2} > (E_{n1} - \Delta E_{j1})$.. An S_1 curve preceded by a jump in reach 1 or an M_3 curve followed by a jump in reach 2, as shown in Figure 4.7, will form.

PROBLEMS

4-24   Two reaches of a rectangular channel are connected by a short transition consisted of a rise of 0.2 m in the channel bottom. The flow conditions in the reaches are $q = 1.0 \, \text{m}^2/\text{s}$; $y_{n1} = 20$ cm; and $y_{n2} = 75$ cm. Is the solution for the water-surface profile in the channel unique?

4-25   Two reaches of a rectangular channel are connected by a short contraction. Is the water-surface profile in the channel unique if $q_1 = 1.0 \, \text{m}^3/\text{s/m}$; $y_{n1} = 0.30$ m; $q_2 = 1.5 \, \text{m}^3/\text{s/m}$; and $y_{n2} = 0.95$ m? Neglect the energy loss in the transition.

4-26   Two reaches of a rectangular channel are connected by a short transition consisted of a rise of 0.35 m in the bottom. The flow conditions in the channel are $q = 1.5 \, \text{m}^3/\text{s/m}$; $y_{n1} = 0.30$ m; and $y_{n2} = 0.50$ m. Sketch the water-surface profile in the channel. Neglect the friction in the transition.

4-6 Profile Analysis for Given Total Head

The procedure for sketching flow profiles described above is based on the assumption that the discharge is known. However, this may not always be the case, as in flow from a reservoir into a channel (Figure 4-9); the discharge in this case depends upon the total head in the reservoir and the flow depth in the control section. Because the discharge is not known, one cannot determine y_n and y_c and consequently the type of bottom slope and the type of control, CDC or NDC. Two cases are considered: one for a long channel with no downstream control except the channel friction and the other for a channel with an additional control at its downstream end.

Flow in a Long Channel

The flow profile in a long mild channel will be different from that in a steep channel, as shown in Figure 4-9. In the case of a steep channel the control is a CDC at the outlet of the reservoir (Figure 4-9a). For the mild channel (Figure 4-9b) the control is an NDC at the upstream end of the channel. As the flow enters the channel, the water surface drops to y_c in the steep channel and to y_n in the mild channel. There exists an S_2 curve downstream from the CDC in the steep channel. The flow is uniform in the mild channel. The discharge and the flow depth are determined by using the energy equation between the water surface in the reservoir and the control section and the Manning equation, as explained below.

   Assume first that the channel is steep. The energy equation between the water surface in the reservoir and the inlet section can be written as

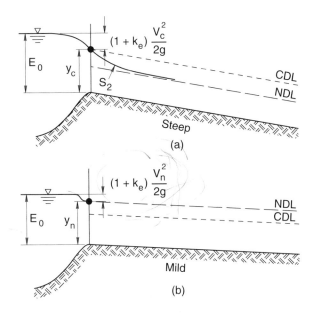

Figure 4-9. Discharge from a reservoir into a channel.

$$E_0 = y_c + \frac{V_c^2}{2g} + k_e \frac{V_c^2}{2g} = y_c + (1 + k_e) \frac{Q^2}{2gA_c^2} \qquad (4\text{-}23)$$

where E_0 is the height of the reservoir surface above the channel bottom at the reservoir outlet; $k_e(V_c^2/2g)$ is entrance loss, where k_e is the entrance loss coefficient, and the subscript c denotes the critical flow. The velocity in the reservoir is assumed negligible. Furthermore, the relation for the critical condition from Eq. (3-30) for $\alpha = 1$ and $\cos\theta = 1$ is

$$\frac{Q^2}{g} = \frac{A_c^3}{B_c} \qquad (4\text{-}24)$$

Equations (4-23) and (4-24) can be solved for Q and y_c. To check whether the assumption of a steep channel is correct, compute the critical slope S_c from the Manning equation for these values of Q and y_c. If $S_0 > S_c$, the channel is steep and the above computations are valid. The normal depth is then obtained from the Manning equation.

If it is found that $S_0 < S_c$, the channel is mild for which the energy equation (4-23) is modified as

$$E_0 = y_n + \frac{V_n^2}{2g} + k_e \frac{V_n^2}{2g} = y_n + (1 + k_e) \frac{Q^2}{2gA_n^2} \qquad (4\text{-}25)$$

where the subscript n denotes the normal flow. The discharge Q and the normal depth y_n can be obtained from Eq. (4-25) and the Manning equation (Example 4-8). The entrance loss in most cases is small and can be neglected.

Example 4-8. A rectangular channel is 3 m in width and draws water from a lake. The channel slope is 0.001 and the Manning n is 0.013. What will be the discharge in the channel when the lake level is 2 m above the high point of the channel bottom? Neglect the inlet losses.

Assume first that the channel is steep. Equations (4-23) and (4-24) are applicable, which for the rectangular channel can be written as

$$E_0 = \tfrac{3}{2} y_c \quad \text{and} \quad q = \sqrt{g y_c^3}.$$

For $E_0 = 2$ m, the critical depth $y_c = \tfrac{2}{3} \times 2 = 1.33$ m, and the unit discharge is $q = \sqrt{9.81 \times 1.33^3} = 4.80$ m^3/s/m. To determine whether the slope is actually steep, compute the critical slope. For $y = 1.33$ m, $A = 3.99$ m^2, $P = 5.66$ m, and $R = 3.99/5.66 = 0.70$ m. The critical slope from the Manning equation is

$$S_c = \left(\frac{nQ}{AR^{2/3}} \right)^2 = \left(\frac{0.013 \times 4.80 \times 3}{3.99 \times 0.70^{2/3}} \right)^2 = 0.0035$$

As this value is greater than channel slope, the channel slope is mild. The Manning equation and Eq. (4-25) are applicable, which can be written as

$$Q = \frac{1}{0.013} \times 3 y_n \left(\frac{3 y_n}{3 + 2 y_n} \right)^{2/3} \sqrt{0.001}$$

$$2 = y_n + \frac{Q^2}{2 \times 9.81 \times (3 y_n)^2}$$

Solve the two equations by the trial-and-error method. The solution is $Q = 11.25$ m^3/s and $y_n = 1.77$ m.

Effect of a Downstream Control

Let there be a downstream control that regulates the water-surface level at the downstream end of the channel. One such control can be a reservoir. The determination of discharge in a channel connecting two reservoirs is sometimes referred to as either the two-lake or the channel-delivery problem. The effects of the water level in the downstream reservoir on the flow profile and the discharge Q in the channel are examined in mild and steep channels. The length and the slope of the channel are L and S_0, respectively.

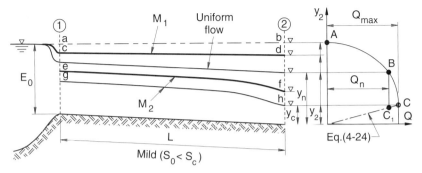

Figure 4-10. Flow between two reservoirs connected by a mild channel.

Mild Channel. The flow profiles for a range of flow depth y_2 at the downstream end are shown in Figure 4-10 and are described as follows:

(i) For $y_2 = E_0 + S_0 L$, the water levels in both upstream and downstream reservoirs are the same; consequently the discharge Q is zero. The flow profile in the channel is a horizontal line ab. The flow conditions are represented by point A in Figure 4-10 wherein the variation of Q with y_2 is shown. As y_2 goes down, so does y_1 until a limiting water-surface profile, as explained below, is reached.

(ii) For $y_2 = y_n$, the flow in the channel is uniform. The flow profile is a straight line along the normal-depth line. The discharge is Q_n, which, as explained earlier, can be determined from Eq. (4-25) and the Manning equation. The flow conditions are represented by point B.

(iii) $E_0 + S_0 L > y_2 > y_n$, the flow profile cd is of the M_1 type. The discharge is between zero and Q_n, as shown by the curve AB. The discharge can be obtained by computing the M_1 profile for various assumed Q values until the energy equation in section 1 is satisfied. The procedures for computing flow profiles are described in the next chapter.

(iv) For $y_2 < y_n$, the flow profile ef is of the M_2 type. The discharge, as shown by the curve BC, is larger than Q_n (why?). The discharge can be determined by computing the M_2 profile for various assumed Q values until the energy equation in section 1 is satisfied. The lower limit of the M_2 profile is shown by the curve gh. The flow at the downstream end is critical, i.e., $y_2 = y_c$ and the discharge is the maximum. The flow conditions are represented by point C, which lies on a curve given by the critical-depth relation (4-24). Note that if the channel is long enough such that the M_2 curve reaches the normal depth before it reaches the upstream reservoir, Q remains unchanged at its uniform flow value along the line BC_1.

(v) For $y_2 < y_c$, the flow profile and the discharge in the channel remain unchanged, as shown by the curve gh and point C (or C_1), respectively.

Steep Channel. The flow profiles for a range of flow depth y_2 at the downstream end are shown in Figure 4-11 and are described as follows:

(i) For $y_2 = E_0 + S_0 L$, the water levels in both upstream and downstream reservoirs are the same; consequently the discharge Q is zero. The flow profile in the channel is a horizontal line ab. The flow conditions are represented by point A

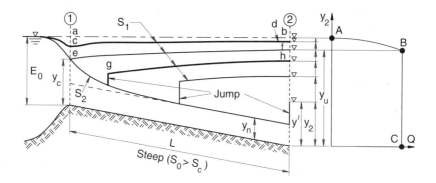

Figure 4-11. Flow between two reservoirs connected by a steep channel.

in Figure 4-11, wherein the variation of Q with y_2 is shown. As y_2 goes down, so does y_1. Define a depth $y_2 = y_u$ such that as $y_2 \to y_u$, $y_1 \to y_c$.

(ii) For $y_2 = y_u$, the flow depth in section 1 is y_c and the discharge Q_c in the channel can be determined from Eqs. (4-23) and (4-24), as explained earlier. The flow conditions are represented by point B.

(iii) For $E_0 + S_0 L > y_2 > y_u$, the flow profile cd is of the S_1 type. The flow depth in section 1 is larger than y_c; the CDC in section 1 is "drowned." The discharge is between zero and Q_c, as shown by the curve AB. The discharge can be obtained by computing S_1 profiles for various assumed Q values until the energy equation in section 1 is satisfied.

(iv) For $y_u > y_2 > y'$, where y' is the conjugate depth of y_n and y_n is the normal depth in the channel at the discharge Q_c, a hydraulic jump develops in the channel and the flow profiles downstream and upstream from the jump are of the S_1 and S_2 types, respectively. As y_2 decreases, the jump moves downstream, first along the S_2 profile and then along the normal-depth line, until it reaches the downstream end of the channel. The flow upstream from the jump, and hence the discharge, is not affected by y_2.

(v) For $y' > y_2 > y_n$, a rapid change in depth, termed "imperfect" jump, occurs at the downstream end.

(vi) For $y_n > y_2$, the flow in the channel is not affected by y_2; the flow from the channel shoots out like a jet.

The flow depth y_u can be easily determined if one neglects the friction losses in the channel. The energy equation between sections 1 and 2 then can be written as

$$E_0 + S_0 L = y_u + \frac{Q_c^2}{2gA_u^2} \tag{4-26}$$

where A_u is the flow area at depth y_u. Equation (4-26) has two positive roots, and y_u is equal to the upper alternate depth. The discharge curve AB in Figure 4-11 also can be determined by neglecting the channel friction.

PROBLEMS

4-27   Water from a reservoir is discharged into a rectangular channel of 4 m in width. The channel is made of concrete ($n = 0.013$) and has a slope of 0.001. Determine the discharge if the reservoir level is 2 m above the high point of the channel bottom. Neglect the entrance loss.

4-28   A rectangular channel 6 m wide and with $n = 0.013$ consists of two reaches of different slopes. The upstream reach with a slope of 0.005 is supplied water from a reservoir with a water surface 2 m above the high point of the channel bottom. The downstream reach with a slope of 0.001 discharges into another reservoir with a water surface 2 m above the low point of the channel bottom. The bottom of the channel at the section where it changes slope is 1.5 m above the channel bottom at the downstream end and is 2 m below the channel bottom at the upstream end. Sketch the profile in the channel.

4-29   A rectangular channel with $B = 6$ m, $S_0 = 0.01$, and $L = 100$ m connects two reservoirs. The channel slope is steep. The water level in the upstream reservoir is 3 m above the channel bed at the reservoir outlet. Determine the maximum water level in the downstream reservoir that will not affect the discharge in the channel. Determine the discharge in the channel if the water level in the downstream reservoir is 5 percent higher than the maximum water level calculated earlier. Neglect channel friction.

4-30   A canal with mild slope connects two reservoirs. For a constant discharge in the canal, sketch a curve showing the relationship between the depths at the upstream end (y_1) and the downstream end (y_2) of the canal. Identify points on the curve where (i) $y_2 = y_c$ and (ii) $y_2 = y_n$ and $y_2 \to \infty$.

4-7 Location of Hydraulic Jump

A hydraulic jump forms when a supercritical flow changes into a subcritical flow. Examples of the flow profiles with a hydraulic jump are presented in Figures 4-3f and 4-3g. The location of the hydraulic jump in these two figures can be determined by computing the length of the M_3 profile in Figure 4-3f and the length of the S_1 profile in Figure 4-3g. The lengths of different profiles can be computed by following the procedures described in the next chapter provided the depths at both ends of the flow profiles are known. The depth at one of the ends located in the control section is always known, such as at point N′ in Figure 4-3f and point N in Figure 4-3g. In some cases, as in Figures 4-3f and 4-3g, the depth at the other end can be easily obtained. The depth at the other end in Figures 4-3 f and 4-3g are equal to the conjugate depth to y_{n2} and y_{n1}, respectively. The conjugate depth can be determined by Eq. (3-17) or (3-18) for rectangular channels and by Eq. (3-21) for nonrectangular channels.

In other cases, as shown in Figure 4-4b, the depth at the other end cannot be determined easily because the depth neither before nor after the jump is known. The controls are a CDC at point C and an ACC at point A′. The profile from point C is an A_2 curve preceded by an H_2 curve, and from point A′ is an H_3 curve followed by an A_3 curve. The figure also includes a conjugate-depth curve HGFE.

The depths on this curve are conjugate to the corresponding depths on the H_3 and A_3 curves. The intersection of the curve A_2 (or H_2) with the conjugate-depth curve at point G establishes the location of the jump. It should be mentioned that the above analysis does not take the length of the jump into consideration. The length of the jump is discussed in Section 10-3.

4-8 Profiles in Compound Channels

The flow profiles in compound channels for flow conditions that yield only one critical depth are same as shown in Figure 4-2 for simple channels. The flow profiles in mild and steep compound channels with three critical depths are shown in Figure 4-12. The flow profiles in Figures 4-12a and 4-12c were discussed briefly by Quintela (1982). The normal-depth line and the three critical-depth lines divide the flow region into five zones. The channel slope is mild when $y_n > y_{c''}$ or $y_{c'} > y_n > y_c$, and it is steep when $y_n < y_c$ or $y_{c'} < y_n < y_{c''}$. Whether the water-surface profile in a zone is a backwater curve ($dy/dx > 0$) or a drawdown curve ($dy/dx < 0$) can be determined from Eqs. (4-8), (4-11), and (4-12), and this information is summarized in Table 4-5. The water-surface slope at the zonal boundaries can be determined as explained earlier for simple channels. With all this information the shapes of the various profiles can be sketched, as shown in Figure 4-12.

In theory, a different classification is needed to designate flow profiles in compound channels as there are five zones instead of three as in simple channels. But in practice the classification for flow profiles in simple channels is also used for compound channels. The flow remains confined to the upper three zones during high flows and to the lower three zones during low flows. The profiles in these three zones are designated as M_1, M_2, and M_3 in mild channels and S_1, S_2, and S_3 in steep channels, as shown in Figure 4-12.

It should be mentioned (Quintela, 1982) that gradually varied flow in the zone $y_{c''} > y > y_{c'}$ cannot be followed by gradually varied flow in the zone $y_{c'} > y > y_c$ and vice versa. Rapidly varied flow must occur when the flow passes from $y > y_{c'}$ to $y < y_{c'}$ or vice versa.

It is instructive to locate the critical- and normal-depth controls and determine the type of curve in a long channel with a compound cross section that originates from a reservoir and terminates in a freefall. Let the cross section of the compound channel be as shown in Figure 3-13 as an inset. The channel carries a discharge of 220 m³/s. The normal depth in the channel depends on the slope of the channel. As can be seen from the specific-energy curve in Figure 3-13, three critical depths, namely, $y_c = 2.31$ m, $y_{c'} = 3.05$ m, and $y_{c''} = 3.27$ m, are possible in the channel. Let the water surface in the reservoir be 4 m above the channel invert at the inlet. Depending upon the channel slope, the following four cases are possible:

Case 1, $y_c > y_n$: The channel slope is steep. The controls are a CDC at the critical depth y_c in the inlet section and an NDC at the downstream end of the channel. The flow depth rapidly changes from the reservoir level to the critical depth y_c in the inlet section, then gradually varies from the critical depth to the normal depth as an S_2 curve, and finally remains constant at the normal depth.

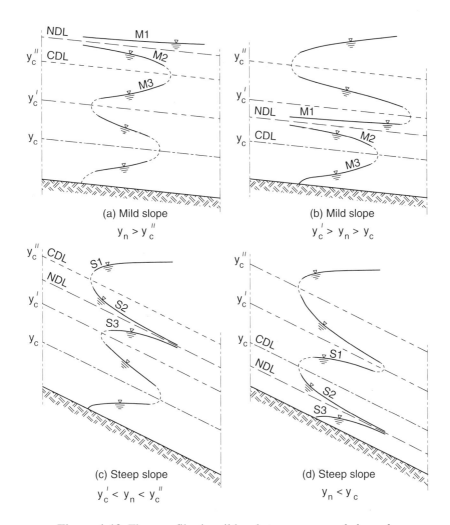

Figure 4-12. Flow profiles in mild and steep compound channels.

Case 2, $y_{c'} > y_n > y_c$: The channel slope is mild. The controls are an NDC at the upstream end of the channel and a CDC at the downstream end of the channel. The flow depth rapidly changes from the reservoir level to the normal depth in the inlet section, then stays at the normal depth in the channel, and finally gradually changes from the normal depth to the critical depth y_c as an M_2 curve near the downstream end of the channel.

Case 3, $y_{c''} > y_n > y_{c'}$: The channel slope is steep. The controls are a CDC at the critical depth $y_{c''}$ in the inlet section and an NDC at the downstream end of the channel. The flow depth rapidly changes from the reservoir level to the critical depth $y_{c''}$ in the inlet section. The flow profile is an S_2 curve between $y_{c''}$ and y_n.

GRADUALLY VARIED FLOW

Table 4-5. Flow characteristics in compound channels

Depth	F	S_f	dy/dx
(a) Mild slope; $y_n > y_{c''}$			
$y > y_n$	<1	$<S_0$	>0
$y_n > y > y_{c''}$	<1	$> S_0$	<0
$y_{c''} > y > y_{c'}$	>1	$> S_0$	>0
$y_{c'} > y > y_c$	<1	$> S_0$	<0
$y_c > y > 0$	>1	$> S_0$	>0
(b) Mild slope; $y_{c'} > y_n > y_c$			
$y > y_{c''}$	<1	$< S_0$	>0
$y_{c''} > y > y_{c'}$	>1	$< S_0$	<0
$y_{c'} > y > y_n$	<1	$< S_0$	>0
$y_n > y > y_c$	<1	$> S_0$	<0
$y_c > y > 0$	>1	$> S_0$	>0
(c) Steep slope; $y_{c'} < y_n < y_{c''}$			
$y > y_{c''}$	<1	$< S_0$	>0
$y_{c''} > y > y_n$	>1	$< S_0$	<0
$y_n > y > y_{c'}$	>1	$> S_0$	>0
$y_{c'} > y > y_c$	<1	$> S_0$	<0
$y_c > y > 0$	>1	$> S_0$	>0
(d) Steep slope; $y_n < y_c$			
$y > y_{c''}$	<1	$< S_0$	>0
$y_{c''} > y > y_{c'}$	>1	$< S_0$	<0
$y_{c'} > y > y_c$	<1	$< S_0$	>0
$y_c > y > y_n$	>1	$< S_0$	<0
$y_n > y > 0$	>1	$< S_0$	>0

Case 4, $y_n > y_{c''}$: The channel slope is mild. The controls are an NDC at the upstream end and a CDC at the downstream end of the channel. The flow depth rapidly changes from the reservoir level to the normal depth in the inlet section, and then stays at the normal depth except near the downstream end of the channel, where the flow profile is an M_2 curve between y_n and $y_{c''}$.

For a given water-surface level in the reservoir the discharge in the compound channel can be determined following the procedure outlined in Section 4-7.

PROBLEMS

4-31   Water flows from a lake into a long channel. The water surface in the lake is 1.2 m above the channel bottom at the lake outlet. The channel cross section is compound and is similar to that shown as an inset in Figure 3-13. The bottom of the main channel is 1 m below the bottom of the overbank channels. The widths and the Manning roughness coefficients for the overbank channels and the main channel are $B_o = 3$ m, $n_o = 0.0144$, $B_m = 1$ m; and $n_m = 0.013$, where the subscripts o and m denote the overbank channel and main channel, respectively.

(a) Using the energy equation at the channel entrance, plot the variation of discharge with depth in the entrance. The energy coefficient α is a function of y and is given by Eq. (3-41).
(b) From the above plot determine the critical depths and the corresponding critical discharges. (Note that the discharge is either a local maximum or a local minimum at the critical condition.)
(c) Determine the critical slopes corresponding to the critical discharges.
(d) Determine the flow profile if the channel slope is (i) 0.01; (ii) 0.005; (iii) 0.002; and (iv) 0.001.

4-32   Determine conveyance K, energy coefficient α, and specific energy E if the flow depth and the discharge in the channel section shown as an inset in Figure 3-13 are 4 m and 500 m^3/s, respectively.

CHAPTER 5

COMPUTATION OF GRADUALLY VARIED FLOW

The procedure for qualitative sketching of flow profiles in gradually varied flow is described in the last chapter. The integration of the governing equation for quantitative determination of the flow profiles is discussed in this chapter. Different methods of integration and different forms of the governing equation are used for prismatic and nonprismatic channels. The governing equation can be written in either of the following forms:

$$\frac{dH}{dx} = -S_f \qquad \text{[Eq. (1-55)]}$$

$$\frac{dy}{dx} = \frac{S_0 - S_f}{1 - F^2} \qquad \text{[Eq. (4-8)]}$$

$$\frac{dE}{dx} = S_0 - S_f \qquad \text{[Eq. (4-13)]}$$

The governing equation is a first-order nonlinear ordinary differential equation, and its solution requires one boundary condition that is generally stated in terms of depth in the control section. The governing equation is integrated either analytically or numerically. The direct integration method, i.e., the analytical integration, is applicable only to prismatic channels. The direction of integration in the direct integration method has no consequence on the solution. Because most natural channels are not prismatic, the governing equation for such channels is solved numerically. As stated in the Chapter 3, the numerical integration of the governing equation is carried out in the upstream direction in subcritical flow and in the downstream direction in supercritical flow.

It may be mentioned that a general-case computer model, called HEC-2, developed by the Hydrological Engineering Center of the U.S. Army Corps of Engineers (1979) is very popular in the United States. This model uses Eq. (1-55) for solution and includes the effects of various obstructions such as bridge piers, weirs, and culverts.

5-1 Direct Integration Method

The direct integration method is applicable to prismatic channels only. This method uses Eq. (4-8) as the governing equation. Using Eqs. (2-17) and (4-10), the channel slope and the friction slope in Eq. (4-8) can be expressed in terms of conveyance K and discharge Q. From these two equations the ratio of the two slopes can be written as

$$\frac{S_f}{S_0} = \left(\frac{K_n}{K}\right)^2 \tag{5-1}$$

Using Eqs. (3-31), (3-34), and (3-35), the Froude number in Eq. (4-8) can be expressed in terms of section factor Z for critical flow as

$$F^2 = \left(\frac{Z_c}{Z}\right)^2 \tag{5-2}$$

Substitution of Eqs. (5-1) and (5-2) into Eq. (4-8) yields

$$\frac{dy}{dx} = S_0 \left[\frac{1-(K_n/K)^2}{1-(Z_c/Z)^2}\right] \tag{5-3}$$

The direct integration of Eq. (5-3) requires that the terms $(K_n/K)^2$ and $(Z_c/Z)^2$ be expressed in terms of y. It is assumed that both K and Z can be expressed as exponential functions of y, i.e.,

$$K^2 \propto y^N \tag{5-4}$$

and

$$Z^2 \propto y^M \tag{5-5}$$

in which N and M are called hydraulic exponents. Substitution of Eqs. (5-4) and (5-5) into Eq. (5-3) gives

$$\frac{dy}{dx} = S_0 \left[\frac{1-(y_n/y)^N}{1-(y_c/y)^M}\right] \tag{5-6}$$

This equation can be expressed for dx as

$$dx = \frac{y_n}{S_0}\left[1 - \frac{1}{1-u^N} + \left(\frac{y_c}{y_n}\right)^M \frac{u^{N-M}}{1-u^N}\right] du \tag{5-7}$$

in which $u = y/y_n$. The integration of Eq. (5-7) requires that N and M should be constant. Channels can be divided into a number of reaches so that N and M are constants for each reach. Under this assumption the integration of Eq. (5-7) yields

$$x = \frac{y_n}{S_0}\left[u - \int_0^u \frac{du}{1-u^N} + \left(\frac{y_c}{y_n}\right)^M \int_0^u \frac{u^{N-M}}{1-u^N} du\right] + \text{constant} \tag{5-8}$$

The first integral

$$\int_0^u \frac{du}{1-u^N} = F(u,N)$$

is known as the varied flow function. The second integral may be expressed in terms of the varied flow function by the substitution $v = u^{N/J}$ and $J = N/(N-M+1)$ so that

$$\int_0^u \frac{u^{N-M}}{1-u^N} du = \frac{J}{N} \int_0^v \frac{dv}{1-v^J} = \frac{J}{N} F(v,J) \qquad (5-9)$$

Equation (5-8) can be written as

$$x = \frac{y_n}{S_0}\left[u - F(u,N) + \left(\frac{y_c}{y_n}\right)^M \frac{J}{N} F(v,J)\right] + \text{constant}$$

The length Δx of the profile between sections 1 and 2 in Figure 5-1 is obtained from the above equation as

$$\Delta x = x_2 - x_1$$

$$= \frac{y_n}{S_o}\left[(u_2 - u_1) - \{F(u_2,N) - F(u_1,N)\} + \left(\frac{y_c}{y_n}\right)^M \frac{J}{N} \{F(v_2,J) - F(v_1,J)\}\right] (5-10)$$

The transformation in Eq. (5-9) was proposed by Chow (1955); the method is, therefore, known as the Chow method. The values of the varied flow function are given in Appendix E. It should be noted that $F(u,N)$ can be expressed in a series form (French, 1985) which can be easily evaluated on a computer.

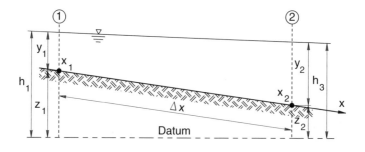

Figure 5-1. Definition sketch.

Hydraulic Exponents

The hydraulic exponents M and N are determined from Eqs. (5-5) and (5-4), respectively. Substituting in Eqs. (5-5) and (5-4) for Z and K from Eqs. (3-34) and (2-18), taking the logarithm of both sides of the equations, differentiating the resulting equations with respect to y, and rearranging the terms in the final equations (Probs. 5-1 and 5-2), one obtains

$$M = \frac{y}{A}\left(3B - \frac{A}{B}\frac{dB}{dy}\right) \tag{5-11}$$

and

$$N = \frac{2y}{3A}\left(5B - 2R\frac{dP}{dy}\right) \tag{5-12}$$

For trapezoidal channels Eqs (5-11) and (5-12) become (Probs. 5-3 and 5-4)

$$M = \frac{3(1+2my/b)}{(1+my/b)} - \frac{2my/b}{(1+2my/b)} \tag{5-13}$$

$$N = \frac{10}{3}\frac{(1+2my/b)}{(1+my/b)} - \frac{8}{3}\frac{(y/b)\sqrt{1+m^2}}{\left[1+2(y/b)\sqrt{1+m^2}\right]} \tag{5-14}$$

The values of M and N respectively are 3 and vary from 2 to $3\frac{1}{3}$ for rectangular channels and 5 and $5\frac{1}{3}$ for triangular channels. For trapezoidal channels M and N vary from 3 to 5 and from 2 to $5\frac{1}{3}$, respectively. The application of the direct integration method is illustrated through the following example.

Example 5-1. A trapezoidal channel with a bottom width of 5 m and side slopes of 1H:1V terminates in a free overfall. The channel is laid on a slope of 0.001 and the Manning n for the channel is 0.013. The normal depth of flow in the channel is 3 m. Compute the length of the water-surface curve extending from the overfall to an upstream section where depth is 1 percent less than the normal depth.

$A_n = (5+1\times3)\times3 = 24\,\text{m}^2; \quad P_n = (5+2\times3\times\sqrt{1+1^2}) = 13.49\,\text{m}; \quad R_n = 24/13.49 = 1.78\,\text{m}$

$$Q = \frac{24\times(1.78)^{2/3}(0.001)^{1/2}}{0.013} = 85.7\,\text{m}^3/s$$

$$\sqrt{\frac{Q^2m^3}{gb^5}} = \sqrt{\frac{85.7^2\times1^3}{9.81\times5^5}} = 0.49$$

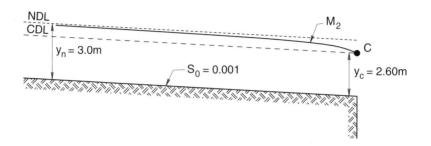

Example 5-1.

From Figure 3-10,

$$\frac{my_c}{b} = 0.52 \quad \text{and} \quad y_c = \frac{0.52 \times 5}{1} = 2.60 \text{ m}$$

Because $y_n > y_c$, the channel slope is mild.

Control: a CDC at point C at the freefall.

Flow profile: M_2 curve between $y_c = 2.60$ m at the fall and depth $= 0.99 \times 3.0 = 2.97$ m in an upstream section.

Because the variation in depth is small, M and N can be computed for the average depth of $(2.60 + 2.97)/2 = 2.79$ m and are assumed constant for the entire length of the curve. The values of M and N from Eqs. (5-13) and (5-14) are $M = 3.54$ and $N = 3.70$:

$$J = N/(N-M+1) = 3.70/(3.70 - 3.54 + 1) = 3.20; \quad v = u^{N/J} = u^{3.70/3.20} = u^{1.16}$$

The length of the M_2 curve is computed as follows. The values of $F(u,.N)$ and $F(v,J)$ were obtained from tables in Appendix E:

y	u	v	F(u,N)	F(v,J)
2.60	0.867	0.848	1.039	1.038
2.97	0.990	0.988	1.787	1.888
Difference	0.123		0.748	0.850

Now
$$\frac{y_n}{S_o} = \frac{3}{0.001} = 3000; \quad \left(\frac{y_c}{y_n}\right)^M \frac{J}{N} = \left(\frac{2.60}{3.00}\right)^{3.54} \frac{3.20}{3.70} = 0.52$$

The length of the M_2 curve from Eq. (5-10) is

$$\Delta x = 3000 \, [(0.123) - (0.748) + 0.52 \, (0.850)] = -549 \text{ m}.$$

The length is negative because section 2 is upstream from section 1.

One of the main advantages of the direct integration method is that there is no constraint on the step length Δx, as long as the variations in M and N are acceptable. Additionally the effect of channel shape on lengths of flow profiles can be easily examined using this method (Jain, 1971). It should be noted that the direct method of integration can be extended to channels with horizontal or adverse slopes and to circular channels in which the hydraulic exponents near the crown vary with depth at high rates (Chow, 1959; French, 1985).

PROBLEMS

5-1   Derive Eq. (5-11).

5-2   Derive Eq. (5-12).

5-3   From Eq. (5-11) derive Eq. (5-13).

5-4   From Eq. (5-12) derive Eq. (5-14).

5-5   Water flows under a sluice gate into a wide rectangular channel that is laid on a slope of 0.001 and has the Manning $n = 0.02$. The sluice gate is regulated to discharge 6 m^3/s per meter width of the channel, and the flow depth at the vena contracta is 0.5 m. Determine the location of the hydraulic jump in the channel.

5-6   A trapezoidal channel with a bottom width of 5 m and side slopes of 2H:1V terminates in a free overfall. The channel is laid on a slope of 0.001 and the Manning n for the channel is 0.013. The normal depth of flow in the channel is 3 m. Compute the length of the water-surface curve extending from the overfall to an upstream section where depth is 1 percent less than the normal depth.

5-7   A trapezoidal channel connects two reservoirs that are 1 km apart. The channel has a base width of 2 m, side slopes of 2H:1V, a Manning n of 0.013, and a bottom slope of 0.0007. The channel is required to convey 60 m^3/s of water. Determine the water level in the upper reservoir above the invert of the channel inlet when the water level in the lower reservoir above the invert of the channel outlet is 3.5 m.

5-8   Derive by direct integration an expression for the water-surface profile in a wide horizontal channel assuming that the Chézy C is a constant.

5-9   Show that the flow profile in a horizontal channel can be expressed as

$$x = \frac{y_c}{S_c}\left(\frac{u^{N-M+1}}{N-M+1} - \frac{u^{N+1}}{N+1}\right) + \text{constant}$$

where $u = y/y_c$.

5-10   Two reservoirs are connected with a wide rectangular channel. The channel carries a unit discharge of 1.8 m^3/s/m. The channel has a slope of 0.001 and a Manning $n = 0.03$. The water levels in the upper and lower reservoirs are 1.5 m above the channel inlet and 0.7 m above the channel outlet, respectively. Determine the length of the channel using the Chow method.

5-11   A wide rectangular channel with Manning $n = 0.025$ and bottom slope $S_0 = 0.01$ carries a discharge of 5 m³/s/m. A backwater curve is created by a dam, and the water depth immediately behind the dam is 2.4 m. Sketch the water-surface profile and compute the length of the backwater curve.

5-12   A wide rectangular channel with bottom slope $S_0 = 0.001$ carries a unit discharge of 2 m³/s/m at a normal depth of 1.0 m. Due to illegal dumping of waste material in the channel, the Manning coefficient for this reach of the channel increased by 100%. The length of the roughened reach of the channel is such that the flow depth at its upstream end differs by 20% from the normal depth for the roughened reach. Sketch the water-surface profile and determine the length of the roughened reach.

5-2 Direct Step Method

The computation of a flow profile by a step method consists of dividing the channel into short reaches and determining reach by reach either the change in depth for a given length of a reach or the length of the reach for a given change in depth. The solution by the former approach, termed the standard step method, involves iterations, while the latter approach, termed the direct step method, yields a direct solution. In natural channels the details of the cross sections in general are measured at locations that are easily accessible. Therefore the lengths of the reaches, which are determined from the measurement locations, are known, and the standard step method is applicable. This method is described in the next section. In prismatic channels where the question of accessibility has no relevance, the computations are made by the direct step method.

The form of the governing equation used in the direct step method is given by Eq. (4-13), which can be written in a finite-difference form for a reach shown in Figure 5-1 as

$$\Delta x = \frac{\Delta E}{S_0 - \overline{S}_f}$$

or

$$x_2 - x_1 = \frac{E_2 - E_1}{S_0 - \overline{S}_f} \tag{5-15}$$

where the subscripts 1 and 2 refer to sections 1 and 2 in Figure 5-1 and $\overline{S}_f$ is the average slope and is given by

$$\overline{S}_f = \tfrac{1}{2}\left(S_{f_1} + S_{f_2}\right) \tag{5-16}$$

It should be noted that the average slope can be defined by equations other than Eq. (5-16), but this definition is commonly used in practice (Chaudhry, 1993). The computational procedure by the direct step method is illustrated by means of the following example.

Example 5-2. A trapezoidal channel with a bottom width of 5 m, side slopes of 1H:1V, and a Manning n of 0.013 carries a discharge of 50 m³/s at a slope of 0.0004.

Compute by the direct step method the backwater profile created by a dam that backs up the water to a depth of 6 m immediately behind the dam. The upstream end of the profile is assumed at a depth equal to 1 percent greater than the normal depth.

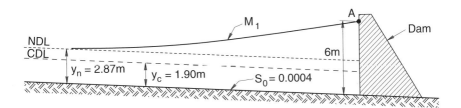

Example 5-2

Normal depth:
$$\frac{nQ}{b^{8/3}S_0^{1/2}} = \frac{0.013 \times 50}{5^{8/3} \times 0.0004^{1/2}} = 0.445$$

From Table D-1 for $m = 1.0$,
$$\frac{y_n}{b} = 0.574; \quad y_n = 2.87 \text{ m}$$

Critical depth: From Eq. (3-32),
$$\frac{y_c'^3(1 + y_c')^3}{2y_c' + 1} = \frac{Q^2 m^3}{gb^5} = \frac{50^2 \times 1^3}{9.81 \times 5^5} = 0.0815.$$

From Figure 3-10 or by trial,
$$y_c' = \frac{my_c}{b} = 0.38; \quad y_c = 1.90 \text{ m}$$

Because $y_n > y_c$, the channel slope is mild.

Control: an ACC at point A at the downstream end of the channel where the depth is 6 m.

Flow profile: The profile lies in zone 1 and therefore it is an M_1 curve. The range of depth is 6 m at the downstream end and $1.01 \times 2.87 = 2.90$ m at the upstream end.

The computations are shown in Table 5-1, which is self-explanatory.

Explanatory Remarks on Table 5-1

In column 1 the flow depth is assigned values from 6.00 to 2.90 m. The decrement in y is large initially and is gradually made smaller as the rate of change of depth with distance becomes small. The values of S_f in column 8 are calculated from the Manning equation, $S_f = n^2 V^2 / R^{4/3}$. The average friction slope in a reach in column 9 is obtained from Eq. (5-16) and the distance Δx in column 11 is computed from Eq. (5-15). The distance L in column 12 is obtained by algebraically adding Δx of column 11.

Table 5-1. Example 5-2 by the Direct Step Method

1	2	3	4	5	6	7	8	9	10	11	12
y m	A m^2	P m	R m	V m/s	E m	ΔE m	S_f ×10^4	$\overline{S}_f$ ×10^4	$S_0 - \overline{S}_f$ ×10^4	Δx m	L m
6.00	66.00	21.97	3.005	0.758	6.029		0.223				0
						−0.491		0.271	2.729	−1,320	
5.50	57.75	20.56	2.808	0.865	5.538		0.318				−1,320
						−0.487		0.394	3.606	−1,350	
5.00	50.00	19.14	2.612	1.000	5.051		0.469				−2,670
						−0.481		0.592	3.408	−1,410	
4.50	42.75	17.73	2.411	1.170	4.570		0.715				−4,080
						−0.472		0.925	3.075	−1,526	
4.00	36.00	16.31	2.207	1.390	4.099		1.135				−5,606
						−0.455		1.505	2.495	−1,750	
3.50	29.75	14.90	1.997	1.680	3.644		1.875				−7,356
						−0.260		2.268	1.732	−1,500	
3.20	26.24	14.05	1.868	1.900	3.384		2.660				−8,856
						−0.082		2.830	1.170	−700	
3.10	25.11	13.77	1.824	1.990	3.303		3.000				−9,556
						−0.080		3.190	0.810	−987	
3.00	24.00	13.49	1.779	2.080	3.222		3.380				−10,543
						−0.057		3.545	0.455	−1,250	
2.90	22.91	13.20	1.735	2.182	3.165		3.710				−11,793

PROBLEMS

5-13　A rectangular channel 6 m wide carries a discharge of 0.8 m^3/s. At a certain section the channel roughness changes from rough to smooth. The normal depths in the rough and the smooth reaches are 0.9 and 0.7 m, respectively. The channel slope is 0.005. Using the single step of the direct step method, estimate the length of the reach of nonuniform flow.

5-14　A wide rectangular channel with Manning $n = 0.025$ is laid with a change in slope from 0.01 to 0.025. The depth of the uniform flow in the downstream reach is 2 m. Determine the location of the hydraulic jump using the single step of the direct step method.

5-15　Solve Problem 5-5 using the direct step method.

5-16　Solve Problem 5-6 using the direct step method.

5-17　Solve Problem 5-12 using the direct step method.

5-3 Standard Step Method

This method is suitable for natural channels where the channel cross sections are available only at predetermined locations, which in turn determines the length of the channel reaches. Because in practice the cross-sectional data, i.e., area A and wetted perimeter P, are expressed in terms of river stage instead of flow depth, the changes in river stage for the given lengths of channel reaches are determined in this method. The form of the governing equation in this method is given by Eq. (1-55), which in reference to sections 1 and 2 in Figure 5-1 can be written as

$$H_2 - H_1 = -\tfrac{1}{2}\left(S_{f_1} + S_{f_2}\right)\left(x_2 - x_1\right) \tag{5-17}$$

in which
$$H_\alpha = y + z + \alpha \frac{V^2}{2g} = h + \alpha \frac{V^2}{2g} \tag{5-18}$$

and $h = y+z =$ river stage. The value of α for compound channels is obtained from Eq. (3-41). Though the procedure for determining the stage in section 2 for the known stage in section 1 is simple, it requires iterations. Assume a value of h_2 and calculate the values of $H_{\alpha 2}$ for the assumed h_2 from Eqs. (5-17) and (5-18). If the two values of $H_{\alpha 2}$ are the same, the assumed value of h_2 is correct; otherwise, try another value of h_2. The solution procedure is illustrated through two examples; one example is for compound channels. Using the Newton-Raphson method for solving nonlinear algebraic equations, the number of iterations in most cases can be minimized to one. Equation (5-17) can be expressed in terms of y_2 as

$$f(y_2) = y_2 + z_2 + \alpha_2 \frac{V_2^2}{2g} - H_1 + \frac{1}{2}\left(S_{f_1} + S_{f_2}\right)\left(x_2 - x_1\right) \tag{5-19}$$

Let y_2^* be an estimate of y_2. Based on the Newton-Raphson method, a better estimate of y_2 is obtained from

$$y_2 = y_2^* - \frac{f\left(y_2^*\right)}{f'\left(y_2^*\right)} \tag{5-20}$$

in which $f' = df/dy_2$, which is obtained by differentiating Eq. (5-19) with respect to y_2 as

$$\frac{df}{dy_2} = 1 + \frac{d}{dy_2}\left\{\alpha_2 \frac{V_2^2}{2g} + \frac{1}{2}S_{f_2}\left(x_2 - x_1\right)\right\} \tag{5-21}$$

On the assumption that the friction slope varies approximately y^{-3}, and P and y are approximately equal to B and R, respectively,

$$\frac{d}{dy_2}\left(\alpha_2\frac{V_2^2}{2g}\right)=\alpha_2\frac{Q^2}{2g}\frac{d}{dy_2}\left(\frac{1}{A_2^2}\right)=-\frac{\alpha_2Q^2B_2}{gA_2^3}\approx-\frac{\alpha_2V_2^2}{gR_2}$$

$$\frac{dS_{f2}}{dy_2}\approx-\frac{3S_{f2}}{y_2}\approx-\frac{3S_{f2}}{R_2}$$

Equation (5-21) can be written as

$$f'(y_2)=1-\alpha_2\frac{V_2^2}{gR_2}-\frac{3}{2}\frac{S_{f2}}{R_2}(x_2-x_1) \tag{5-22}$$

The computational procedure by the standard step method is illustrated by means of examples. Example 5-3 is for single channels and Example 5-4 is for compound channels.

Example 5-3. Find the depth of flow in a section 1000 m upstream from the dam in Example 5-2.

The calculations are done in tabular form and are shown in Table 5-2, which is self-explanatory.

Table 5-2. Example 5-3 by the Standard Step Method

1	2	3	4	5	6	7	8	9	10	11	12
Δx	y	z	A	P	R	V	H_α	S_f	S_f	$\bar{S_f}\Delta x$	H_α
m	m	m	m^2	m	m	m/s	m	$\times10^4$	$\times10^4$	m	m
0	6.00	0	66.00	21.97	3.005	0.758	6.029	0.223			
−1000	5.50	0.4	57.75	20.56	2.808	0.865	5.938	0.318	0.271	−0.027	6.056
−1000	5.62	0.4	59.68	20.90	2.856	0.838	6.056	0.293	0.258	−0.026	6.055

Explanatory Remarks on Table 5-2

In this example the calculations are done in terms of depth instead of river stage, and the energy coefficient α is assumed to be unity. A depth $y = 5.5$ m in column 2 is assumed at $\Delta x = -1000$ m, i.e., in section 2. To measure elevation head, the datum is assumed at the channel bottom in the section at $\Delta x = 0$, i.e., in section 1. The elevation head z for section 2 in column 3 is 0.4 m. The values of total head H_α in section 2 are computed using Eqs. (5-17) and (5-18) and are listed in columns 12 and 8, respectively. For the assumed value of $y = 5.5$ m the two values of the total head in section 2, 5.938 m in column 8 and 6.056 m in column 12, are not the same. Another value of $y = 5.62$ m, which is determined from Eq. (5-20), is tried, and it yields the same value of the total head in section 2. The new trial value of y_2 equal to 5.62 m is computed as follows.

For $y_2^* = 5.5$ m, $f\left(y_2^*\right)$ from Eq. (5-19) is

$$f(y_2^*) = H(y_2^*) - H_1 + \overline{S}_f \Delta x = 5.938 - 6.029 - 0.027 = -0.118 \text{ m}$$

$f'(y_2^*)$ from Eq. (5-22) is

$$f'(y_2^*) = 1 - \frac{0.865^2}{9.81 \times 2.808} - \frac{3 \times 0.318 \times 10^{-4} \times (-1000)}{2 \times 2.808} = 0.990$$

and y_2 from Eq. (5-20) is

$$y_2 = 5.50 - \frac{(-0.118)}{0.990} = 5.62 \text{ m}$$

Example 5-4. Determine the change in the river stage in a 500-m-long river reach for a flow of 3000 m³/s. The cross section of the river in this reach consists of two subsections, a main-channel section and an overbank-channel section, both of which are approximately rectangular in section. The properties of each subsection at the downstream end of the reach are $B_m = 150$ m; $z_m = 15.0$ m; $B_o = 300$ m; and $z_o = 18.0$ m; and at the upstream end of the reach are $B_m = 170$ m; $z_m = 15.2$ m; $B_o = 250$ m; and $z_o = 18.5$ m, where the subscripts m and o refer to the main channel and overbank channel, respectively. The values of the Manning n for the main and overbank channels are 0.03 and 0.05, respectively. The river stage at the downstream section is 20.5 m.

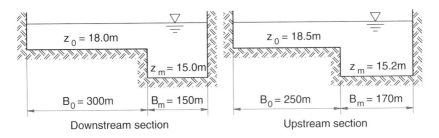

Example 5-4

The computations are summarized in Table 5-3.

Explanatory Remarks on Table 5-3

The flow parameters are computed separately for each subsection. A trial stage $h = 21.00$ m in column 3 is assumed at $L = -500$ m, i.e., in section 2. The K values in column 7 are computed from Eq. (2-20). The energy coefficient α in column 9 is obtained from Eq. (3-41). The values of total head H in section 2 are computed using Eqs. (5-17) and (5-18) and are listed in columns 11 and 16, respectively. For the assumed value of $h = 21.00$ m the two values of the total head in section 2, 21.26 m in column 11 and 21.13 m in column 16, are not the same. Another value of $h = 20.87$ m, which is determined from Eq. (5-20), is tried, and it yields the same value of the total head in section 2. The new trial value is computed as follows.

For $h_2^* = 5.5$ m, $f(y_2^*)$ from Eq. (5-19) is

Table 5-3. Example 5-4 by the Standard Step Method

1	2	3	4	5	6	7	8	9	10	11	12	13	14	15	16
L	Sub sec.*	h	A	P	R	K	K^3/A^2	α	V	H_α	S_f	$\bar{S}_f$	Δx	$S_f \Delta x$	H_α
m		m	m^2	m	m	$\times 10^{-4}$	$\times 10^{-8}$		m/s	m	$\times 10^3$	$\times 10^3$	m	m	m
0	M	20.50	825	158.5	5.205	8.26	8.28								
	O	20.50	750	302.5	2.479	2.75	0.37								
	Total		1575			11.01	8.65	1.61	1.90	20.80	0.742				
−500	M	21.00	986	179.1	5.505	10.25	11.08								
	O	21.00	625	252.5	2.475	2.29	0.31								
	Total		1611	431.6	3.733	12.54	11.39	1.50	1.86	21.26	0.572	0.657	−500	−0.33	21.13
−500	M	20.87	964	179.0	5.385	9.87	10.35								
	O	20.87	593	252.4	2.349	2.10	0.26								
	Total		1557	431.4	3.609	11.97	10.61	1.50	1.93	21.15	0.628	0.685	−500	−0.34	21.14

*The notations M and O in this column refer to the main channel and overbank channel, respectively.

$$f(y_2^*) = H_\alpha(y_2^*) - H_{\alpha 1} + \bar{S}_f \, \Delta x = 21.26 - 20.80 - 0.33 = 0.13 \, \text{m}$$

$f'(y_2^*)$ from Eq. (5-22) is

$$f'(y_2^*) = 1 - 1.5 \times \frac{1.86^2}{9.81 \times 3.733} - \frac{3 \times 0.572 \times 10^{-3} \times (-500)}{2 \times 3.733} = 0.973$$

The value of R_2 in Eq. (5-22) is obtained by dividing the total area A in column 4 by the total wetted perimeter P in column 5. The value of h_2 from Eq. (5-20) is

$$h_2 = 21.00 - \frac{0.13}{0.973} = 20.87 \, \text{m}$$

PROBLEMS

5-18   Determine the flow depth in the section 100 m upstream from the free overfall in Problem 5-6.

5-19   Rework Example 5-4 if the length of the river reach is 700 m.

5-20   Below is shown a trial calculation to determine the river stage at kilometer 20.0. The cross-sectional shape of the river is approximately trapezoidal with a bottom width of 10 m and the side slopes are 2H:1V. The river discharge is 30 m^3/s and the Manning n is 0.013. The river slope is 0.001 and the elevation of the channel bottom at kilometer 19.0 is 0.0.

1	2	3	4	5	6	7	8	9	10	11	12
x	y	z	A	P	R	V	H	S_f	$\bar{S}_f$	$\bar{S}_f \Delta x$	H
km	m	m	m^2	m	m	m/s	m	$\times 10^5$	$\times 10^5$	m	m
19.0	5.00	0.0	100.00	32.36	3.09	0.30	5.01	0.379			
20.0	3.95	1.0	70.71	27.66	2.56	0.42	4.96	0.870	0.625	−0.006	5.01

Using the results of this trial, determine the next trial value that should be close to the correct value, and check whether the next trial value is correct or not.

5-4 The Ezra Method

This method (Ezra, 1954) solves the governing equation (1-55) graphically and is recommended when flow profiles are required for several initial stages and/or discharges. Referring to Figure 5-1, Eq. (1-55) can be written as

$$h_2 + \alpha_2 \frac{V_2^2}{2g} = h_1 + \alpha_1 \frac{V_1^2}{2g} - \frac{1}{2}(S_{f_1} + S_{f_2})(x_2 - x_1) \tag{5-23}$$

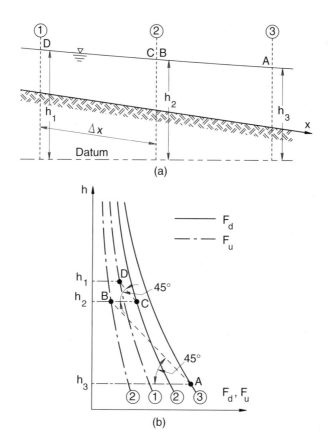

Figure 5-2. Solution by the Ezra method.

This equation can be written in the following form

$$h_2 + F_d(h_2) = h_1 + F_u(h_1)$$ (5-24)

where

$$F_u(h) = \alpha \frac{V^2}{2g} - \frac{1}{2} S_f \Delta x$$ (5-25)

$$F_d(h) = \alpha \frac{V^2}{2g} + \frac{1}{2} S_f \Delta x$$ (5-26)

$\Delta x = x_2 - x_1$ and the subscripts u and d denote the upstream and downstream sections, respectively. For each chosen section in the channel under consideration, the values of the functions $F_u(h)$ and $F_d(h)$ for a given discharge Q_0 are determined for a range of stages. These functions are then plotted, as shown schematically in Figure 5-2 for three sections, sections 1, 2, and 3. It should be noted that two curves are required for each intermediate section, such as section 2; and only one curve is needed for each of the two end sections, such as sections 1 and 3.

The solution of Eq. (5-24) lies along straight lines at an angle θ, where $\tan \theta = -1$, i.e., at 45° to the horizontal axis, as shown by the dotted lines in Figure 5-2 wherein both h and F are plotted using the same scale. The solution for subcritical flow proceeds in the upstream direction from the known initial stage at section 3, which is represented by point A on the F_d curve for section 3. A straight line is drawn through point A at 45°. At point B it intersects the F_u curve for section 2. The stage in section 2 is represented by point B. A horizontal line is drawn through point B. At point C it intersects the F_d curve for section 2. A straight line at 45° is drawn through point C. It intersects the F_u curve for section 1 at point D. The stage in section 1 is represented by point D. Solutions for different initial stages can be obtained in a similar fashion. For supercritical flow the solution proceeds in the downstream direction from the known initial stage at an upstream section. The curves in Figure 5-2 are drawn for a discharge Q_0. These curves can be used to determine water-surface profiles for another discharge Q in the following manner. Note that functions F_u and F_d are proportional to the square of the velocities or the discharges. These functions for Q can be obtained by multiplying the corresponding functions for Q_0 by a factor $(Q/Q_0)^2$. Equation (5-24) for a discharge Q can be written as

$$h_2 + r F_d (h_2) = h_1 + r F_u (h_1) \qquad (5\text{-}27)$$

where $r = (Q/Q_0)^2$ and the functions F_u and F_d are for discharge Q_0. The solution of Eq. (5-27) lies along a straight line inclined at an angle θ to the horizontal axis, where

$$\tan \theta = -r \qquad (5\text{-}28)$$

The water-surface profile for a discharge Q is obtained following the procedure described above for Q_0, except that the 45° lines are drawn at an angle θ given by Eq. (5-28). Note that for $Q = Q_0$, $r = 1$ and $\tan \theta = -1$.

For prismatic channels the computations can be minimized by keeping Δx constant. Each subreach is then represented by the same pair of curves, but adjacent pairs are separated by a vertical distance $S_0 \Delta x$.

PROBLEMS

5-21 Two reservoirs are connected by a wide rectangular channel which carries a unit discharge of 1 m³/s/m. The length, slope, and Manning roughness coefficient of the channel are 200 m, 0.001, and 0.02, respectively. The water level in the lower reservoir is 1.0 m above the invert of the channel at its downstream end.

(a) Determine the type of flow profile in the channel.

(b) Determine by the Ezra method the water level in the upper reservoir:

(c) Determine the water level in the upper reservoir if the unit discharge in the channel is reduced to 0.8 m^3/s/m.

5-22   Solve Example 5-4 by the Ezra method.

5-23   Explain how the Ezra method can be used for estimating the discharge in a channel reach by measuring the stages at the upstream and downstream ends of the reach.

5-5 Inclusion of Form Losses

The previous discussion has assumed that the energy losses in channel reaches are due to channel friction that can be accounted for by the Manning equation. However, there can be additional energy losses caused by local features such as bends, bridge piers, and other structures that produce sudden changes in cross sections. These additional energy losses are termed form or eddy losses and are estimated empirically in most cases. The form losses are expressed as

$$h_e = C_L \frac{V^2}{2g} \qquad (5-29)$$

where h_e is the eddy loss and C_L is the loss coefficient. The loss coefficient depends on the shape and size of the transition. In subcritical flow the form losses produce an increase in upstream water-surface levels. In the case of bridge piers, the following approaches can be used to estimate the rise in water levels upstream from the bridge piers.

One approach would be to apply the energy equation between sections 1 and 2 and the momentum equation between sections 2 and 3, where the sections are identified in Figure 5.3. Sections 1 and 3 are upstream and downstream from the. piers, respectively, and section 2 is within the constricted reach. In the energy equation the energy losses between sections 1 and 2 are neglected, and in the momentum equation the pressure on the rear face of the pier is assumed to be the same as in section 2 (Prob. 5-24). The energy loss between sections 1 and 2 are negligible only for the small contraction ratio $\sigma = b_2/b_1$. The other approach is to apply the momentum equation between sections 1 and 3. But the momentum equation includes a term that represents the drag force on the pier and cannot be easily estimated for different pier shapes. For a blunt-nosed pier, the drag coefficient of 2 can be assumed

The third approach is empirical. The experiments on the flow through bridge piers by Yarnell (1934) have resulted in the empirical relation

$$\frac{\Delta y}{y_3} = KF_3^2 \left(K + 5F_3^2 - 0.6 \right)\left(\alpha + 15\alpha^4 \right) \qquad (5-30)$$

in which Δy is the rise in water level due to piers = $y_1 - y_3$; K is a function of the pier shape and is given in Table 5-4; F_3 is the Froude number in section 3; σ is the contraction ratio = b_2/b_1, and $\alpha = 1-\sigma$. .

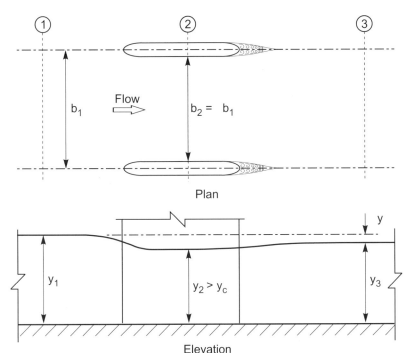

Figure 5-3. Flow between bridge piers.

Table 5-4. Variation of K with Pier Shape

Pier Shape	K
Semicircular nose and tail	0.90
Lens-shaped* nose and tail	0.90
Twin-cylinder piers with connecting diaphragm	0.95
Twin-cylinders without diaphragm	1.05
90° triangular nose and tail	1.05
Square nose and tail	1.25

*A lens-shaped nose or tail is formed from two circular curves, each of radius equal to twice the pier width and each tangential to the pier face.

Equation (5-30) is valid if the contraction does not choke the flow and create the critical condition in section 2. The contraction ratio that causes critical flow in section 2 can be obtained under two different assumptions. The assumption that $E_1 = E_2$ yields (Prob. 5-20)

$$\sigma^2 = \frac{27F_1^2}{2 + F_1^2} \qquad (5\text{-}31)$$

and if $F_2 = F_3$, then (Prob. 5-21)

$$\sigma = \frac{(2+1/\sigma)^3 F_3^4}{(1+2F_3^2)^3}$$

(5-32)

where E_1 and E_2 are the specific energies in sections 1 and 2 and F_2 and F_3 are the specific-force functions in sections 2 and 3, respectively. Though Yarnell used Eq. (5-31) to determine the critical contraction ratio, Eq. (5-32) is more likely to be correct as it does not depend on the energy conservation assumption.

PROBLEMS

5-24 A channel of rectangular cross section 3 m in width carries 10 m³/s of water at a normal depth of 1.5 m. Determine the change in upstream depth that would result from the installation of a centrally located pier 1 m wide and 2 m long with a well-rounded nose.

5-25 Derive Eq. (5-31).

5-26 Derive Eq. (5-32).

5-6 Flow in Parallel Channels

An island in rivers divides the flow into two parallel channels, as shown in Figure 5-4. The division of discharge between two channels and the flow profiles in the two channels are determined as follows. Considering the usual case of subcritical flow in rivers, the control section is located at the downstream end at point B. A certain discharge Q_1 is assumed in the channel. The discharge in channel 2 is obtained from the continuity equation, i.e., $Q_2 = Q - Q_1$. Starting from point B to point A, the water-surface profiles are computed in channel 1 for Q_1 and in channel 2 for Q_2. If the elevations of the water surface at A are the same for both the channels, the assumed distribution of discharge is correct; otherwise a new trial value of Q_1 is required.

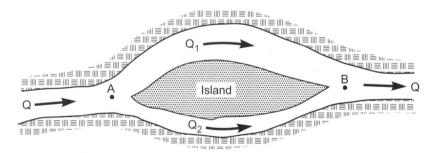

Figure 5-4. Flow past an island.

APPENDIX E

TABLE OF THE VARIED-FLOW FUNCTION

Table E-1. Values of $F(u,N)$

N u	2.6	3.0	3.4	3.8	4.2	4.6	5.0	5.4	5.8
0.00	0.000	0.000	0.000	0.000	0.000	0.000	0.000	0.000	0.000
0.10	0.100	0.100	0.100	0.100	0.100	0.100	0.100	0.100	0.100
0.20	0.201	0.200	0.200	0.200	0.200	0.200	0.200	0.200	0.200
0.30	0.304	0.302	0.301	0.301	0.300	0.300	0.300	0.300	0.300
0.40	0.411	0.407	0.404	0.403	0.402	0.401	0.401	0.400	0.400
0.50	0.525	0.517	0.511	0.508	0.505	0.504	0.503	0.502	0.501
0.54	0.574	0.563	0.556	0.551	0.548	0.546	0.544	0.543	0.542
0.58	0.626	0.612	0.603	0.596	0.592	0.589	0.587	0.585	0.583
0.62	0.680	0.663	0.651	0.643	0.637	0.633	0.630	0.628	0.626
0.66	0.738	0.717	0.703	0.692	0.685	0.679	0.675	0.672	0.669
0.70	0.802	0.776	0.757	0.744	0.735	0.727	0.722	0.717	0.714
0.72	0.836	0.807	0.786	0.772	0.761	0.752	0.746	0.741	0.737
0.74	0.868	0.840	0.817	0.800	0.788	0.779	0.771	0.766	0.761
0.76	0.909	0.874	0.849	0.830	0.817	0.806	0.798	0.791	0.786
0.78	0.950	0.911	0.883	0.862	0.847	0.834	0.825	0.817	0.811
0.80	0.994	0.950	0.919	0.896	0.878	0.865	0.854	0.845	0.838
0.81	1.017	0.971	0.938	0.914	0.895	0.881	0.869	0.860	0.852
0.82	1.041	0.993	0.958	0.932	0.913	0.897	0.885	0.875	0.866
0.83	1.067	1.016	0.979	0.952	0.931	0.914	0.901	0.890	0.881
0.84	1.094	1.040	1.001	0.972	0.949	0.932	0.918	0.906	0.897
0.85	1.121	1.065	1.024	0.993	0.969	0.950	0.935	0.923	0.912
0.86	1.153	1.092	1.048	1.015	0.990	0.970	0.954	0.940	0.930
0.87	1.182	1.120	1.074	1.039	1.012	0.990	0.973	0.959	0.947
0.88	1.228	1.151	1.101	1.064	1.035	1.012	0.994	0.978	0.966
0.89	1.255	1.183	1.131	1.091	1.060	1.035	1.015	0.999	0.986
0.90	1.294	1.218	1.163	1.120	1.087	1.060	1.039	1.021	1.007
0.91	1.338	1.257	1.197	1.152	1.116	1.088	1.064	1.045	1.029
0.92	1.351	1.300	1.236	1.187	1.148	1.117	1.092	1.072	1.054
0.93	1.435	1.348	1.279	1.226	1.184	1.151	1.123	1.101	1.081
0.94	1.504	1.403	1.328	1.270	1.225	1.188	1.158	1.134	1.113
0.95	1.582	1.467	1.385	1.322	1.272	1.232	1.199	1.172	1.148
0.96	1.665	1.545	1.454	1.385	1.329	1.285	1.248	1.217	1.188
0.97	1.780	1.644	1.543	1.464	1.402	1.351	1.310	1.275	1.246
0.98	1.946	1.783	1.666	1.575	1.502	1.443	1.395	1.354	1.339
0.99	2.212	2.017	1.873	1.761	1.671	1.598	1.537	1.487	1.444

Table E-1. (continued)

N u	2.6	3.0	3.4	3.8	4.2	4.6	5.0	5.4	5.8
0.995	2.478	2.250	2.070	1.945	1.838	1.751	1.678	1.617	1.565
1.000	∞	∞	∞	∞	∞	∞	∞	∞	∞
1.005	2.022	1.679	1.384	1.188	1.036	0.915	0.817	0.737	0.669
1.01	1.757	1.419	1.182	1.007	0.873	0.766	0.681	0.610	0.551
1.02	1.493	1.191	0.982	0.828	0.711	0.620	0.546	0.486	0.436
1.03	1.340	1.060	0.866	0.725	0.618	0.535	0.469	0.415	0.370
1.04	1.232	0.967	0.785	0.653	0.554	0.477	0.415	0.365	0.324
1.05	1.150	0.896	0.723	0.598	0.504	0.432	0.374	0.328	0.289
1.06	1.082	0.838	0.672	0.553	0.464	0.396	0.342	0.298	0.262
1.07	1.026	0.790	0.630	0.516	0.431	0.366	0.315	0.273	0.239
1.08	0.978	0.749	0.595	0.485	0.403	0.341	0.292	0.252	0.220
1.09	0.935	0.713	0.563	0.457	0.379	0.319	0.272	0.234	0.204
1.10	0.897	0.681	0.536	0.433	0.357	0.299	0.254	0.218	0.189
1.11	0.864	0.652	0.511	0.411	0.338	0.282	0.239	0.204	0.176
1.12	0.833	0.626	0.488	0.392	0.321	0.267	0.225	0.192	0.165
1.13	0.805	0.602	0.468	0.374	0.305	0.253	0.212	0.181	0.155
1.14	0.780	0.581	0.450	0.358	0.291	0.240	0.201	0.170	0.146
1.15	0.756	0.561	0.432	0.343	0.278	0.229	0.191	0.161	0.137
1.16	0.734	0.542	0.417	0.329	0.266	0.218	0.181	0.153	0.130
1.17	0.713	0.525	0.402	0.317	0.255	0.208	0.173	0.145	0.123
1.18	0.694	0.509	0.388	0.305	0.244	0.199	0.165	0.138	0.116
1.19	0.676	0.494	0.375	0.294	0.235	0.191	0.157	0.131	0.110
1.20	0.659	0.480	0.363	0.283	0.226	0.183	0.150	0.125	0.105
1.22	0.628	0.454	0.341	0.264	0.209	0.168	0.138	0.114	0.095
1.24	0.600	0.431	0.322	0.248	0.195	0.156	0.127	0.104	0.086
1.26	0.574	0.410	0.304	0.233	0.182	0.145	0.117	0.095	0.079
1.28	0.551	0.391	0.288	0.219	0.170	0.135	0.108	0.088	0.072
1.30	0.530	0.373	0.274	0.207	0.160	0.126	0.100	0.081	0.066
1.32	0.510	0.357	0.260	0.196	0.150	0.118	0.093	0.075	0.061
1.34	0.492	0.342	0.248	0.185	0.142	0.110	0.087	0.069	0.056
1.36	0.475	0.329	0.237	0.176	0.134	0.103	0.081	0.064	0.052
1.38	0.459	0.316	0.226	0.167	0.127	0.097	0.076	0.060	0.048
1.40	0.444	0.304	0.217	0.159	0.120	0.092	0.071	0.056	0.044
1.42	0.431	0.293	0.208	0.152	0.114	0.087	0.067	0.052	0.041
1.44	0.417	0.282	0.199	0.145	0.108	0.082	0.063	0.049	0.038
1.46	0.405	0.273	0.191	0.139	0.103	0.077	0.059	0.046	0.036
1.48	0.394	0.263	0.184	0.133	0.098	0.073	0.056	0.043	0.033
1.50	0.383	0.255	0.177	0.127	0.093	0.069	0.053	0.040	0.031
1.60	0.335	0.218	0.148	0.103	0.074	0.054	0.040	0.030	0.023
1.70	0.298	0.189	0.125	0.086	0.060	0.043	0.031	0.023	0.016
1.80	0.267	0.166	0.108	0.072	0.049	0.034	0.024	0.017	0.012

Table E-1. (continued)

N u	2.6	3.0	3.4	3.8	4.2	4.6	5.0	5.4	5.8
1.90	0.242	0.147	0.0.94	0.062	0.041	0.028	0.020	0.014	0.010
2.00	0.221	0.132	0.082	0.053	0.035	0.023	0.016	0.011	0.007
2.20	0.186	0.107	0.065	0.040	0.025	0.016	0.011	0.007	0.005
2.40	0.160	0.089	0.052	0.031	0.019	0.012	0.008	0.005	0.003
2.60	0.140	0.076	0.043	0.025	0.015	0.009	0.005	0.003	0.002
2.80	0.124	0.065	0.036	0.020	0.012	0.007	0.004	0.002	0.001
3.0	0.110	0.056	0.030	0.017	0.009	0.005	0.003	0.002	0.001
3.5	0.085	0.041	0.021	0.011	0.006	0.003	0.002	0.001	0.001
4.0	0.069	0.031	0.015	0.007	0.004	0.002	0.001	0.000	0.000
4.5	0.057	0.025	0.011	0.005	0.003	0.001	0.001	0.000	0.000
5.0	0.048	0.020	0.009	0.004	0.002	0.001	0.000	0.000	0.000
7.0	0.028	0.010	0.004	0.002	0.001	0.000	0.000	0.000	0.000
10.0	0.016	0.005	0.002	0.001	0.000	0.000	0.000	0.000	0.000
20.0	0.011	0.002	0.001	0.000	0.000	0.000	0.000	0.000	0.000

CHAPTER 6

SPATIALLY VARIED FLOW

Spatially varied flow in a channel reach, as defined earlier in Section 1-8, is a gradually varied flow in which the flow is continuously added or subtracted along the channel reach. The discharge in the channel reach is a function of the longitudinal distance x. Such flows are not uncommon. Examples of spatially varied flow are flow in roof gutters, flow in side-channel spillways for dams, and flow in channels with withdrawal of flow through a side-discharge weir, a bottom rack, and a permeable boundary in the form of seepage. The first two examples pertain to lateral inflow, and the remaining examples are related to lateral outflow.

The governing equations for spatially varied flow are presented in Section 1-8 wherein it is shown that the momentum and energy equations for spatially varied flow are different and a choice between these two equations must be made. It is explained therein that the energy equation is more appropriate for lateral outflow and the momentum equation is more suitable for lateral inflow. Because the governing equations are different for the two types of spatially varied flow, the two types are discussed separately.

6-1 Lateral Outflow

The energy equation is more appropriate for lateral-outflow problems, because the longitudinal velocity V_ℓ of lateral outflow in the momentum equation cannot be easily estimated. Moreover it is reasonable to assume that the energy slope in the energy equation for lateral outflow is equal to the frictional slope that can be estimated from the Manning equation.

Governing Equations

The governing equations for lateral outflow as derived in Section 1-8 are as follows:

Continuity equation: $\qquad \dfrac{dQ}{dx} + q_\ell = 0 \qquad$ [Eq. (1-56)]

Energy equation: $\qquad \dfrac{dH_\alpha}{dx} + \dfrac{\overline{U}_\ell^{\,2} - \alpha V^2}{2gQ} q_\ell = -S_f \qquad$ [Eq. (1-58)]

in which the use of Eq. (1-49) and the assumption that $S_e = S_f$ have been made. An estimate of total velocity $\overline{U}_\ell$ of lateral outflow is needed to solve the governing equations. It is reasonable to assume that the total velocities of the lateral and main flows are the same, i.e., $\overline{U}_\ell = V$. The second term on the left-hand side of

180

Eq. (1-58) vanishes, and Eq. (1-58) can be written as

$$\frac{d}{dx}\left(\frac{V^2}{2g}\right) + \cos\theta\frac{dy}{dx} + \frac{dG_0}{dx} = -S_f \tag{6-1}$$

An expression for the first term on the left-hand side of Eq. (6-1) can be derived as follows:

$$\frac{d}{dx}\left(\frac{V^2}{2g}\right) = \frac{1}{2g}\frac{d}{dx}\left(\frac{Q^2}{A^2}\right) = \frac{1}{2g}\left(\frac{2Q}{A^2}\frac{dQ}{dx} - \frac{2Q^2}{A^3}\frac{dA}{dy}\frac{dy}{dx}\right)$$
$$= \frac{Q}{gA^2}\frac{dQ}{dx} - \frac{Q^2 B}{gA^3}\frac{dy}{dx} \tag{6-2}$$

Substituting Eq. (6-2) into Eq. (6-1), noting that $dG_0/dx = -S_0$ and $F^2 = (Q^2 B/gA^3)$ for $\cos\theta = 1$, and rearranging the terms in the resulting equation, one obtains

$$\frac{dy}{dx} = \frac{S_0 - S_f - \left(Q/gA^2\right)\left(dQ/dx\right)}{1 - F^2} \tag{6-3}$$

Equation (6-3) can be integrated either analytically or numerically. However, only the simplified form of Eq. (6-3) can be solved analytically.

Analytical Solutions

The analytical solutions of Eq. (6-3) are presented for two cases: (i) flow through a side-discharge weir, which is installed along the side of a channel for diverting excess flow, and (ii) flow through a bottom rack. The racks are installed in the bottom of mountain streams as one of the components of "water intakes" for removing gravel from the streams. The length of a side-discharge weir or of a bottom rack generally is short; consequently the effect of channel and friction slopes on the flow profile can be assumed to be negligible. On the assumption that $S_0 = 0$ and $S_f = 0$, Eq. (6-3) reduces to

$$\frac{dy}{dx} = \frac{-\left(Q/gA^2\right)\left(dQ/dx\right)}{1 - F^2} \tag{6-4}$$

As both $S_0 = 0$ and $S_f = 0$, the specific energy E in the transition is constant, as can be seen from Eq. (4-13). The flow profile in the channel reach with lateral outflow, referred to hereinafter as transition reach, can be either a backwater curve or a drawdown curve depending on the sign of the term dy/dx in Eq. (6-4). The numerator in the term on the right-hand side is positive, because dQ/dx is negative for lateral outflow. For subcritical flow, i.e., for $F < 1$, dy/dx is positive and therefore the flow profile in the transition reach is a backwater curve. On the other hand, for supercritical flow, i.e., for $F > 1$, dy/dx is negative and the profile is a drawdown curve.

Several types of flow profiles are possible in the upstream reach (referred to as reach 1), in the transition reach, and in the downstream reach (referred to as reach 2). The types of flow profiles depend on the flow conditions in reach 1 and reach 2, which in turn prescribe the controls in sections 1 and 2, which are respectively located at the upstream and downstream ends of the transition reach. The possible controls are either a normal-depth control or a critical-depth control. A CDC is possible only in section 1, because the minimum head required to permit the flow is larger in section 1 than section 2. However, the CDC in section 1 is not active if the slope of reach 1 is steep; in that case there exists an NDC in section 1.

Possible flow profiles are shown in Figure 6-1. Though the flow profiles in this figure are drawn for a side-discharge weir, these flow profiles also are applicable for the case where the lateral outflow occurs through a bottom rack. The characteristics of the flow profiles for various flow conditions in reaches 1 and 2 are summarized in Table 6-1, which is self-explanatory. As an example consider the flow conditions given in row 1 of Table 6-1. Reaches 1 and 2 are mild; therefore the possible controls are (i) the NDCs at the upstream end of the mild channels, i.e., one NDC at point N in section 2 and another NDC at the upstream end of reach 1 (neither listed in the table nor shown in Figure 6-1a), and (ii) the CDC at point C in section 1. Let the flow conditions be such that $E_{n2} > E_{n1} > E_{c1}$. Of the two controls, the NDC is active because $E_{n2} > E_{c1}$. The flow in the transition is subcritical as it is at the control N. The flow profile in the transition from Eq. (6-4) is a backwater (BW) curve. The flow profile in reach 1 is M_1 because $E_{n2} > E_{n1}$. The flow in reach 2 is uniform because there is no other control in the reach where it is controlled by the channel slope and the friction.

Let us examine the effect on the flow profile of the normal depth in reach 2. Let y_{n1}, and hence E_{n1}, be kept constant and y_{n2}, and consequently E_{n2}, be allowed to decrease such that $E_{n1} > E_{n2} > E_{c1}$. This flow condition is given in row 2 of Table 6-1 and in Figure 6-1b. Of the two controls, the NDC is still active because $E_{n2} > E_{c1}$. The flow profile is similar to that in Figure 6-1a, except in reach 1, where the flow profile is an M_2 curve because $E_{n1} > E_{n2}$. Let y_{n2} be further lowered so that $E_{n2} < E_{c1}$, as given in row 3 of Table 6-1. Of the two controls, the CDC becomes active as $E_{c1} > E_{n2}$. The flow downstream from the CDC is supercritical, but the flow in reach 2 is subcritical. The flow changes from supercritical to subcritical through a hydraulic jump, as shown in Figure 6-1c. The location of the jump depends on y_{n2}. The jump moves downstream as y_{n2} decreases. The energy loss in the jump increases as the jump moves from section 1 to section 2. Let ΔE_{j2} be the energy loss in the jump when it forms in section 2. The specific energy immediately downstream from the jump is $E_{n2} = (E_{c1} - \Delta E_{j2})$. If $E_{n2} > (E_{c1} - \Delta E_{j2})$, the jump will form between sections 1 and 2. For this flow condition the NDC also is an active control. The depth in reach 1 changes from y_{n1} to y_{c1}; hence the flow profile in reach 1 is an M_2 curve. The flow downstream from the CDC is supercritical; hence the flow profile there from Eq. (6-4) is a drawdown (DD) curve. The flow upstream from the NDC is subcritical; hence the flow profile there

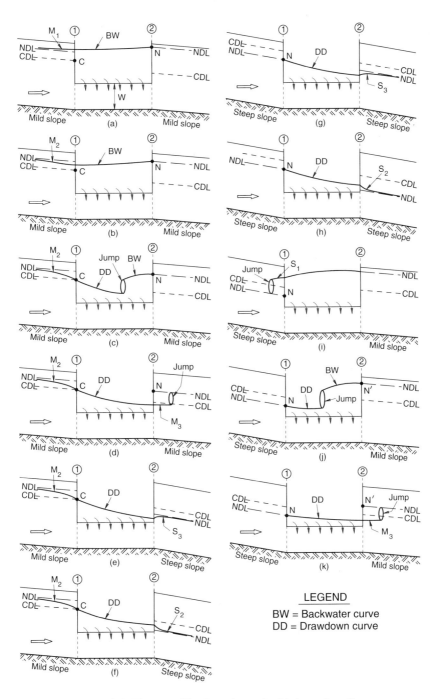

Figure 6-1. Flow profiles in a channel with lateral outflow.

Table 6-1. Types of Flow Profiles in a Channel with Lateral Outflow

Reach 1 Slope	Reach 2 Slope	Possible Control(s)	Figure No.	Relative Magnitudes of Specific Energies	Active Control(s)	Flow Profiles Reach 1	Flow Profiles Transition	Flow Profiles Reach 2
Mild	Mild	C in section 1 N in section 2	6-1a	$E_{n2} > E_{n1} > E_{c1}$	N in section 2	M_1	BW	UF
			6-1b	$E_{n1} > E_{n2} > E_{c1}$	N in section 2	M_2	BW	UF
			6-1c	$E_{c1} > E_{n2} > E_{c1} - \Delta E_{j2}$	C in section 1 N in section 2	M_2	DD, J, & BW	UF
			6-1d	$E_{c1} - \Delta E_{j2} > E_{n2}$	C in section 1	M_2	DD	M_3 & J
Mild	Steep	C in section 1	6-1e	$E_{n1} > E_{c1} > E_{n2}$	C in section 1	M_2	DD	S_3
			6-1f	$E_{n1} > E_{n2} > E_{c1}$	C in section 1	M_2	DD	S_2
Steep	Steep	N in section 1	6-1g	$E_{n1} > E_{n2}$	N in section 1	UF	DD	S_3
			6-1h	$E_{n2} > E_{n1}$	N in section 1	UF	DD	S_2
Steep	Mild	N in section 1 N' in section 2	6-1i	$E_{n1} - \Delta E_{j2} > E_{n2}$	N in section 1	UF	DD	M_3 & J
			6-1j	$E_{n2} > E_{n1} - \Delta E_{j1}$	N in section 2	J & S_1	BW	UF
			6-1k	$E_{n1} - \Delta E_{j1} > E_{n2} > E_{n1} - \Delta E_{j2}$	N in section 1 N' in section 2	UF	DD, J, & BW	UF

Legend: C = CDC at C; N = NDC at N; DD = Drawdown curve; BW = Backwater curve; J = Jump; UF = Uniform flow.

is a BW curve. There exists a hydraulic jump between the DD and BW curves. The flow in reach 2 is uniform because there is no other control in the reach and is controlled by the channel slope and friction. If y_{n2} is further decreased so that $E_{n2} < (E_{c1} - \Delta E_{j2})$ as given in row 4 of Table 6-1, the jump then forms in reach 2 and is preceded by an M_3 curve, as shown in Figure 6-1d. The flow profile in the transition is a DD curve. The flow profile in reach 1 remains unchanged as an M_2 curve. The other profile given in Figure 6-1 and Table 6-1 can be analyzed in a similar fashion.

Side-Discharge Weir. The integration of Eq. (6-4) requires an outflow-head relationship for side weirs. Such relations for different weirs are discussed in a later chapter. An outflow-head relation for a side weir is given by

$$-\frac{dQ}{dx} = C_s \sqrt{2g} \left(y - W \right)^{3/2} \tag{6-5}$$

where C_s is the discharge coefficient for the side weir and W is the height of the weir crest above the channel bottom (see Figure 6-1a). As both $S_0 = 0$ and $S_f = 0$, the specific energy E in the transition is constant. Additional assumptions are needed to solve Eq. (6-4) analytically. These assumptions are (i) the channel is rectangular and (ii) the discharge coefficient C_s is constant. Using the definition of specific energy, discharge Q in a rectangular channel can be expressed in terms of E and y as

$$Q = by\sqrt{2g(E - y)} \tag{6-6}$$

Substituting into Eq. (6-4) for dQ/dx and Q from Eqs. (6-5) and (6-6), noting that $F^2 = Q^2 b/g(by)^3$ in Eq. (6-4), and rearranging the terms in the resulting equation, one obtains

$$\frac{dy}{dx} = \frac{2C_s}{b} \frac{\sqrt{(E - y)(y - W)^3}}{(3y - 2E)} \tag{6-7}$$

Equation (6-7) was integrated by de Marchi (1934), and the result is

$$\frac{xC_s}{b} = \frac{2E - 3W}{E - W} \sqrt{\frac{E - y}{y - W}} - 3\sin^{-1}\sqrt{\frac{E - y}{E - W}} + \text{constant} \tag{6-8}$$

The constant in Eq. (6-8) is determined from the conditions at the control. If the control is in section 1 where the flow conditions are known, the determination of the constant is straightforward. However, if the control is in section 2 (the control in section 2 is an NDC, as shown in Table 6-1), the determination of the constant requires a trial-and-error procedure because the discharge Q_2 in section 2 is not known. Such a trial-and-error procedure is as follows: (i) Guess a trial value of Q_2. (ii) Using the Manning equation, calculate y_{n2} for the assumed Q_2. (iii) Calculate E_{n2}. (iv) Determine the constant in Eq. (6-8) for $E = E_{n2}$, $y = y_{n2}$, and $x = L$. (v) Because E is constant in the transition, calculate y_1 for $E_1 = E_{n2}$ and the given Q_1. (vi) Check whether Eq. (6-8) is satisfied in section 1 for $y = y_1$ and $x = 0$. If Eq. (6-8) is not satisfied in section 1, try another value of Q_2.

Note that the discharge over the side weir is a function of x because the flow depth, and consequently the head over the weir crest, is a function of x. However, if the width of the channel in the transition reach decreases linearly, it can be shown from Eqs. (6-5) and (6-6) that the discharge over the weir is a constant (Prob. 6-7). The experimental result of a side weir of this type is presented by Jain and Fischer (1982).

Bottom Screen. The flow in a channel with a lateral outflow through a bottom screen is shown in Figure 6-2. An outflow-head relation for flow through the screen is given by

$$-\frac{dQ}{dx} = \varepsilon C_r b\sqrt{2gE} \qquad (6-9)$$

in which b is the channel width, ε is the ratio of the opening area to the total area of the screen surface, and C_r is the coefficient of discharge for the screen. Substituting into Eq. (6-4) for dQ/dx from Eq. (6-9) and for Q from Eq. (6-6) and rearranging the terms in the resulting equation, one obtains

$$\frac{dy}{dx} = \frac{2\varepsilon C_r \sqrt{E(E-y)}}{3y-2E} \qquad (6-10)$$

Integration of Eq. (6-10) yields

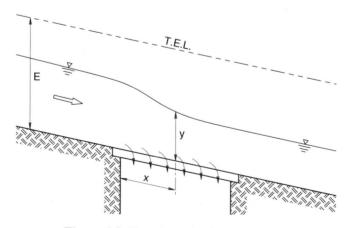

Figure 6-2. Flow through a bottom screen.

$$x = -\frac{y}{\varepsilon C_r}\sqrt{1-\frac{y}{E}} + \text{constant} \tag{6-11}$$

The constant of integration can be determined in a manner similar to that described above for the side-discharge weir. It should be noted that Eq. (6-11) can also be derived using only Eqs. (6-6) and (6-9), i.e., without using Eq. (6-4) (Prob. 6-4).

Numerical Integration

In problems where the terms S_0 and S_f in the governing equation cannot be neglected, the solution of the governing equation is obtained by the numerical integration method. Equation (6-1) is used to derive the finite-difference form of the governing equation. Assuming $\cos\theta = 1$, this equation can be written as

$$\frac{dy}{dx} = S_0 - S_f - \frac{V}{g}\frac{dV}{dx} \tag{6-12}$$

Equation (6-12) can be written in a finite-difference form as

$$\Delta y = \left(S_0 - \overline{S}_f\right)\Delta x - \frac{\overline{Q}}{g\overline{A}}\Delta V \tag{6-13}$$

where the bar over a term denotes the average value of that term within the finite increment Δx of the channel length. The average discharge and area can be written as

$$\overline{Q} = \frac{Q_1 + Q_2}{2}; \quad \overline{V} = \frac{V_1 + V_2}{2}; \quad \overline{A} = \frac{\overline{Q}}{\overline{V}} = \frac{Q_1 + Q_2}{V_1 + V_2} \tag{6-14}$$

where the subscripts 1 and 2 refer to sections 1 and 2. Substitution for the average discharge and area from Eq. (6-14) into Eq. (6-13) yields

$$\Delta y = \left(S_0 - \overline{S}_f\right)\Delta x - \frac{Q_1(V_1 + V_2)}{g(Q_1 + Q_2)}\left(1 + \frac{\Delta Q}{2Q_1}\right)\Delta V \tag{6-15}$$

The procedure for solving Eq. (6-15) is similar to that illustrated in the next section for the equation for lateral inflow. The computations proceed from the control section, which in the case of $S_0 = 0$ and $S_f = 0$ is in either section 1 or section 2, as described above. Is a control section possible within the transition reach if S_0 and S_f are not zero? If the answer is positive, the control is a CDC because an NDC is not possible in the transition reach. The F is unity at a CDC, and dy/dx at the CDC from Eq. (6-3) is either infinite if the numerator is nonzero or finite if the numerator is zero. The former condition implies a negative hydraulic jump because the flow upstream and downstream from a CDC is subcritical and supercritical, respectively. It can be shown that there should be a gain of energy in a negative hydraulic jump, which is at variance with the principle of conservation of energy; therefore the possibility of a negative jump is ruled out. The latter condition implies that

$$S_0 = S_f + \frac{Q}{gA^2}\frac{dQ}{dx} \tag{6-16}$$

Equation (6-16) can be solved by a trial-and-error procedure that is similar to that illustrated in the next section for lateral inflow.

PROBLEMS

6-1   The width of a horizontal rectangular channel linearly decreases from 3 to 1.2 m and then the channel terminates in a free overfall. The water is withdrawn through a perforated screen located in the bottom of the tapered section of the channel. The rate of withdrawal is such that the unit discharge in the channel varies linearly. The discharge in the channel upstream from the tapered section is 6 m^3/s and is controlled by a sluice gate with an opening of 0.4 m. Determine the flow depths in sections (i) upstream from the gate, (ii) at the beginning of the tapered reach, and (iii) at the end of the tapered reach if the discharge through the perforated screen is (a) 3.6 m^3/s, (b) 2.0 m^3/s, and (c) 1.5 m^3/s. Assume that the coefficient of discharge for the gate is unity.

6-2   Sketch the flow profile in a channel with lateral outflow when
$y_{c1} = 1.5 y_{c2} = 2.0 y_{n1} = y_{n2}$.

6-3   Consider the flow of water in a long channel of rectangular section 1-m-wide and with a 2-m-long perforated screen in the bottom. The Manning roughness coefficient for the channel is 0.03 and the channel slope is 0.008. The discharge in the channel upstream from the screen is 2 m^3/s. The ratio of the opening area to the total area of the screen is 0.5 and the coefficient of discharge for the screen is 0.6. Neglecting the effect of channel slope and friction in the transition reach, sketch the flow profile in the channel

6-4   Derive Eq. (6-10) from Eqs. (6-6) and (6-9) if E is constant within the transition reach.

6-5   A long rectangular channel includes a side-discharge weir which comes into operation when the flow in the channel reaches 0.6 m^3/s. The width, slope, and Manning n for the channel are 2 m, 0.001, and 0.013, respectively. The coefficient of discharge for the weir is 0.52. Determine the length of the weir and height of the weir crest if the total discharge over the weir is 0.2 m^3/s when the upstream flow is 0.8 m^3/s.

6-6   A bottom rack, consisting of parallel bars oriented in the flow direction, is located on the bed of a 3-m-wide rectangular channel. The rack extends all the way across the channel, and the bars are spaced evenly in such a way that one-half of the total rack area is taken up by the bars. The discharge coefficient for the rack is 0.5. Upstream from the rack the discharge is 15 m³/s, and the flow is subcritical. It may be assumed that the flow passes through the critical depth at the upstream end of the rack. What will be the depth at the downstream end of the rack if the rack is just long enough to withdraw one-half of the flow? How long must this rack be? The effect of resistance and channel slope may be neglected.

6-7   Show that the discharge over a side weir is constant if the width of the channel in the transition reach decreases linearly.

6-8   Consider the flow of water in a long rectangular channel with a side-discharge weir. Sketch the flow profile in the entire channel (upstream from, within, and downstream from the weir) for the following flow conditions: width of the channel = 6 m; bottom slope of the channel = 0.005; discharge in the channel upstream from the weir = 12 m³/s; Manning's n for the channel upstream from the weir = 0.015; discharge in the channel downstream from the weir = 6 m³/s; depth of flow in the channel at the downstream edge of the weir = 0.25 m; and Manning's n for the channel downstream from the weir = 0.03.

6-9   A long flume (Manning n = 0.014) is laid on a slope of 0.001 and has a rectangular section 3 m wide. An open grille with ε = 0.25 and C_r = 0.5 is inserted in the bed of the flume. Assuming that the control is in the section immediately upstream of the grille, determine the flow through the grille if the discharge in the flume upstream from the grille is 8 m³/s. Check whether the assumption about the location of the control is correct. For what value of the upstream discharge would the entire discharge pass through the grille?

6-10   A 3-m-wide rectangular channel carries a discharge of 3.6 m³/s at a normal depth of 0.70 m. A side-discharge weir that carries a discharge of 1.5 m³/s is installed in one of the sidewalls of the channel. The normal depth in the channel downstream from the weir is 0.45 m. Identify the possible controls and their locations. Which control (or controls) is (or are) active and why? Sketch the water-surface profile in the channel. Neglect S_0 and S_f in the transition. Determine the length of the weir if the coefficient of discharge for the weir is 0.62 and the height of the weir is 15 cm.

6-2 Lateral Inflow

The momentum equation is more appropriate for lateral inflows, because it is difficult in the energy equation to account for the energy dissipation due to additional turbulence generated by the mixing of the lateral inflow with the main flow.

Governing Equations

The governing equations for lateral inflow as derived in Section 1-8 are as follows:

Continuity equation: $$\frac{dQ}{dx} + q_\ell = 0 \qquad (1\text{-}56)$$

Momentum equation: $$\frac{dH}{dx} + \frac{V_\ell - V}{gA} q_\ell = -S_f \qquad (1\text{-}57)$$

in which Eq. (1-48) has been used. Equation (1-57) is solved for a side-channel spillway (see Figure 6-3), wherein the lateral inflow enters the channel normal to the main flow in the channel, i.e., $V_\ell = 0$. If $\cos\theta$ is assumed unity, Eq. (1-57) can be written as

$$\frac{d}{dx}\left(\frac{V^2}{2g}\right) + \cos\theta\,\frac{dy}{dx} + \frac{dG_0}{dx} - \frac{Q}{gA^2}q_\ell = -S_f \qquad (6\text{-}17)$$

Combining Eq. (6-2) with Eq. (6-17), noting that $dG_0/dx = -S_0$, $F^2 = (Q^2B/gA^3)$, and $q_\ell = -dQ/dx$, and rearranging the terms in the resulting equation, one obtains

$$\frac{dy}{dx} = \frac{S_0 - S_f - \left(2Q/gA^2\right)\!\left(dQ/dx\right)}{1 - F^2} \qquad (6\text{-}18)$$

Equation (6-18) can be integrated either analytically or numerically. However, only the simplified form of Eq. (6-18) can be solved analytically.

Analytical Solution

The analytical solution of Eq. (6-18) is presented for the case where the length of the channel reach with lateral inflow is short so that the effect of the channel slope and friction can be neglected. On the assumption that $S_0 = 0$ and $S_f = 0$, Eq. (6-18)

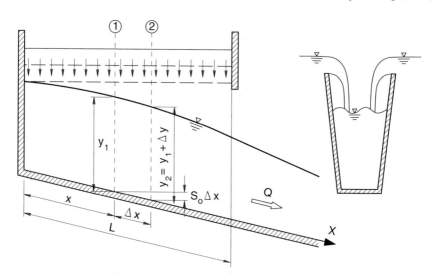

Figure 6-3. Flow profile in a side channel.

reduces to

$$\frac{dy}{dx} = \frac{-\left(2Q/gA^2\right)\left(dQ/dx\right)}{1-F^2} \tag{6-19}$$

Note that Eq. (6-19) is similar to Eq. (6-4), except that the denominator of the right-hand side of the former is twice that of the latter. The flow profile in the channel reach with lateral inflow, referred to as transition reach, can be either a backwater curve or a drawdown curve depending on the sign of the term dy/dx in Eq. (6-19). The numerator in the term on the right-hand side is negative, because dQ/dx is positive for lateral inflow. For subcritical flow, i.e., $F<1$, dy/dx is negative and therefore the flow profile in the transition reach is a drawdown curve. On the other hand, for supercritical flow, the term dy/dx is positive and the profile is a backwater curve.

Similar to the lateral-outflow case, several types of flow profiles in the lateral-inflow case also are possible in the upstream, the transition, and the downstream reaches. The possible controls are either an NDC or a CDC. The CDC in the lateral-inflow case is possible only in section 2, because the minimum head required to permit the flow is larger in section 2 than in section 1. However, the CDC in section 2 is not possible if the slope of reach 2 is mild; in that case there exists an NDC in section 2.

Additional assumptions are needed to solve Eq. (6-19) analytically. These assumptions are (i) the channel is rectangular, (ii) the rate of lateral inflow per unit channel length is constant, and (iii) the discharge in reach 1 is zero. The discharge in the transition reach can be expressed as

$$Q_x \equiv \frac{dQ}{dx} = \frac{Q}{x} \tag{6-20}$$

where x is the distance from section 1 along the transition reach. Substituting Eq. (6-20) into Eq. (6-19), noting that $F^2 = (Q^2B/gA^3)$, and rearranging the terms in the resulting equation, one obtains

$$\frac{d\left(x^2\right)}{dy} - \frac{x^2}{y} = -\frac{gb^2y^2}{Q_x^2} \tag{6-21}$$

where b is the channel width. Equation (6-21) is a linear ordinary differential equation in x^2; its solution is

$$x^2 = -\frac{gb^2y^3}{2Q_x^2} + Ay \tag{6-22}$$

where A is a constant of integration. The control in this case is in section 2 and the constant A can be determined from the known value of y in section 2 (Prob. 6-10).

Numerical Integration

In problems where the terms S_0 and S_f in the governing equation cannot be neglected, the solution of the governing equation is obtained by the numerical integration method. Equation (6-17) is used to derive the finite-difference form of the governing equation. This equation for $\cos \theta = 1$ can be written as

$$\frac{dy}{dx} = S_0 - S_f - \frac{Q}{gA^2}\frac{dQ}{dx} - \frac{V}{g}\frac{dV}{dx} \tag{6-23}$$

Equation (6-23) can be written in a finite-difference form as

$$\Delta y = \left(S_0 - \overline{S}_f\right)\Delta x - \frac{1}{g\overline{A}}\left(\overline{V}\Delta Q + \overline{Q}\Delta V\right) \tag{6-24}$$

where the bar over a term denotes the average value of that term within the finite increment Δx of the channel length. The average velocity, discharge, and area can be written as

$$\overline{V} = \frac{V_1 + V_2}{2}; \quad \overline{Q} = \frac{Q_1 + Q_2}{2}; \quad \overline{A} = \frac{\overline{Q}}{\overline{V}} = \frac{Q_1 + Q_2}{V_1 + V_2} \tag{6-25}$$

where the subscripts 1 and 2 refer to sections 1 and 2 in Figure 6-3. Substitution from Eq. (6-25) into Eq. (6-24) yields

$$\Delta y = \left(S_0 - \overline{S}_f\right)\Delta x - \frac{Q_1\left(V_1 + V_2\right)}{g\left(Q_1 + Q_2\right)}\left(\Delta V + \frac{V_2}{Q_1}\Delta Q\right) \tag{6-26}$$

The procedure for solving Eq. (6-26) is illustrated through an example, but first the possibility of a CDC within the transition reach is examined. As explained earlier, the numerator of Eq. (6-18) must be zero at the CDC, i.e.,

$$S_0 = S_f + \frac{2Q}{gA^2}\frac{dQ}{dx} \tag{6-27}$$

Equation (6-27) is solved by a trial-and-error procedure as follows: (i) Guess a value of x where the CDC is located; (ii) find Q at this section from Eq. (6-20); (iii) find the critical depth from the critical-depth relation; (iv) compute A, P, and R at the critical depth; (v) compute S_f from the Manning equation; (vi) check whether Eq. (6-27) is satisfied; and (vii) if not, try another value of x. The CDC is within the transition reach if x is less than the length L of the transition reach. Otherwise the control is outside the transition reach where the flow is a gradually varied flow, and the location of the control can be determined by following the procedure outlined in Chapter 4.

For $S_0 = 0$ and $S_f = 0$, the numerator of Eq. (6-18) is nonzero; hence a CDC within the transition reach does not exist. Equation (6-27) has the following explicit solution for wide rectangular channels if the Chézy equation (2-4) is used to compute the friction slope (Prob. 6-11):

$$x = \frac{8Q_x^2}{gB^2\left(S_0 - gP/C^2 B\right)^3} \tag{6-28}$$

where $Q_x = dQ/dx = Q/x$ and C is the Chézy coefficient.

Example 6-1. Determine the flow profile in a trapezoidal channel of a side-channel spillway for the following data: $b = 3$ m; $n = 0.015$; $m = 0.5$; $S_0 = 0.15$; $L = 122$ m; and $dQ/dx = 4.0$ m³/s/m.

The first step in the solution is to examine the possibility of a CDC within the transition reach. Equation (6-27) is solved by a trial-and-error procedure described earlier. The computations are shown in Table 6-2, which is self-explanatory.

Table 6-2. Location of the CDC

1	2	3	4	5	6	7	8	9	10	11
x	Q	$\sqrt{\dfrac{Q^2 m^3}{gb^5}}$	$\dfrac{my_c}{b}$	y_c	A	P	R	V	S_f	$S_f + \dfrac{2Q}{gA^2}\dfrac{dQ}{dx}$
m	m³/s			m	m²	m	m	m		
50	200	1.45	0.90	5.40	30.78	15.07	2.04	6.50	0.0037	0.1758
58	232	1.68	1.01	6.06	36.54	16.55	2.21	6.35	0.0032	0.1449
57	228	1.65	1.00	6.00	36.00	16.42	2.19	6.33	0.0032	0.1468
55	220	1.59	0.98	5.88	34.93	16.15	2.16	6.30	0.0032	0.1502 ok

Explanatory Remark on Table 6-2

A trial value of $x = 50$ m in column 1 is assumed. The critical depth in column 5 is obtained on solving Eq. (3-32) which involves two parameters that are listed in columns 3 and 4. The friction slope in column 10 is computed from the Manning equation. The right-hand side of Eq. (6-27) computed in column 11 is not equal to the left-hand side of Eq. (6-27), i.e., S_0. Another trial value of x is needed. Finally the value of $x = 55$ m yields a value of 0.1502 in column 11 that is very close to the channel slope. There exists a CDC at $x = 55$ m.

With the CDC section located, the flow profiles both upstream and downstream from the control section are computed solving Eq. (6-26) by a trial-and-error procedure. The computations are shown in Table 6-3.

Table 6-3. Computation of the Flow Profile

1 x	2 Δx	3 y	4 Δy	5 A	6 P	7 R	8 Q	9 V	10 S_f ×10³	11 $\bar{S}_f$ ×10³	12 Q_1+Q_2	13 V_1+V_2	14 ΔQ	15 ΔV	16 Δy
m	m	m	m	m²	m	m	m³/s	m/s			m³/s	m/s	m³/s	m/s	m
(a) Upstream Subcritical Flow Profile															
55		5.88		34.93	16.15	2.16	220	6.30	3.2						
25	−30	5.00	−0.88	27.50	14.48	1.94	100	3.64	1.2	2.2	320	9.94	−120	−2.66	−1.20
		4.70	−1.18	25.15	13.51	1.86	100	3.98	1.6	2.4	320	10.28	−120	−2.32	−1.19 ok
0	−25	2.60	−2.10*												
(b) Downstream Supercritical Flow Profile															
55		5.88		34.93	16.15	2.16	220	6.30	3.2						
90	35	6.60	0.72	41.58	17.76	2.34	360	8.66	5.4	4.3	580	14.96	140	2.36	0.51
		6.70	0.82	42.55	17.98	2.37	360	8.46	5.1	4.1	580	14.76	140	2.16	0.77
		6.74	0.86	42.93	18.07	2.38	360	8.39	5.0	4.1	580	14.69	140	2.09	0.86 ok
122	32	7.50	0.76	50.63	19.77	2.56	488	9.64	6.0	5.5	848	18.03	128	1.25	0.97
		7.40	0.66	49.58	19.55	2.54	488	9.84	6.3	5.7	848	18.23	128	1.45	0.71
		7.38	0.64	49.37	19.50	2.53	488	9.88	6.4	5.7	848	18.27	128	1.49	0.65 ok

*Because $Q = 0$, $V = 0$, and $S_f = 0$ at $x = 0$, Eq. (6-26) reduces to $\Delta y = (S_0 - S_{f1})\Delta x + \dfrac{V_1^2}{g}$, where S_{f1} and V_1 are the friction slope and velocity, respectively, at $x = 25$ m.

Explanatory Remark on Table 6-3

The longitudinal distances between the section of computation and the beginning of the channel are listed in column 1. A trial depth $y = 5.00$ m at $x = 25$ m is assumed, and it is listed in column 3. The friction slope in column 10 is computed from the Manning equation. For the assumed value of $y = 5.00$ m the two values of Δy, $- 0.88$ m in column 4 and -1.20 m in column 16, are not the same. Another value of $y = 4.70$ m is tried, and it yields almost the same value of Δy in columns 4 and 16. The computations are repeated for other sections.

PROBLEMS

6-11   Rework Example 6-1 for $dQ/dx = 3$ m^3/s/m.

6-12   A channel of 2-m-wide rectangular section drains rainwater from a parking lot. The flow from the parking lot enters the drain at a uniform rate of 10 liters per second per unit length of the channel. The channel is 50 m in length and terminates in a free overfall. Determine the flow depths at both ends of the channel if the channel slope and resistance are negligible.

6-13   Derive Eq. (6-28). Note that at the CDC the Froude number $F = \sqrt{Q^2 B/gA^3} = 1$.

CHAPTER 7

UNSTEADY FLOW I

The techniques for solving the governing equations for unsteady flow caused by the movement of flood waves, the propagation of tides, the operation of the control gates, the failure of a dam, etc., are described in this and the next chapter. Though the complete governing equations, which include the continuity and momentum equations, can be solved numerically on computers, the meaningful, analytical solutions of the simplified governing equations can be obtained with much less effort.

One of the simplest cases of unsteady flow is *uniformly progressive flow*. This type of flow has a stable wave profile of various forms that does not change in shape as the wave travels in the channel. It can be reduced to steady flow by using a coordinate system that is moving with the speed of the wave profile as described in Section 3.1. One of the forms of wave configuration is the monoclinal wave that resembles the rising limb of a flood wave in rivers. The speed of the monoclinal wave is found to be a good approximation of the speed of the flood wave in rivers. The technique used in analyzing the monoclinal wave cannot be applied to solve other unsteady flow problems, as they cannot be reduced to steady flow. The analysis of the monoclinal wave is included in Appendix F.

For purposes of discussion, unsteady flow is classified into gradually varied unsteady flow and rapidly varied unsteady flow. In the former flow the change in depth is gradual; consequently. the effect of streamline curvature is not significant and the momentum equation derived in Section 1-8 is applicable. Examples of gradually varied unsteady flow are flood waves, tidal waves, and waves due to slow operation of controlling structures. In the rapidly varied unsteady flow the change in depth is large and leads to the formation of surges. The momentum equation is not valid in a zone in the vicinity of a surge; but this zone is very narrow. The momentum equation is applicable to both the upstream and downstream regions from this narrow zone. It is possible to assume that the two regions are linked by the surge compatibility equation. Therefore, the rapidly varied unsteady flow problems also are solved using the momentum equation. Examples of rapidly varied unsteady flow are surges of various kinds developed by rapid operation of controlling structures.

The momentum equation can be simplified by neglecting terms that are smaller than the other terms. The magnitudes of the various terms in the momentum equation depend on the nature of the unsteady flow. Two simplified forms of the momentum equation will be examined: first, in which friction-slope and the channel-slope terms are neglected; second, in which the acceleration terms and the $\partial y/\partial x$ term are neglected. There are many practical cases in which the length of the channel reach under consideration is rather short; in such cases the effect of channel and friction slopes in the short reach can be neglected as a first approximation. An example is the sudden release of water due to the collapse of a dam. In this case the time of arrival of the disturbance at a town located a few kilometers downstream of the dam can be estimated by neglecting S_0 and S_f. Such problems are discussed in the present chapter. On the other hand, the movement of a flood wave down a channel can be predicted with a reasonable accuracy by neglecting the inertia and the $\partial y/\partial x$ terms if the channel is steep. This type of problem is dealt with in the next chapter.

7-1 Governing Equations

The continuity and the momentum equations for unsteady flow without lateral flow as given in Section 1-8 are

$$\frac{\partial Q}{\partial x}+\frac{\partial A}{\partial t}=0 \qquad \text{[Eq. (1-63)]}$$

$$\frac{1}{g}\frac{\partial V}{\partial t}+\frac{\partial H}{\partial x}=-S_f \qquad \text{[Eq. (1-64)]}$$

where
$$H=\frac{V^2}{2g}+y\cos\theta+G_0 \qquad \text{[Eq. (1-54)]}$$

Substituting the relations $Q=VA$, $D=A/B$, and $dA=B\,dy$ in Eq. (1-63), the definition of H from Eq. (1-54) into Eq. (1-64) with $\cos\theta=1$, the above two equations can be written as

$$\frac{\partial y}{\partial t}+D\frac{\partial V}{\partial x}+V\frac{\partial y}{\partial x}=0 \tag{7-1}$$

$$\frac{1}{g}\frac{\partial V}{\partial t}+\frac{V}{g}\frac{\partial V}{\partial x}+\frac{\partial y}{\partial x}=S_0-S_f \tag{7-2}$$

Equations (7-1) and (7-2) are nonlinear partial differential equations and known as de Saint-Venant equations. Theoretically it should be possible to solve these equations for the two unknown quantities, namely, V and y, for the given initial and boundary conditions.

7-2 Characteristic Equations

The system of equations given in Eqs. (7-1) and (7-2), which contains differentiation in two directions, namely, x and t directions, can be transformed to another system of equations that contains differentiation only in one direction, termed the characteristic direction. In other words partial differential equations can be converted to ordinary differential equations, which are called the *characteristic differential equations*. The characteristic equations are helpful in understanding the requirements of the initial and boundary conditions and the progression of the solutions. Moreover the characteristic equations for some important problems can be solved analytically. The characteristic equations can be derived as follows.

Addition of Eq. (7-1) to Eq. (7-2) after multiplying it by an unknown variable λ and rearrangement of the terms in the resulting equation yield

$$\left[\frac{\partial V}{\partial t} + (V + \lambda D)\frac{\partial V}{\partial x}\right] + \lambda\left[\frac{\partial y}{\partial t} + \left(V + \frac{g}{\lambda}\right)\frac{\partial y}{\partial x}\right] = g\left(S_0 - S_f\right) \tag{7-3}$$

The total derivatives of $V = V(x,t)$ and $y = y(x,t)$ can be written as

$$\frac{DV}{Dt} = \frac{\partial V}{\partial t} + \frac{\partial V}{\partial x}\frac{dx}{dt} \tag{7-4}$$

$$\frac{Dy}{Dt} = \frac{\partial y}{\partial t} + \frac{\partial y}{\partial x}\frac{dx}{dt} \tag{7-5}$$

A comparison of Eq. (7-3) to Eqs. (7-4) and (7-5) shows that the terms inside the brackets in Eq. (7-3) can be expressed as total derivatives if

$$\frac{dx}{dt} = (V + \lambda D) = \left(V + \frac{g}{\lambda}\right) \tag{7-6}$$

An expression for λ is obtained from Eq. (7-6) as

$$\lambda = \pm\sqrt{\frac{g}{D}} \tag{7-7}$$

Substitution for λ from Eq. (7-7) into Eq. (7-6) gives

$$\frac{dx}{dt} = V \pm c \tag{7-8}$$

where c is a celerity of a small disturbance defined in Eq. (3-13) as

$$c = \sqrt{gD} \tag{3-13}$$

Using Eqs. (7-4)–(7-8), Eq. (7-3) can be written (Prob. 7-1) as

$$\frac{D}{Dt}(V + \omega) = g\left(S_0 - S_f\right) \tag{7-9}$$

if
$$\frac{dx}{dt} = V + c \qquad (7\text{-}10)$$

and
$$\frac{D}{Dt}(V - \omega) = g(S_0 - S_f) \qquad (7\text{-}11)$$

if
$$\frac{dx}{dt} = V - c \qquad (7\text{-}12)$$

where
$$\omega = \int_0^y \lambda\, dy \quad \text{or} \quad d\omega = \lambda\, dy \qquad (7\text{-}13)$$

and ω is known as Escoffier's stage variable. Using Eqs. (7-7) and (3-13), Eq. (7-13) can be written for rectangular channels (Prob. 7-2) as

$$\omega = 2c \qquad (7\text{-}14)$$

Note that from Eq. (3-13) c for rectangular channels is equal to $\sqrt{gy}$. Thus for any point moving through the fluid Eqs. (7-9) and (7-11) are valid along the curves in the x-t plane given by Eqs. (7-10) and (7-12), respectively. These curves are referred to as characteristics: the positive characteristic, C+, given by Eq. (7-10), and the negative characteristic, C–, given by Eq. (7-12). Substituting for ω from Eq. (7-14) into Eqs. (7-9) and (7-11), the characteristic equations for rectangular channels are obtained as

$$\frac{D}{Dt}(V + 2c) = g(S_0 - S_f) \qquad (7\text{-}15)$$

along the C+ characteristics given by Eq. (7-10) and

$$\frac{D}{Dt}(V - 2c) = g(S_0 - S_f) \qquad (7\text{-}16)$$

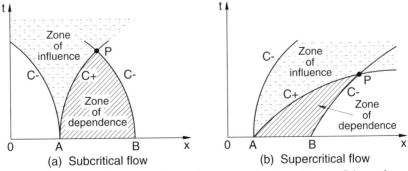

Figure 7-1. Zone of influence of disturbance at point A and zone of dependence of point P.

along the C– characteristics given by Eq. (7-12).

In subcritical flow $V < c$; therefore dx/dt for the C+ and C– characteristics from Eqs. (7-10) and (7-12) is positive and negative, respectively, as shown in Figure 7-1a. In supercritical flow $V > c$; consequently dx/dt for both characteristics is positive, as shown in Figure 7-1b. In critical flow dx/dt for the C– characteristic is zero. The characteristics are the paths along which a small disturbance propagates at velocities $V \pm c$. The perturbation that initiates, say at point A in Figures 7-1a and 7-1b, propagates along curves C+ and C–, and influences the flow conditions in the speckled region. This region is defined as the *zone of influence* of disturbance at point A. Flow conditions in the speckled region are influenced by the flow conditions at point A. Similarly a region, called the *zone of dependence* and shown as the hatched area in Figure 7-1, can be delineated by the two characteristics meeting at point P. The flow conditions at point P depend upon the flow conditions in the zone of dependence. As points A and B in Figure 7-1a are upstream and downstream of point P, respectively, the flow conditions at point P for subcritical flow are influenced from both upstream and downstream flow conditions. As both points A and B in Figure 7-1b are upstream of point P, the flow conditions at point P for supercritical flow are not affected from the downstream conditions and depend only on the upstream flow conditions.

PROBLEMS

7-1 Derive Eqs. (7-9) and (7-11) from Eq. (7-3).

7-2 Derive Eq. (7-14) from Eq. (7-13).

7-3 Initial and Boundary Conditions

For determining the locations and the number of the initial and boundary conditions required to solve an unsteady-flow problem, consider a channel reach AB, shown in Figure 7-2. Also consider that the flow conditions at time $t = t_1$ in the channel reach, say at points A_1, P, and B_1, are to be determined. The locations of points A_1 and B_1 are selected at the upstream and downstream sections of the channel reach, respectively, and point P is located inside the channel reach. It should be possible in theory to delineate the zone of dependence for each of these points, as shown in Figure 7-2a for subcritical flow and Figure 7-2b for supercritical flow. The zone of dependence for point P at time $t = t_1 > t_0$ lies within points 3 and 4. The flow conditions at point P at time t_1 can be determined by solving Eqs. (7-9) and (7-11) for V and y (y can be obtained from ω), but the solution requires that the values of V and y be given at points 3 and 4 at time t_0. In other words two conditions in terms of V and y at some initial time, called initial conditions, must be specified along the channel reach.

A portion of the zone of dependence for points A_1 and B_1 in Figure 7-2a and the whole zone of dependence for point A_1 in Figure 7-2b lie outside the channel reach, where the flow conditions are not known. Therefore, the flow conditions at these points cannot be determined from the flow conditions within the channel

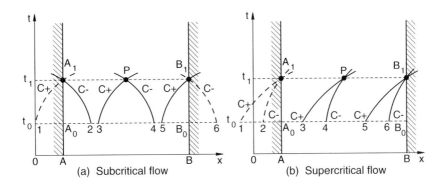

Figure 7-2. Illustration of the initial and boundary data requirements.

reach only. For points A_1 and B_1 in Figure 7-2a for subcritical flow only one of the characteristics is within the computation domain; consequently only one equation, either Eq. (7-9) or Eq. (7-11), can be utilized. The two unknowns at these points in subcritical flow can be determined only if V or y, or a relation between V and y, termed boundary conditions, is prescribed at both the upstream and the downstream sections of the channel reach. For point A_1 in Figure 7-2b for supercritical flow both characteristics lie outside the computation domain and none of Eqs. (7-9) and (7-11) can be utilized; consequently two boundary conditions in terms of V and y must be specified at the upstream section of the channel reach. Both characteristics for point B_1 in Figure 7-2b are within the computation domain; the flow conditions at point B_1 can be obtained from Eqs. (7-9) and (7-11), similar to that for point P. The boundary conditions at the downstream section of the channel reach are not needed in the case of supercritical flow.

7-4 Simple-Wave Problem

A simple-wave problem deals with the propagation of a disturbance in a long horizontal frictionless channel $S_0 = 0$ and $S_f = 0$ in which the flow initially is steady and uniform. Useful solutions of practical problems, in which the flow changes so fast that the acceleration terms in the momentum equation are large in comparison to the bed and friction slope terms, can be obtained by neglecting the latter terms. Assuming $S_0 = 0$ and $S_f = 0$, Eqs. (7-15) and (7-16) reduce to

$$\frac{D}{Dt}(V + 2c) = 0 \tag{7-17}$$

and

$$\frac{D}{Dt}(V - 2c) = 0 \tag{7-18}$$

The integration of Eqs. (7-17) and (7-18) yields

$$V + 2c = \text{constant} \quad \text{(along C+ characteristics)} \qquad (7\text{-}19)$$

and
$$V - 2c = \text{constant} \quad \text{(along C− characteristics)} \qquad (7\text{-}20)$$

where the constants can be obtained from the initial and boundary conditions. The requirements for the boundary conditions are different for subcritical and supercritical flow: in the former one boundary condition at both the upstream and downstream sections of the channel reach, and in the latter two boundary conditions at the upstream section of the channel reach. Therefore, the simple-wave problem for the two types of flows is dealt with separately.

Subcritical Flow

Initial Conditions. Consider flow in a channel of length L. The channel initially carries the uniform flow at a depth y_0 and a velocity V_0, i.e., the initial conditions in the channel are

$$V(x,0) = V_0, \quad -L \le x \le 0$$

$$(7\text{-}21)$$

$$c(x,0) = c_0, \quad -L \le x \le 0$$

where $c_0 = \sqrt{gy_0}$ and the origin of x is at the downstream end of the channel and $V_0 < c_0$.

Boundary Conditions. A disturbance that takes the form of a prescribed variation of c or V with the time t is introduced at the origin. Assume that the effect of the disturbance on the flow conditions in the channel is studied for $t < T$ and the length of the channel is long enough so that the time taken by the disturbance to reach the upstream end of the channel is longer than T. The flow conditions at the upstream end, therefore, remain unchanged for $t < T$. The boundary conditions for the above-described flow are

$$V(0,t) = f(t) \qquad (7\text{-}22a)$$

or
$$c(0,t) = g(t) \qquad (7\text{-}22b)$$

and
$$V(-L,t) = V_0, \quad t < T \qquad (7\text{-}23a)$$

or
$$c(-L,t) = c_0, \quad t < T \qquad (7\text{-}23b)$$

Solution. The solution is obtained by delineating the C+ and C− characteristics in the x-t plane. The equation of the C− characteristic that originates from the origin is obtained from Eq. (7-12) as

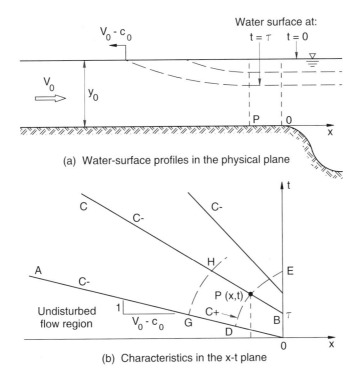

(a) Water-surface profiles in the physical plane

(b) Characteristics in the x-t plane

Figure 7-3. The simple-wave problem.

$$\frac{dx}{dt} = V_0 - c_0 \tag{7-24}$$

In Figure 7-3b Eq. (7-24) is plotted in the x-t plane as a straight line OA with an inverse slope of $(V_0 - c_0)$ which is negative because the flow is subcritical, i.e., $V_0 < c_0$. In physical terms the line OA divides the disturbed and undisturbed flow regions: the undisturbed region below OA and the disturbed region above OA. It is shown later in Example 7-1 that if any member of the C+ or C− family of characteristics is a straight line, then so are all other members of the same family, and the members of the other family are not straight lines. Therefore all C− characteristics in this simple-wave problem are straight lines and C+ characteristics are curved, as shown in Figure 7-3b. It should be pointed out that for supercritical flow the above statement is not always valid in the entire x-t plane, as shown later in this section.

In order to determine the flow condition at any point P(x,t) in the x-t plane, consider the C− and C+ characteristics passing through point P, shown respectively by BC and DE in Figure 7-3. The C− characteristic BC is issued from point B on the t-axis where $t = \tau$. Both terms, $(V - 2c)$ and $(V - c)$, are constant along the

C– characteristics: the former according to Eq. (7-20) and the latter according to Eq. (7-12) and the fact the C– characteristics are straight lines, i.e., dx/dt is constant along the characteristics. Consequently, both V and c are constant along the C– characteristics, i.e., the velocity and celerity at point P are given by $V = V(\tau)$ and $c = c(\tau)$ where $V(\tau)$ and $c(\tau)$ are the values of V and c at point B on the t-axis at $t = \tau$. One of the two variables V and c along the t-axis is prescribed by Eqs. (7-22a) or (7-22b); the other variable along the t-axis is determined from Eq. (7-26), which is given below and is derived as follows. The C+ characteristic DE through point P intersects the initial C– characteristic OA at point D where $V = V_0$ and $c = c_0$. From Eq. (7-19), which is valid along the C+ characteristic DE, one obtains

$$V + 2c = V_0 + 2c_0 \qquad (7\text{-}25)$$

According to Eq. (7-25) the term $(V+2c)$ is constant over the entire x-t plane, including the t-axis. Hence Eq. (7-25) can be written as

$$V(\tau) + 2c(\tau) = V_0 + 2c_0 \qquad (7\text{-}26)$$

The value of τ, i.e., the location of point B, is obtained as follows. The inverse slope of the characteristic BC from Eq. (7-12) is

$$\frac{dx}{dt} = V(\tau) - c(\tau) \qquad (7\text{-}27)$$

From Eqs. (7-26) and (7-27) the inverse slope can be written in terms of either $V(\tau)$ or $c(\tau)$ and is given by

$$\frac{dx}{dt} = \tfrac{3}{2}V(\tau) - \tfrac{1}{2}V_0 - c_0 \qquad (7\text{-}28)$$

or

$$\frac{dx}{dt} = V_0 + 2c_0 - 3c(\tau) \qquad (7\text{-}29)$$

Also the slope of the straight line BC that passes through points $P(x,t)$ and $B(0,\tau)$ is

$$\frac{dx}{dt} = \frac{x}{t - \tau} \qquad (7\text{-}30)$$

For known values of x and t for point P, τ can be obtained either from Eqs. (7-28) and (7-30) or Eqs. (7-29) and (7-30). Substituting $c(\tau) = \sqrt{gy(\tau)}$ in Eq. (7-29), the flow profile at any time t can be obtained from Eqs. (7-29) and (7-30) as

$$\frac{x}{t - \tau} = V_0 + 2c_0 - 3\sqrt{gy(\tau)} \qquad (7\text{-}31)$$

If the boundary condition at the downstream end is prescribed in terms of $q(\tau)$, it can be transformed in terms of $y(\tau)$ or $V(\tau)$ by using Eq. (7-26). For the boundary condition as a step function, all C− characteristics in the nonuniform flow region originate from the origin (Prob. 7-6). For a certain boundary condition a C− characteristic that originates at the downstream end of the channel can coincide with the t axis along which $V = c$. The disturbance prescribed at the downstream end then cannot propagate upstream in the channel and the downstream end acts as a CDC.

In this simple-wave problem, the parameter $(V+2c)$ is constant, not only along C+ characteristics but also in the entire x-t plane [see Eq. (7-25)]. It means $(V+2c)$ appears constant to all observers everywhere, and it is possible only if a disturbance of zero amplitude propagates along C+ characteristics. The propagation of such a disturbance cannot be called a wave motion. Disturbances in a simple-wave problem propagate along one of the characteristics only.

The above solution is presented for the channel reach wherein the disturbance is introduced at the downstream end of the reach. A similar solution can be obtained for the channel reach wherein the disturbance is introduced at the upstream end of the reach. The disturbances in this case propagate along C+ characteristics. The equation of the initial C+ characteristic is obtained from Eq. (7-10) as

$$\frac{dx}{dt} = V_0 + c_0 \tag{7-32}$$

The initial C+ characteristic is a straight line; hence all C+ characteristics are straight lines. The difference between Eqs. (7-24) and (7-32) is in the sign of the celerity term. Equations (7-24)–(7-31) after changing the sign of the celerity terms are valid for flows where the disturbances travel downstream along C+ characteristics (see Example 7-2). Note that $(V+2c)$ is constant across a disturbance traveling at $(V- c)$, and $(V- 2c)$ is constant across a disturbance traveling at $(V+c)$.

Example 7-1. Show that if any curve of the C+ or C− family of characteristics in a simple-wave problem is a straight line, then all other curves of the same family are also straight lines, and the members of the other family are not straight lines.

(a) The first statement is proved for the C− family of characteristics, and a similar proof can be obtained for the C+ family of characteristics. Consider in Figure 7-3b two C− characteristics OA and BC and two C+ characteristics DP and GH. Let OA be a straight line. Both $(V- c)$ and $(V- 2c)$ are constant along OA, and it is possible only if both V and c are constant along OA. It implies that

$$V_D = V_G; \quad c_D = c_G$$

Because $(V+2c)$ is constant along C+ characteristics, one can write

$$V_D + 2c_D = V_P + 2c_P; \quad V_G + 2c_G = V_H + 2c_H$$

whence $\qquad\qquad\qquad V_H + 2c_H = V_P + 2c_P \tag{1}$

Because BC is a C– characteristic along which $(V-2c)$ is constant, one can write

$$V_H - 2c_H = V_P - 2c_P \tag{2}$$

Equations (1) and (2) can be satisfied only if

$$V_H = V_P; \quad c_H = c_P$$

i.e., if PH, hence BC, is a straight line.

  (b) Along C+ characteristics

$$\frac{dx}{dt} = V + c \tag{3}$$

In a simple-wave problem

$$V + 2c = V_0 + 2c_0 \tag{4}$$

Elimination of V between Eqs. (3) and (4) gives

$$\frac{dx}{dt} = V_0 + 2c_0 - c \tag{5}$$

The slope (dx/dt) of the C+ characteristics is not constant unless c is constant, which is a trivial case; whence C+ characteristics are not straight lines.

Example 7-2. An 80-km-long rectangular channel connects two reservoirs. Initially the flow in the channel is steady and uniform at a depth 1.50 m and velocity 1.0 m/s. The water levels in the upstream and the downstream reservoirs start falling at a rate of 0.30 m/h at time $t = 0$ and 0.15 m/h at time $t = 2$ h, respectively. At what time and in which channel section will the depth of flow be ultimately affected by lowering the water levels in the reservoirs? At what distance from the upstream reservoir will the depth of flow be 0.6 m at that time?

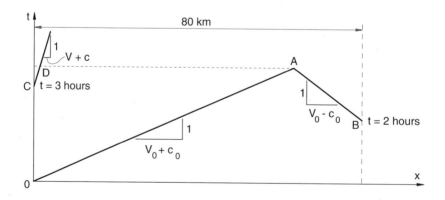

Example 7-2

The problem is solved by drawing characteristics in the x-t plane (see figure above). The upstream of the channel is located at $x = 0$, and the downstream reservoir is at $x = 80$ km. The first disturbance from the upstream reservoir travels downstream in the channel along the C+ characteristic OA that originates at point A(0,0) and the first disturbance from the downstream reservoir travels upstream in the channel along the C– characteristic BA that starts at point B(80,2).

The initial depth and velocity in the channel are $y_0 = 1.5$ m and $V_0 = 1.0$ m/s $= 3.60$ km/h. The initial celerity is

$$c_0 = \sqrt{1.5g} = 3.83 \text{ m/s} = 13.81 \text{ km/h}$$

Here, dx/dt is equal to $(V_0 + c_0)$ along OA and $(V_0 - c_0)$ along BA:

$$\left(\frac{dx}{dt}\right)_{OA} = \frac{x_A - 0}{t_A - 0} = V_0 + c_0 = 17.41 \text{ km/h}$$

$$\left(\frac{dx}{dt}\right)_{BA} = \frac{x_A - 80}{t_A - 2} = V_0 - c_0 = -10.21 \text{ km/h}$$

where (x_A, t_A) are the coordinates of point A. Solving the above two equations, one obtains

$$x_A = 63.37 \text{ km}; \quad t_A = 3.64 \text{ h}$$

Now consider the C+ characteristic CD, along which c is constant and equal to $\sqrt{0.6g}$; it originates at point C(0,3) representing the depth of 0.6 m at the upstream end of the channel, i.e., a drop of 0.9 m in the water level in the upstream reservoir, which occurs at $t = 3$ h. The slope of CD from Eq. (7-29) after changing the sign of the celerity terms is

$$\left(\frac{dx}{dt}\right)_{CD} = V_0 - 2c_0 + 3c(\tau) = 3.60 - 2\times13.81 + 3\sqrt{0.6g} = 2.23 \text{ km/h}$$

The x-coordinate of point D is

$$x_D = (3.64 - 3.00)\times 2.23 = 1.43 \text{ km}$$

Surge Inception. The C– characteristics in Figure 7-3 are shown diverging, i.e., the inverse slope of the C– characteristics increases with time. It follows from Eq. (7-29) that the celerity, hence depth, at the boundary $x = 0$ decreases with time t. Such a disturbance is termed negative disturbance or wave. For a disturbance where the depth increases with time, the C– characteristics will be converging and therefore must eventually intersect, as shown in Figure 7-4. It means that the solution has a discontinuity beyond points of intersections of the C– characteristics, and the equations presented in this section are valid up to points of intersections. The points of intersections form an envelope as indicated by the thick line in Figure 7-4. In physical terms the discontinuity represents the formation of a surge. Such a disturbance is termed a positive disturbance or wave. The propagation of surges in a horizontal channel without resistance was described in Chapter 3.

A method of obtaining an envelope of a family of curves is as follows. Let the equation of the family of curves be represented by

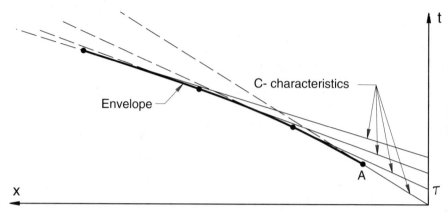

Figure 7-4. Converging characteristics forming an envelope.

$$f(x,t,\tau) = 0 \qquad (7-33)$$

The slope of the envelope at a point of contact with a curve of the family is the same as the slope of this curve at that point. By equating the two slopes, it can be shown that on the envelope

$$\frac{\partial f}{\partial \tau} = 0 \qquad (7-34)$$

The equation of the envelope is obtained by eliminating τ between Eqs. (7-33) and (7-34). The C– characteristics in Figure 7-4 are represented by Eq. (7-31), which can be written in the form of Eq. (7-33) as

$$f(x,t,\tau) = x - (t - \tau)\left[V_0 + 2c_0 - 3\sqrt{gy(\tau)}\right] = 0 \qquad (7-35)$$

Differentiation of Eq. (7-35) with respect to τ and its substitution in Eq. (7-34) yield

$$\frac{\partial f}{\partial \tau} = -\left[V_0 + 2c_0 - 3\sqrt{gy(\tau)}\right] + 3(t - \tau)\frac{\partial\left(\sqrt{gy(\tau)}\right)}{\partial \tau} = 0 \qquad (7-36)$$

Elimination of τ between Eqs. (7-35) and (7-36) may not be easy. Instead, the coordinates of the envelope for various values of τ are obtained from Eqs. (7-35) and (7-36) as

$$x = \frac{\left[V_0 + 2c_0 - 3\sqrt{gy(\tau)}\right]^2}{3\dfrac{\partial\left(\sqrt{gy(\tau)}\right)}{\partial \tau}} \qquad (7-37)$$

and
$$t = \tau + \frac{x}{\left[V_0 + 2c_0 - 3\sqrt{gy(\tau)}\right]} \qquad (7\text{-}38)$$

For a specified increase in depth, $y = y(\tau)$, the envelope can be traced from Eqs. (7-37) and (7-38) (see Prob. 7-8). The location where the surge first develops is indicated by point A in the x-t plane in Figure 7-4. The coordinates of point A can be obtained from Eqs. (7-37) and (7-38) by substituting $\tau = 0$.

Equations (7-37) and (7-38) are valid for the disturbance propagating upstream. Similar equations for the disturbance propagating downstream are obtained by changing the sign of the celerity terms in these two equations, as explained earlier.

Example 7-3. Water flows at a uniform depth of 1.5 m and velocity of 1 m/s in a rectangular channel that discharges into a large estuary. Initially the water level in the estuary is the same as in the channel. At a certain time the estuary level rises at the rate of 0.5 m/h. When, where, and at what depth will a surge first develop.

The initial depth and velocity in the channel are $y_0 = 1.5$ m and $V_0 = 1$ m/s $= 3.60$ km/h. The initial celerity is $c_0 = \sqrt{1.5g}$) $= 3.84$ m/s $= 13.81$ km/h. The depth in the channel where it meets the estuary is given by

$$y(t) = 1.5 + 0.5t \qquad (1)$$

From Eq. (1)

$$\frac{d\sqrt{y(t)}}{dt} = \frac{1}{2} \times 0.5 \left(1.5 + 0.5t\right)^{-1/2} \qquad (2)$$

From Eq. (2)

$$\left. \frac{d\sqrt{y(t)}}{dt} \right|_{t=0} = \frac{1}{2} \times 0.5 (1.5)^{-1/2} = 0.20 \ \text{m}^{1/2}/\text{h}$$

The location where the surge first develops is given by Eq. (7-37) upon substituting $\tau = 0$ as

$$x = -\frac{\left(3.6 + 2 \times 13.81 - 3\sqrt{(9.81 \times 3600^2 / 1000)(1.5/1000)}\right)^2}{3\sqrt{9.81 \times 3600^2 / 1000}(0.20 / \sqrt{1000})} = -15.41 \ \text{km}$$

The time when the surge first develops is obtained from Eq. (7-38) by substituting $\tau = 0$ as

$$t = \frac{-15.41}{\left(3.6 + 2 \times 13.81 - 3\sqrt{(9.81 \times 3600^2 / 1000)(1.5/1000)}\right)} = 1.51 \ \text{h}$$

Effect of Resistance. Henderson (1966) analyzed the effect of resistance on dispersion of small disturbances. He linearized the governing equations by neglecting higher order terms. The conclusions of the analysis are: for $F_0 < 2$, where subscript 0 indicates the initial uniform flow condition, resistance makes positive disturbance more dispersive and delays surge formation; and resistance makes negative disturbance less dispersive, but it is doubtful that resistance can lead to surge formation.

Supercritical Flow

If all members of one family of characteristics are straight lines, the solution of unsteady problems can be obtained easily, as shown above for a simple-wave problem in subcritical flow. However, the solution of a simple-wave problem for supercritical flow is not so simple because all members of one family of characteristics are not straight lines in the entire *x-t* plane, as shown below. In supercritical flow the slopes of all members of both families of characteristics are positive, as shown in Figure 7-5. The initial conditions in the channel are steady and uniform at depth y_0 and velocity V_0. The C+ and C− characteristics originating from the origin are OA and OF, respectively. Let C+ characteristic OA be a straight line at an inverse slope of $(V_0 + c_0)$. Consider two additional C− characteristics GH and DE and another C+ characteristic BC. The C− characteristic GH originates from the *t*-axis. The C− characteristic DE intersects the initial C+ characteristic OA at D. The C− characteristic OF intersects the C+ characteristic BC at point J. Both $(V + c)$ and $(V + 2c)$ are constant along OA, and it is possible only if both V and c are constant along OA. It implies that

$$V_O = V_D; \quad c_O = c_D \tag{7-39}$$

Because $(V+2c)$ is constant along C+ characteristics, one can write

$$V_H + 2c_H = V_J + 2c_J = V_E + 2c_E \tag{7-40}$$

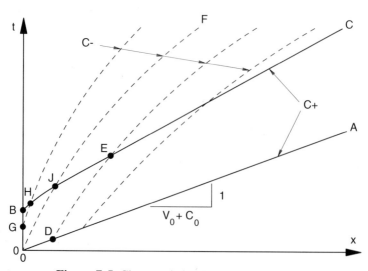

Figure 7-5. Characteristics for supercritical flow.

Because $(V-2c)$ is constant along C$-$ characteristics, one can write

$$V_H - 2c_H = V_G - 2c_G;\ V_J - 2c_J = V_O - 2c_O;\ V_E - 2c_E = V_D - 2c_D \quad (7\text{-}41)$$

Because $V_G \neq V_O$ and $c_G \neq c_O$, one can see from Eqs. (7-39), (7-40), and (7-41) that

$$V_J = V_E;\ c_J = c_E;\ \text{and}\ V_J \neq V_H;\ c_J \neq c_H \qquad\qquad (7\text{-}42)$$

These relations imply that JE, hence JC, is a straight line, and HJ, hence BJ, is not a straight line. A problem of determinacy arises if one assumes that BJ is a straight line; because then one has three equations to solve for V_J and c_J; both $(V+2c)$ = constant and $(V+c)$ = constant along the C+ characteristic BJ; and $(V-2c)$ = constant along the C$-$ characteristics OJ.

If the C$-$ characteristic OF that originates from the origin is assumed to be a straight line, it can be shown that only the C$-$ characteristics originating from the x-axis are straight lines and the C$-$ characteristics originating from the t-axis are not straight lines. Therefore, an analytical solution of a general simple-wave problem for supercritical flow is not available. However, if the boundary conditions are such that either all C$-$ characteristics or C+ characteristics in the nonuniform region start from the origin, then these characteristics are straight lines as in the case of a sudden change in gate opening; such problems can be solved analytically. The problem of gate operation is discussed in Section 7-6.

PROBLEMS

7-3   Plot the flow profile at $t = 3.0$ h in Example 7-2.

7-4   Determine the velocity and depth at $t = 2$ h in the section 50 km from the upstream end of the channel in Example 7-2.

7-5   Water flows at a uniform depth of 0.9 m and velocity of 1.2 m/s in a rectangular channel. The depth at the downstream end of the channel begins to fall at a rate of 0.1 m/h for 4 h and then stays steady at 0.5 m. Plot the water-surface profile at $t = 4.5$ h. Determine the rate at which the region of nonuniform depth is lengthening.

7-6   A long rectangular channel discharges into an estuary. The flow in the channel is steady and uniform at a depth of 2.4 m and a velocity of 1.5 m/s. Water level in the estuary initially is the same as in the river. The estuary level suddenly falls by 1.2 m at time $t = 0$ and then remains constant forever. Determine the length of the nonuniform-flow reach at time $t = 10$ minutes. Neglect bed slope and resistance in the channel.

7-7   A 70-km-long rectangular channel has a reservoir at the upstream end and a sluice gate at the downstream end. Flow in the channel is steady and uniform at a depth of 3 m and velocity of 1.2 m/s. Water-surface level in the reservoir begins to fall at a rate of 0.3 m/h at time $t = 0$, and the gate is suddenly closed at time $t = 1$ h. At what time and where will the depth of flow be last affected? What will be the flow depth at that time in a section 10 km downstream from the reservoir? Neglect S_0 and S_f.

7-8   Water flows in a rectangular channel at a uniform depth of 0.9 m and velocity of 1.2 m/s. The depth at the downstream end of the channel increases at a rate of 0.2 m/h for 3 h and then remains steady. Plot the envelope of the C– characteristics. Determine the location where the surge will first develop.

7-9   Flow in a rectangular channel is initially uniform at velocity 3 m/s and depth 5.4 m. During the slow closure of a gate at the downstream end of the channel the velocity behind the gate is observed to be

$$V = V_0(1 - t/30) \quad \text{for} \quad 0 < t < 30 \text{ s}$$

Assuming $S_0 = S_f = 0$, find the depth behind the gate at $t = 15$ s.

7-10   A long, rectangular canal of width 6 m carries flow of 30 m^3/s at uniform depth 2.5 m. The flow to the canal at its upstream end is suddenly cut off by closing a sluice gate. Determine the depth of the stationary water downstream of the gate and the speed with which the leading and trailing edges of the resulting negative wave travel downstream. Neglect channel slope and resistance.

7-5 Dam-Break Problem

Consider a dam that retains water to a depth y_0 in a rectangular channel as shown in Figure 7-6a. The dam is represented in the figure by a vertical plate. The failure of the dam can be simulated by the sudden removal of the plate. Two cases with different depths in the downstream channel will be considered: first, with zero depth and, second, with a finite depth.

Dry Downstream Channel Bed

Consider first that the plate is gradually accelerated to a speed w. The path of the plate in the x-t plane is shown by the line OBEH in Figure 7-6b. The speed of the plate is given by inverse slope of the line. With increasing velocity of the plate from point O to point E, the inverse slope of the line gradually increases. The inverse slope becomes constant beyond point E as the plate begins to move at a constant speed w. The velocity of water in contact with the plate is equal to the velocity of the plate. The boundary condition in this problem is, therefore, prescribed in terms of velocity along OBEH, instead of in terms of depth along the t-axis as in Example 7-2.

A number of C– characteristics issuing from the line OBEH can be drawn as shown in Figure 7-6b. The slope of the C– characteristics can be obtained from Eq. (7-12). The inverse slope of the first C– characteristic OA that divides the undisturbed and disturbed zones is $-c_0$ as $V_0 = 0$ in the present problem. The undisturbed zone is identified by Zone I in Figure 7-6. The slope of the C– characteristic BC issuing from point B can be written as

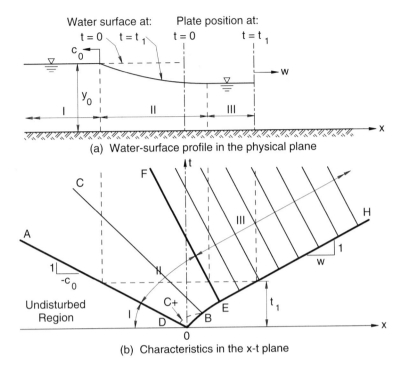

Figure 7-6. Flow development by movement of a plate.

$$\frac{dx}{dt} = V_B - c_B \tag{7-43}$$

A C+ characteristic BD can be drawn through point B to the initial C− characteristic OA. From Eq. (7-19), which is valid along a C+ characteristic such as BD in Figure 7-6b

,

$$V_B + 2c_B = V_D + 2c_D = 2c_0 \tag{7-44}$$

because $V_D = V_0 = 0$ and $c_D = c_0$. Note that Eq. (7-44) is valid in the entire x-t plane and can be written as

$$V + 2c = 2c_0 \tag{7-45}$$

From Eqs. (7-43) and (7-44) the slope of the characteristic BC can be expressed in terms of either V_B or c_B as

$$\frac{dx}{dt} = \frac{3}{2}V_B - c_0 \tag{7-46}$$

or
$$\frac{dx}{dt} = 2c_0 - 3c_B \qquad (7\text{-}47)$$

Because V_B increases with time until $V_B = w$ at point E, the inverse slope given by Eq. (7-46) of the C– characteristics issuing from the line OBE increases as t increases along OBE. Consequently, these characteristics diverge in the zone marked II in Figure 7-6, and the depth in a section in Zone II decreases with time. The characteristics issuing from the straight line EH in the zone marked III are parallel, as plate velocity is constant along EH. The flow conditions, i.e., V and y, in Zone III are, therefore, constant. The flow conditions are variable in Zone II, which connects Zones I and III, wherein the flow conditions are constant. The constant depth y_E in Zone III can be obtained by substituting

$$V = w \quad \text{and} \quad c = \sqrt{g y_E}$$

in Eq. (7-45) as

$$y_E = \frac{(c_0 - w/2)^2}{g} \qquad (7\text{-}48)$$

It is seen from Eq. (7-48) that the depth decreases as w increases and becomes zero when $w = 2c_0$. The inverse slope of the characteristics in Zone III from Eq. (7-46) for $V_E = w = 2c_0$ is w. This means all characteristics in Zone III coalesce into the straight line EH; Zone III disappears, as shown in Figure 7-7. The water-

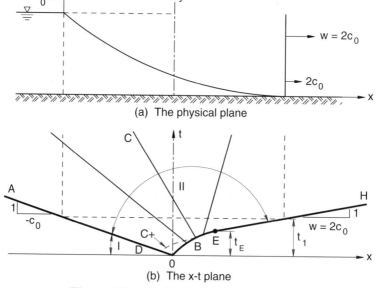

(a) The physical plane

(b) The x-t plane

Figure 7-7. A limiting case wherein $w = 2c_0$.

surface profile in Figure 7-7a is shown for $t > t_E$, where t_E is the period in which the plate is accelerated from speed zero to $w = 2c_0$. The depth at the leading edge of the wave is zero. The leading edge that is in contact with the plate is moving with a velocity $2c_0$. It can be easily seen from Figure 7-7 that if w exceeds $2c_0$, the plate loses contact with the water behind it. In other words the flow field is not affected by the removal of the plate.

The sudden removal of the plate can be simulated by accelerating the plate with an infinite acceleration to a velocity equal to or greater than $2c_0$. In that case point E coincides with point O at the origin and the line EH originates from the origin, as shown in Figure 7-8 by line OH. The flow condition at any point P $(x,.t)$ can be obtained from Eqs. (7-46) or (7-47), as the slope dx/dt of the C− characteristic passing through point P is equal to x/t. The water-surface profile at any time t is given by Eq. (7-47) as

$$\frac{x}{t} = 2\sqrt{gy_0} - 3\sqrt{gy} \qquad (7-49)$$

At any instant the water-surface profile is a parabola. The leading edge of the profile advances at a speed $2c_0$ and the trailing edge recedes at speed c_0. It should be noted that one of the C− characteristics coincides with the t-axis; its inverse slope is zero, and $V = c$ along the t-axis, i.e., the flow is critical at $x = 0$. It can be easily shown that the depth and velocity at $x = 0$ are constant and equal to $4y_0/9$ and $2c_0/3$, respectively (Prob. 7-11). The slopes of the C− characteristics in the

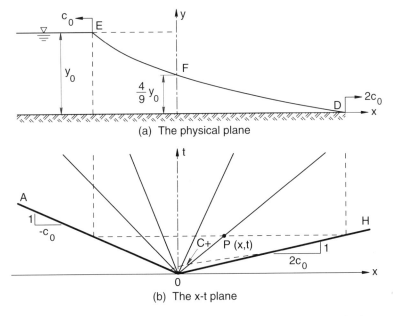

(a) The physical plane

(b) The x-t plane

Figure 7-8. Sudden failure of the dam with dry downstream bed.

first quadrant are positive, i.e., the flow there is supercritical. Note that the sloping wavefront EF is a Type C wavefront as it is propagating upstream. The sloping wavefront FD is moving downstream and therefore is of Type B. Though both wavefronts are propagating at velocity $(V - c)$, V is smaller than c in the former and V is greater than c in the latter.

Finite Depth in the Downstream Channel

The problem for the case of finite depth in the channel downstream from the dam is not very different from the problem described above. The downstream water limits the extent of Zone II in Figure 7-8, without affecting the relationship existing within this zone. The solution obtained for the dry bed produces a discontinuity in velocity at point B (see Figure 7-9) where the parabola meets the water surface in the downstream channel. Such a discontinuity in velocity is possible across a surge of Type A. The surge moves at a velocity V_w, as shown in Figure 7-9. The depth and velocity downstream of the surge in Zone IV are constant, so are the depth and velocity upstream of the surge in Zone III. The continuity and the momentum equations across the surge [Eqs. (3-1) and (3-2)] for $V_4 = 0$ can be written as

$$y_3(V_3 - V_w) = -y_4 V_w \qquad (7-50)$$

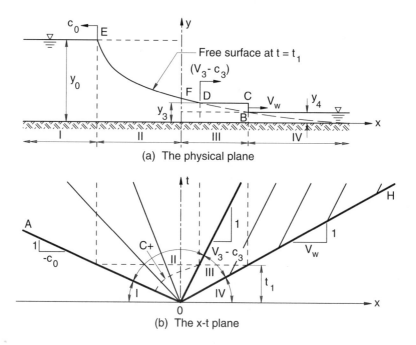

(a) The physical plane

(b) The x-t plane

Figure 7-9. Sudden failure of a dam with a finite depth in the downstream channel; $0.138y_0 > y_4 > 0$.

$$V_w = \sqrt{\frac{g}{2}\frac{y_3}{y_4}(y_3 + y_4)} \qquad (7\text{-}51)$$

where the subscripts refer to the zones. Equation (7-51) can also be obtained from Eq. (3-10); the negative sign before the radical is taken as V_w is positive. From Eq. (7-45) along the C+ characteristic in Zone II (one of these is shown by a dotted curve in Figure 7-9b), one gets

$$V_3 + 2c_3 = 2c_0 \qquad (7\text{-}52)$$

The unknown quantities, namely, y_3, V_3, and V_w, can be determined by solving Eqs. (7-50)–(7-52). Substitution into Eq. (7-50) for V_3 from Eq. (7-52) and for V_w from Eq. (7-51) yields

$$\sqrt{\frac{g}{2}\frac{y_3}{y_4}(y_3 + y_4)} = \frac{y_3\left(2\sqrt{gy_0} - 2\sqrt{gy_3}\right)}{(y_3 - y_4)} \qquad (7\text{-}53)$$

Equation (7-53) can be solved for y_3 and then V_w and V_3 can be determined from Eqs. (7-51) and (7-52), respectively.

The slope of the C– characteristic in Zone III in Figure 7-9 is positive, which implies that the flow in Zone III is supercritical. One of the C– characteristics in Zone II is in coincidence with the t-axis; the flow is critical in section F at $x = 0$ and therefore the depth and velocity in section F is $4y_0/9$ and $2c_0/3$, respectively. The discharge at $x = 0$ is $8c_0y_0/27$ and independent of both the time and depth y_4. It can be shown from the above equations that V_3 decreases with increasing y_4, and V_3 becomes equal to c_3 (critical flow) when y_4 is about $0.138y_0$ (Prob. 7-12). Figure 7-9 is, therefore, valid only for $0 < y_4 < 0.138y_0$. Note that the sloping wavefront EF is of Type C, the sloping wavefront FD is of Type B, and the steep wavefront BC is of Type A.

It can be shown that the slope of the C– characteristics in Zone III for y_4 larger than about $0.138y_0$ is negative, as shown in Figure 7-10. The flow in Zone III is, therefore, subcritical; section D moves upstream. The t-axis lies in Zone III, where the flow conditions are steady. The discharge at $x = 0$ continues to be independent of the time but now is a function of depth y_4 and is less than the maximum of $8c_0y_0/27$. Note that the inverse slope of the C– characteristics in Zone III for y_4 equal to about $0.138y_0$ is zero; the flow in Zone III then is critical, and section D remains stationary at $x = 0$.

A comparison of the above results with experimental data (Schoklitsch, 1917; Dressler, 1954; U. S. Army Corps of Engineers, 1960,1961) shows a good agreement except near the leading edge where the resistance effect is more pronounced. In experiments for dry channel the wave front was rounded and it propagated with a velocity that was roughly one half of the theoretical value.

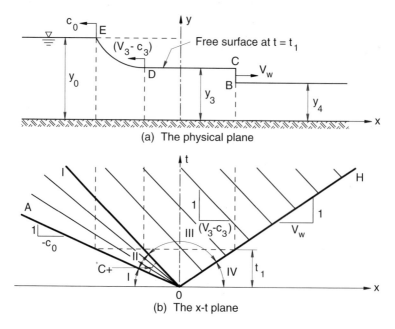

Figure 7-10. Sudden failure of a dam with a finite depth in the downstream channel; $y_0 > y_4 > 0.138y_0$.

Example 7-4. A dam retaining water to a depth of 25.0 m collapsed suddenly. The bed downstream of the dam at the time of failure was dry. Determine (i) the speed of the leading and receding edges of the water-surface profile; (ii) the depth and velocity at the dam; and (iii) depth and velocity in a section 10 km downstream of the dam half an hour after the dam failure.

$$c_0 = \sqrt{gy_0} = \sqrt{9.81 \times 25} = 15.66 \text{m/s}$$

Speed of the leading edge $= 2c_0 = 31.32$ m/s

Speed of the receding edge $= c_0 = 15.66$ m/s

$$\text{Depth at the dam} = \frac{4}{9} y_0 = \frac{4 \times 25}{9} = 11.11 \text{ m}$$

$$\text{Velocity at the dam} = \frac{2}{3} c_0 = \frac{2 \times 15.66}{3} = 10.44 \text{ m/s}$$

The depth y in the section at $x = 10$ km and $t = 30$ min can be determined from Eq. (7 - 49) as

$$\frac{10 \times 1000}{30 \times 60} = 2 \times 15.66 - 3\sqrt{9.81y}$$

or
$$y = 2.74 \text{ m}$$

The velocity in the section can be obtained from Eq. (7 - 45) as

$$V + 2\sqrt{9.81 \times 2.74} = 2 \times 15.66$$

$$V = 20.85 \text{ m/s}$$

PROBLEMS

7-11   Show that the flow depth and velocity at $x = 0$ in Figure 7-8 are $4y_0/9$ and $2c_0/3$, respectively.

7-12   Show that if the depth in the channel downstream of a dam is about $0.138y_0$, then the point D in Figures 7-9 and 7-10 remains stationary at the original dam position.

7-13   To determine flow conditions in a lock resulting by opening the lock gates, consider the following idealized conditions. The lock is rectangular in cross section and its bottom is horizontal. The depths of still water upstream and downstream of the lock gates are 3 and 0.3 m, respectively. The lock gates are suddenly opened. Determine (i) the height and speed of the resulting surge; (ii) the rate of lengthening of the region of constant depth behind the surge; and (iii) the depth and velocity in a section 100 m upstream from the lock gates at 10 s after opening the lock gates.

7-14   A dam impounding a 30-m depth of water fails suddenly. The depth of water in the channel downstream is 1.5 m. Estimate the height and velocity of the surge that moves down the channel following the failure of the dam. Neglect the effect of the channel slope and resistance. Predict the depth at a point 2 km downstream from the dam 10 minutes after the failure.

7-6 Sluice-Gate Operation

Analytical solution of a general simple-wave problem in supercritical flow is not available, as explained in Section 7-4. The flow downstream of a gate is always supercritical as long the gate acts as a control. The changes in flow profiles due to the gate operation can be analyzed analytically only if the gate opening is changed instantaneously. Several cases of sudden opening and closure of a sluice gate are considered. The case of sudden gate opening is considered first as it is similar to the dam-break problem.

Sudden Complete Opening

The problem of sudden complete opening of a sluice gate from the initial closed position is identical to the dam-break problem. The difference between the dam-break problem and the problem of a sudden complete gate opening from an initial opened position is in the initial upstream and downstream velocities, which are zero in the former and nonzero in the latter. Figure 7-9 can be easily modified on superimposing nonzero initial velocities V_0 upstream of the gate and V_4 downstream of the gate. The flow profile is presented in Figure 7-11. The three unknown quantities, namely, y_3, V_3, and V_w, can be determined by solving the following three equations:

Continuity and momentum equations across the surge:

$$y_3(V_3 - V_w) = y_4(V_4 - V_w) \qquad (7\text{-}54)$$

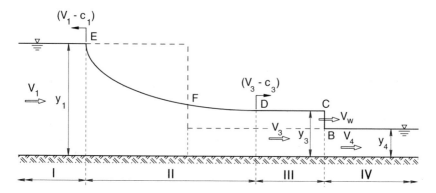

Figure 7-11. Sudden complete opening of a gate.

$$V_w = V_4 + \sqrt{\frac{g}{2}\frac{y_3}{y_4}(y_3 + y_4)}$$ (7-55)

Relation across the negative wave in Zone II:

$$V_1 + 2c_1 = V_3 + 2c_3$$ (7-56)

Sudden Partial Opening

In the case of sudden partial opening of the sluice gate, the negative-wave profile in Zone II in Figure 7-11 must develop a discontinuity at the gate, and the zones of the final steady flow conditions must advance upstream and downstream of the gate. The flow profile is presented in Figure 7-12. The wavefronts EH, GD, and BC are of Types C, B, and A, respectively. Similar types of wavefronts developed in the dam-break problem. There are 11 physical quantities y_1, V_1, y_3, V_3, y_4, V_4, y_6, V_6, y_7, V_7, and V_w and there are six equations available to solve for six unknown quantities. The six equations are as follows:

Continuity and energy equations across the gate:

$$q = V_3 y_3 = V_4 y_4$$ (7-57)

$$y_3 + \frac{q^2}{2gy_3^2} = y_4 + \frac{q^2}{2gy_4^2}$$ (7-58)

Continuity and momentum equations across the surge:

$$y_6(V_6 - V_w) = y_7(V_7 - V_w)$$ (7-59)

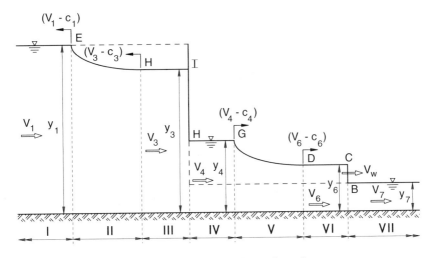

Figure 7-12. Sudden partial opening of a gate.

$$V_w = V_7 + \sqrt{\frac{g}{2}\frac{y_6}{y_7}(y_6 + y_7)}$$
(7-60)

Relations across the negative waves in Zones II and V:

$$V_1 + 2c_1 = V_3 + 2c_3$$
(7-61)

$$V_4 + 2c_4 = V_6 + 2c_6$$
(7-62)

Note that the Types B and C negative waves propagate at velocity $(V-c)$; therefore $(V+2c)$ is constant across the negative waves. The problem can be solved if five of the physical quantities are known. In most cases the flow conditions before increasing the gate opening, i.e., y_1, V_1, y_7, and V_7, and the final gate opening y_4 are known (Example 7-5).

Intuitively a surge of Type A, not the profile shown in Figure 7-12, is expected downstream of the gate due to the sudden increase in the gate opening. The reason why a surge does not form is as follows. The solution of unsteady supercritical flow requires two initial conditions along the channel reach and two boundary conditions at the upstream section of the channel reach, i.e., at the gate. A surge of Type A is characterized by five parameters, namely, depths and velocities upstream and downstream of the surge and the surge velocity. The prescribed values of four of the five parameters in general are not compatible with the continuity and momentum equations (3-1) and (3-2), and such a flow cannot be represented by a simple surge. Montuori (1968, 1993) showed that the flow downstream of the gate for a sudden increase in gate opening is represented by the flow profile shown in Figure 7-12.

Sudden Partial Closure

A sudden partial closure of a gate produces a negative wave BC of Type D and a surge DE of Type E that move downstream and a surge HI of Type F that moves upstream, as shown in Figure 7-13. The reasoning for not having only a surge of Type E on the downstream of the gate is the same as given above for Type A surge in the case of sudden gate opening. Montuori (1968, 1993) showed that the flow downstream of the gate for a sudden decrease in gate opening is represented by the flow profile shown in Figure 7-13. The flow upstream of the gate is subcritical and only one boundary condition at the downstream end is required for solution. The two initial conditions and one boundary condition are compatible with Eqs. (3-1) and (3-2), and the flow upstream can be represented by a simple surge.

The flow conditions before decreasing the gate opening, i.e., y_1, V_1, y_6, and V_6, and the final gate opening y_3 are known. There are seven unknown quantities, y_2, V_2, V_{wu}, V_3, y_4, V_4, and V_{wd}. The seven equations to solve for the seven unknowns are the following (see Example 7-6):

Continuity and energy equations across the gate:

$$q = V_2 y_2 = V_3 y_3 \tag{7-63}$$

$$y_2 + \frac{q^2}{2gy_2^2} = y_3 + \frac{q^2}{2gy_3^2} \tag{7-64}$$

Continuity and momentum equations across the upstream surge:

$$y_1(V_1 - V_{wu}) = y_2(V_2 - V_{wu}) \tag{7-65}$$

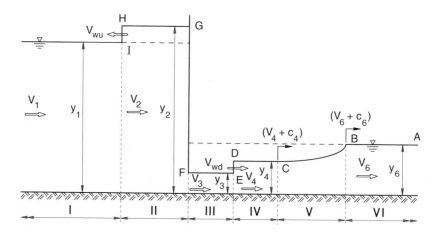

Figure 7-13. Sudden partial closure of a gate.

$$V_{wu} = V_1 - \sqrt{\frac{g}{2}\frac{y_2}{y_1}(y_2 + y_1)} \qquad (7\text{-}66)$$

where V_{wu} is the velocity of the upstream surge and is negative.

Continuity and momentum equations across the downstream surge:

$$y_3(V_3 - V_{wd}) = y_4(V_4 - V_{wd}) \qquad (7\text{-}67)$$

$$V_{wd} = V_3 - \sqrt{\frac{g}{2}\frac{y_4}{y_3}(y_4 + y_3)} \qquad (7\text{-}68)$$

where V_{wd} is the velocity of the downstream surge.

Relation across the negative wave in Zone V:

$$V_4 - 2c_4 = V_6 - 2c_6 \qquad (7\text{-}69)$$

Note that the Type D negative wave in Zone V propagates at velocity $(V+c)$; therefore $(V-2c)$ is constant across the negative wave.

Sudden Complete Closure

The flow conditions due to sudden complete closure of a gate are shown in Figure 7-14, which does not display a surge on the downstream side of the gate. It can be shown that the incorporation of a surge on the downstream side results in a problem of indeterminacy as the number of unknown quantities exceeds the number of equations (Prob. 7-20). The problems upstream and downstream of the gate can be considered independent of each other. On the upstream side there are

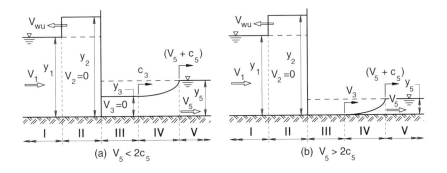

Figure 7-14. Sudden complete closure of a gate.

two unknown quantities, y_2 and V_w ($V_2 = 0$), which can be determined by solving the continuity and momentum equation across the surge. On the downstream side there is only one unknown quantity y_3 or V_3 ; the other one is zero. The unknown quantity can be determined from the relation ($V - 2c$) = constant in Zone IV. It can be shown (Prob. 7-19) that $V_3 = 0$ and $y_3 > 0$ for $V_5 < 2c_5$ (Figure 7-14a); $V_3 = 0$ and $y_3 = 0$ for $V_5 = 2c_5$; and $y_3 = 0$ and $V_3 > 0$ for $V_5 > 2c_5$ (Figure 7-14b).

Example 7-5. The initial depths upstream and downstream of a sluice gate are 4 and 1.0 m, respectively. The gate opening is suddenly increased to make the downstream depth equal to 1.5 m. Calculate the new discharge and the flow conditions in the different zones shown in Figure 7-12.

See Figure 7-12 for different zones. Initial conditions: $y_1 = 4.0$ m; $y_7 = 1.0$ m.

$$c_1 = \sqrt{9.81 \times 4} = 6.26 \text{ m/s}$$

The energy equation between upstream and downstream of the gate yields

$$y_1 + \frac{q^2}{2gy_1^2} = y_7 + \frac{q^2}{2gy_7^2}$$

or
$$q = \frac{\sqrt{2g\, y_1 y_7}}{\sqrt{y_1 + y_7}} = \frac{\sqrt{2 \times 9.81 \times 4 \times 1}}{\sqrt{4+1}} = 7.92 \text{ m}^3/\text{s/m}$$

The continuity equation between upstream and downstream of the gate yields

$$V_1 = \frac{7.92}{4} = 1.98 \text{ m/s}; \quad V_7 = \frac{7.92}{1} = 7.92 \text{ m/s}$$

Final condition : $y_4 = 1.5$ m.

Flow conditions upstream of the gate
The continuity and the energy equations across the gate:

$$q = V_3 y_3 = V_4 y_4 = 1.5 V_4 \tag{1}$$

$$q = \frac{\sqrt{2g\, y_3 y_4}}{\sqrt{y_3 + y_4}} = \frac{6.64 y_3}{\sqrt{y_3 + 1.5}} \tag{2}$$

The relation across the negative wave in Zone II gives

$$V_3 + 2c_3 = V_1 + 2c_1$$

or
$$V_3 + 2\sqrt{9.81 y_3} = 1.98 + 2 \times 6.26 = 14.5 \text{ m/s} \tag{3}$$

Solving Eqs. (1), (2), and (3) for y_3, V_3, and V_4, one obtains

$$y_3 = 3.39 \text{ m}; \quad V_3 = 3.01 \text{ m/s}; \quad V_4 = 6.79 \text{ m/s}$$

$$q = V_3 y_3 = 3.01 \times 3.39 = 10.20 \text{ m}^3/\text{s/m}$$

$$c_4 = \sqrt{9.81 \times 1.5} = 3.84 \text{ m/s}$$

Flow conditions downstream from the gate
The relation across the negative wave in Zone V gives

$$V_6 + 2c_6 = V_4 + 2c_4$$

or
$$V_6 + 2\sqrt{9.81 y_6} = 6.79 + 2 \times 3.84 = 14.47 \text{ m/s} \tag{4}$$

The continuity and momentum equations across the surge give

$$(V_6 - V_w)y_6 = (V_7 - V_w)y_7 = (7.92 - V_w) \times 1 \tag{5}$$

$$V_w = V_7 + \sqrt{\frac{g}{2}\frac{y_6}{y_7}(y_6 + y_7)} = 7.92 + 2.21\sqrt{y_6(y_6 + 1)} \tag{6}$$

Solving Eqs. (4), (5), and (6), one gets

$$V_w = 11.08 \text{ m/s}; \quad V_6 = 8.06 \text{ m/s}; \quad y_6 = 1.05 \text{ m}$$

Example 7-6. The initial conditions are the same as in Example 7-4. The gate opening is suddenly decreased so that the downstream depth is equal to 0.6 m. Calculate the new flow conditions.

See Figure 7-13 for different flow zones. The initial conditions from Example 7-5 are

$$y_1 = 4.0 \text{ m}; \ V_1 = 1.98 \text{ m/s}; \ y_6 = 1.0 \text{ m}; \ V_6 = 7.92 \text{ m/s}$$

$$\text{Final condition} : y_3 = 0.7 \text{ m}$$

The continuity and the energy equations across the gate yield

$$q = V_2 y_2 = V_3 y_3 = 0.7 y_3 \tag{1}$$

$$q = \frac{\sqrt{2g}\, y_2 y_3}{\sqrt{y_2 + y_3}} = \frac{3.10 y_2}{\sqrt{y_2 + 0.7}} \tag{2}$$

Flow conditions upstream from the gate:
The continuity and the energy equations across the upstream surge yield

$$(V_2 - V_{wu})y_2 = (V_1 - V_{wu})y_1 = 7.92 - 4V_{wu} \tag{3}$$

$$V_{wu} = V_1 - \sqrt{\frac{g}{2}\frac{y_2}{y_1}(y_1 + y_2)} = 1.98 - 1.11\sqrt{y_2(4 + y_2)} \tag{4}$$

Solving Eqs. (1)–(4) by trial for V_2, y_2, V_{wu}, and V_3, one obtains

$$V_2 = 1.37 \text{ m/s}; \quad y_2 = 4.40 \text{ m}; \quad V_{wu} = -4.75 \text{ m/s}; \quad V_3 = 8.63 \text{ m/s}$$

Flow conditions downstream from the gate :

The continuity and the momentum equations across the downstream surge yield

$$(V_4 - V_{wd})y_4 = (V_3 - V_{wd})y_3 = 6.04 - 0.7V_{wd} \tag{5}$$

$$V_{wd} = V_3 - \sqrt{\frac{g}{2}\frac{y_4}{y_3}(y_3 + y_4)} = 8.63 - 2.65\sqrt{y_4(0.7 + y_4)} \tag{6}$$

The relation across the negative wave in Zone V yields

$$V_4 - 2\sqrt{gy_4} = V_6 - 2\sqrt{gy_6} = 7.92 - 2\sqrt{9.81 \times 1} = 1.66 \tag{7}$$

Solving Eqs. (5), (6), and (7) by trial for y_4, V_4, and V_{wd}, one gets

$$y_4 = 0.95 \text{ m}; \quad V_4 = 7.77 \text{ m/s}; \quad \text{and} \quad V_{wd} = 5.34 \text{ m/s}$$

PROBLEMS

7-15 Initial depths upstream and downstream of a sluice gate in a rectangular channel are 3.2 and 0.8 m, respectively. The sluice gate is suddenly raised clear of the water. Determine the flow conditions in the different zone shown in Figure 7-11. Is the discharge at the location of the gate steady?

7-16 The gate opening in Problem 7-15 is suddenly increased so that the depth downstream from the gate changes from 0.8 to 1.0 m. Determine the new flow conditions immediately upstream and downstream of the gate. Find the speeds of the front and the tail of the resultant negative wave downstream.

7-17   The gate opening in Problem 7-15 is suddenly decreased to 0.5 m. Determine the heights and velocities of the resulting surges. How much is the new discharge?

7-18   The gate in Problem 7-15 is suddenly closed completely. Determine the height and velocity of the resultant upstream surge. Find the speed of the leading edge of the resultant negative wave downstream. How much is the water depth immediately downstream of the gate?

7-19   Show that the downstream face of a sluice gate will remain high and dry on its sudden complete closure if the initial downstream Froude number is larger than 2.

7-20   Show that for the case of sudden complete closure of a sluice gate the incorporation of a surge on the downstream side results in more number of unknown quantities than the number of equations.

7-21   A sluice gate is placed in a channel of a rectangular section, and the initial depths upstream and downstream from the gate are 2.4 and 1.2 m, respectively. If the sluice gate is instantaneously and completely closed, find (i) the speed with which the front of the negative wave moves downstream; (ii) the depth of water next to the gate on the downstream side; and (iii) the water-surface profile on the downstream side.

7-22   Draw a set of characteristics in the x-t plane for the water-surface profile sketched in Figures 7-12 and 7-13.

APPENDIX F

MONOCLINAL WAVE

The monoclinal wave is a uniformly progressive wave that travels down the channel at a constant speed V_w from an upstream region of uniform flow to a downstream region of uniform flow, as shown in Figure F-1a. The depth of the wavefront gradually varies from a higher depth y_1 to a lower depth y_2. The monoclinal wave can be brought to rest by superimposing a velocity equal and opposite to V_w in Figure F-1b. The continuity equation for the steady flow can be written as

$$(V_w - V)A = Q_r, \text{ a constant} \tag{F-1}$$

or

$$V_w A - Q = Q_r \tag{F-2}$$

where Q_r is termed as "overrun" and is the discharge left behind by the wave. Differentiation of Eq. (F-2) with respect to y gives an expression for wave velocity as

$$V_w = \frac{1}{B}\frac{dQ}{dy} \tag{F-3}$$

It is shown in the next chapter that the velocity of the monoclinal wave given by Eq. (F-3) is equal to the velocity of the flood wave in rivers. If the Manning equation is used to evaluate dQ/dy, it can be shown that for a wide rectangular channel the wave velocity is equal to five-thirds of the flow velocity, i.e.,

$$V_w = \tfrac{5}{3}V \tag{F-4}$$

The differential equation for y can be obtained by eliminating V in Eq. (7-2). Differentiation of Eq. (F-1) with respect to x yields

$$\frac{\partial V}{\partial x} = \frac{BQ_r}{A^2}\frac{\partial y}{\partial x} \tag{F-5}$$

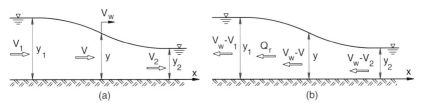

Figure F-1. Monoclinal wave.

As the wave travels at a constant speed, the total derivative of flow velocity is zero, i.e.,

$$\frac{DV}{Dt} = \frac{\partial V}{\partial t} + \frac{\partial V}{\partial x}\frac{dx}{dt} = \frac{\partial V}{\partial t} + V_w\frac{\partial V}{\partial x} = 0$$

or

$$\frac{\partial V}{\partial t} = -V_w\frac{\partial V}{\partial x} \tag{F-6}$$

Substituting from Eqs. (F-6) and (F-5) for $\partial V/\partial t$ and $\partial V/\partial x$ in Eq. (7-2) and rearranging the terms, one obtains

$$\frac{\partial y}{\partial x} = \frac{S_0 - S_f}{1 - BQ_r^2/gA^3} \tag{F-7}$$

Since

$$S_f = \frac{Q^2}{K^2} = \frac{V^2 A^2}{K^2} = \frac{(V_w A - Q_r)^2}{K^2}$$

Eq. (F-7) becomes

$$\frac{dy}{dx} = \frac{S_0 - (V_w A - Q_r)^2/K^2}{1 - (BQ_r^2)/(gA^3)} \tag{F-8}$$

As the wave profile is not a function of time t in the moving coordinate system, the partial derivative sign in Eq. (F-7) has been replaced by the total derivative sign in Eq. (F-8). Equation (F-8) can be easily integrated for a wide rectangular channel and using the Che′zy formula with Che′zy coefficient C as a constant. From the Che′zy formula $K^2 = C^2 B^2 y^3$. For the unit width of a wide channel, Eq. (F-8) can be written as

$$\frac{dy}{dx} = S_0\frac{y^3 - (V_w y - q_r)^2/C^2 S_0}{y^3 - q_r^2/g} \tag{F-9}$$

where $q_r = Q_r/B$. The numerator in Eq. (F-9) is cubic in y and can be written as $(y - y_1)(y - y_2)(y - y_3)$, which on substitution in Eq. (F-9) yields

$$\frac{dy}{dx} = S_0\frac{(y - y_1)(y - y_2)(y - y_3)}{y^3 - y_c^3} \tag{F-10}$$

where $y_c^3 = q_r^2/g$. With the given values of S_0, C, V_w, and Q_r, Eq. (F-10) represents a number of possible wave profiles. One of these profiles is a monoclinal wave having an initial stage y_1 and a final stage y_2, as shown in Figure F-1.

CHAPTER 8

UNSTEADY FLOW II

This chapter deals with the problem of flood routing, which is the process of tracing by computations the course of a flood wave. As a flood wave travels downstream in a river, its spread becomes larger, and consequently its peak becomes lower (i.e., $L_2 > L_1$ and $A_2 < A_1$ in Figure 9-1). The increase in the spread of a flood wave is a consequence of its front travelling faster than its rear, which in turn is primarily due to the $\partial y/\partial x$ term and secondarily due to the acceleration terms in the momentum equation. In some cases all or some of these terms in the momentum equation can be neglected. The decrease in the peak can also be due to storage effects, which are dominant in the case of a flood wave advancing through a reservoir. In such a case the flood routing, termed as reservoir routing, can be carried out using only the continuity equation. There also are methods of flood routing in rivers that are based on the continuity equation alone. Such methods require records of past floods to develop relations between channel storage and discharge. The most widely used of these methods is the Muskingum method.

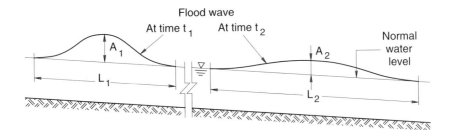

Figure 8-1. Spreading and attenuation of a flood wave.

8-1 Reservoir Routing

The continuity equation for unsteady flow is

$$\frac{\partial Q}{\partial x} + \frac{\partial A}{\partial t} = 0 \tag{8-1}$$

Integration of Eq. (8-1) with respect to x from x_1 to x_2 yields

$$Q(x_2) - Q(x_1) + \frac{d}{dt}\int_{x_1}^{x_2} A\,dx = 0 \tag{8-2}$$

where x_1 and x_2 are longitudinal coordinates of the upstream and downstream ends of the reservoir, respectively. By denoting $Q(x_2)$ as the outflow rate O, $Q(x_1)$ as the inflow rate I, and the integral term as storage volume $\forall$, Eq. (8-2) can be written as

$$I - O = \frac{d\forall}{dt} \tag{8-3}$$

It may be noted that Eq. (8-3) can be written directly from the continuity principle. For a given geometry of the reservoir and on the assumption of a level water surface in the reservoir, the storage volume $\forall$ is a function of water level in the reservoir. For a given discharge structure such as a spillway and/or an outlet, the outflow rate O also is a function of water level in the reservoir. Therefore it should be possible, at least in theory, to express O as a function of $\forall$ as

$$O = R\forall^a \tag{8-4}$$

where R and a are constants dependent on the geometry of the reservoir and the discharge structure. Substitution for O from Eq. (8-4) into Eq. (8-3) yields

$$\frac{d\forall}{dt} + R\forall^a = I \tag{8-5}$$

This equation can be solved analytically or numerically for $\forall$ if the inflow rate $I = f(t)$ is given, and then O can be obtained from Eq. (8-4) (Basha, 1995; Yevjevich, 1959). The solution of Eq. (8-5) for the linear case $a = 1$ is

$$O = R\forall = Re^{-Rt} \int I(t) e^{Rt} dt \tag{8-6}$$

However, Eq. (8-6) is not of much practical use as $I(t)$ cannot be easily expressed in equation form and the relationship between outflow and storage is usually nonlinear. For most practical problems Eq. (8-3) is solved numerically. A finite-difference approximation of Eq. (8-3) is

$$\overline{I} - \overline{O} = \frac{\Delta\forall}{\Delta t} \tag{8-7}$$

in which $\overline{I}$ and $\overline{O}$ are the average values during the time interval Δt. A discretization of Eq. (8-7) leads to

$$\frac{I_1 + I_2}{2} - \frac{O_1 + O_2}{2} = \frac{\forall_2 - \forall_1}{\Delta t} \tag{8-8}$$

in which I_1, O_1, and $\forall_1$ are the inflow rate, outflow rate, and storage volume, respectively, at time level 1; and I_2, O_2, and $\forall_2$ are the inflow rate, outflow rate, and storage volume at time level 2. On collecting the unknown quantities (O_2 and $\forall_2$) on one side of the equation and known quantities (I_1, I_2, O_1, and $\forall_1$) on the other side, Eq. (8-8) becomes

$$N_2 = N_1 + \tfrac{1}{2}\left(I_1 + I_2\right) - O_1 \tag{8-9}$$

in which

$$N = \frac{\forall}{\Delta t} + \frac{O}{2} \tag{8-10}$$

An additional relation between O and N is required to solve Eq. (8-9) for the two unknown quantities. This relationship is obtained from the elevation-storage and elevation-outflow relations. The former relation is determined from the topographic information of the reservoir. The latter relation is based on the hydraulic characteristics of the outlet structure. Since both O and $\forall$ are functions of the water surface elevation in the reservoir, O can be functionally related to N, as shown in Example 8-1 below. Once the O-N relation is obtained, Eq. (8-9) is used to perform the reservoir routing. The procedure is as follows. From the known initial value of O at time level 1 the corresponding value of N_1 is determined from the O-N relation. The value of N_2 is found from Eq. (8-9) and the corresponding value of O at time level 2 is obtained from the O-N relation. This completes one step. The procedure is repeated for subsequent steps and is illustrated in Example 8-1.

Example 8-1. Route the following inflow hydrograph through a reservoir created by a dam and determine the maximum reservoir elevation reached during the passage of the hydrograph. The flow from the reservoir passes over a spillway with a crest elevation of 122 m. The outflow O (m³/s) over the spillway is given by $O = 15H^{3\,*}$, where H is the head at the spillway crest in meters. Above the spillway crest the reservoir has almost vertical banks and a surface area of 65 hectares. The base flow is 10 m³/s.

Time (h)	0	2	4	6	8	10	12	14	16	18	20
Inflow (m³/s)	10	100	180	210	160	90	60	46	30	16	10

*Generally the outflow over a spillway varies as three-halves power of the head over the spillway. The spillway in the example is such that the width B at the free surface varies as three-halves power of the head over the spillway; consequently, the outflow varies as H^3.

The calculations for the O-N relation are shown in Table 8-1, which is self-explanatory. The storage volume in column 4 is calculated above the spillway crest (any other convenient reference level could have been used), and it is equal to the surface area multiplied by the head above the spillway crest. A time interval $\Delta t = 2$ h is used to compute values in column 5. The O-N relation is plotted in Figure a below.

The calculations for routing the inflow hydrograph are presented in Table 8-2. The calculations for the first time interval in the last three columns proceed as indicated by the arrows. The outflow at $t = 0$ is known and is equal to 10 m³/s. The value of N for

Table 8-1. The *O-N* Relation

1	2	3	4	5	6
Reservoir Level m	Head over Spillway, H m	Outflow Rate, O m^3/s	Storage $\forall$ m^3	$\forall/\Delta t$ m^3/s	N m^3/s
122	0	0	0	0	0
123	1	15	650,000	90.28	97.78
124	2	120	1300,000	180.56	240.56
125	3	405	1950,000	270.84	473.34
126	4	960	2600,000	361.12	841.12

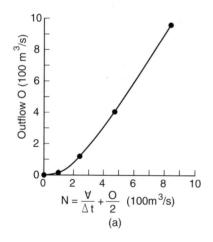

$$N = \frac{\forall}{\Delta t} + \frac{O}{2} \quad (100\text{m}^3/\text{s})$$

(a)

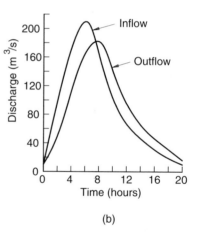

(b)

Example 8-1

$O = 10$ m^3/s is obtained from the *O-N* relations as $N = 84$ m^3/s. The value 45 of $\overline{I} - O$ in column 5 is obtained by subtracting $O = 10$ m^3/s in column 6 from $\overline{I} = 55$ m^3/s in column 3. The value 129 of N at time $t = 2$ h is obtained by adding $\overline{I} - O_1 = 45$ m^3/s in column 5 to $N = 84$ m^3/s at $t = 0$ h according to Eq. (8-9). The outflow $O = 40$ m^3/s for $N = 84$ m^3/s is obtained from the *O-N* relation. The procedure was repeated for subsequent time intervals. The inflow and outflow hydrographs are plotted in Figure b. The maximum outflow is about 183 m^3/s, and the corresponding maximum head over the spillway crest from the outflow relation is $H = 2.30$ m, which gives the maximum reservoir elevation of 124.3 m.

Table 8-2. Storage Routing

1	2	3	4	5	6
Time t h	Inflow I m³/s	$\bar{I}$ m³/s	N m³/s	$\bar{I}-O_1$ m³/s	Outflow O m³/s
0	10		84 ←		10
		55 —→	↓ —→	→ 45	
2	100		129 —→		→ 40
		140		100	
4	180		229		115
		195		80	
6	210		309		172
		185		13	
8	160		322		182
		125		−57	
10	90		265		138
		75		−63	
12	60		202		95
		53		−42	
14	46		160		60
		38		−22	
16	30		138		47
		23		−24	
18	16		113		33
		13		−20	
20	10		93		14

PROBLEMS

8-1   For a certain reservoir the water-surface area and the outflow rate at different water-surface elevations in the reservoir are as follows:

water-surface elevation (m)	160.00	158.00	156.00
water-surface area (km²)	2.16	1.88	1.68
Outflow rate (m³/s)	45.60	42.50	40.10

Determine the time required for the reservoir elevation to fall from 160 to 156 m if there is a steady inflow at a rate of 25.3 m³/s.

8-2   Extend Table 8-2 in Example 8-1 to $t = 22$ h if the inflow rate at that time is 10 m³/s.

8-3   When a flood wave passes through an impounded reach of river and the outflow discharge is not regulated, the inflow and outflow discharge hydrographs must intersect at the peak of the outflow hydrograph. Prove this statement.

8-4   A flood with the inflow hydrograph given below enters a reservoir. The surface area
       of the reservoir is 800,000 m^2 and can be assumed independent of water-surface
       elevation in the reservoir. Outflow in cubic meter per second takes place over a
       spillway and is given by $O = 115H^{3/2}$, where H is the head in meters over the
       spillway crest. An additional discharge of 14 m^3/s passes through a hydropower
       station. Determine the outflow hydrograph if the reservoir level is at the elevation of
       the spillway crest before arrival of the flood:

Time (h)	0	2	4	6	8	10	12
Inflow (m^3/s)	14	30	78	84	58	28	14

8-5   Sketch the inflow and outflow hydrographs for a typical flood through a reservoir
       and show how the rate at which reservoir storage changes with time ($d\forall/dt$) can be
       obtained graphically.

8-6   A reservoir inflow hydrograph for a flood is triangular in shape and superimposed
       on a steady base flow of 110 m^3/s. The inflow rate begins rising above the base flow
       at $t = 1$ day, peaks at 1000 m^3/s at $t = 2$ days, and reverts to the base flow at
       $t = 4$ days. The storage available in the reservoir at the beginning of the flood is
       500 m^3/s-days. Estimate the minimum peak outflow that can be obtained with an
       uncontrolled discharge, say through a conduit. For simplicity assume that the
       outflow hydrograph, like the inflow hydrograph, is a triangle superimposed on a
       steady base flow. Draw the outflow hydrograph. When would the outflow revert
       back to the base flow?

8-2 The Muskingum Method

The Muskingum method of flood routing in rivers was first developed in the 1930s
for flood control schemes in the Muskingum River Basin, Ohio. Similar to the
reservoir routing method, this method, which is also based on the continuity
equation (8-3), requires a relation between storage and outflow. Such a relation is a
single-valued function for reservoirs, but it takes the form of a loop for rivers (see
Section 8-7) and complicates the problem of flood routing through rivers. The
problem would be simplified if the loop curve could be transformed to a single-
valued relationship. Such a transformation is the basis of the Muskingum method,
which uses the following linear algebraic relationship between the channel storage
and both inflow and outflow:

$$\forall = \hat{K}O + \hat{K}X(I - O) = \hat{K}[XI + (1 - X)O] \qquad (8\text{-}11)$$

in which $\hat{K}$ and X are the routing parameters.

       The parameter $\hat{K}$ has a dimension of time, and its value is approximately
equal to the travel time of the flood wave through the river reach. The parameter X
is a dimensionless weighting factor that expresses the relative influence on storage
of inflow and outflow and ranges between 0 and 0.5. It accounts for attenuation of
the flood wave. The parameters $\hat{K}$ and X for a given river reach are found by trial

and error using the past flood data as explained later. The first term $\hat{K}O$ and the second term $\hat{K}X(I-O)$ in Eq. (8-11) are sometimes referred to as prism storage and wedge storage, respectively, and are depicted in Figure 8-2. It may be noted that the wedge storage becomes negative during the falling stage of flood when $O > I$, and Eq. (8-11) for $X = 0$ reduces to Eq. (8-4) with $a = 1$. Also note that Eq. (8-11) is based on the hypothesis that discharge is a unique function of flow depth y. Substituting for $\forall$ from Eq. (8-11) into the discretized form of the continuity equation (8-8) and solving for O_2, one obtains

$$O_2 = C_0 I_2 + C_1 I_1 + C_2 O_1 \tag{8-12}$$

in which C_0, C_1, and C_2 are the routing coefficients and defined as

$$
\begin{aligned}
C_0 &= \frac{\left(\Delta t / \hat{K}\right) - 2X}{2(1 - X) + \left(\Delta t / \hat{K}\right)} \\
C_1 &= \frac{\left(\Delta t / \hat{K}\right) + 2X}{2(1 - X) + \left(\Delta t / \hat{K}\right)} \\
C_2 &= \frac{2(1 - X) - \left(\Delta t / \hat{K}\right)}{2(1 - X) + \left(\Delta t / \hat{K}\right)}
\end{aligned}
\tag{8-13}
$$

Note that $C_0 + C_1 + C_2 = 1$. For a given inflow hydrograph, the outflow hydrograph can be obtained from Eq. (8-12) provided $\hat{K}$ and X are known. It should be noted that Δt must be equal to or larger than $2\hat{K}X$; otherwise the coefficient C_0 would then become negative, which in turn leads to an irrational relationship of decreasing O_2 with increasing I_2 in Eq. (8-12). Values of X greater than 0.5 increase the flood peak, which is in variance with reality. For $\hat{K} = \Delta t$ and $X = 0.5$, the coefficients are $C_0 = 0$, $C_1 = 1$, and $C_2 = 0$, and Eq. (8-12) yields $O_2 = I_1$, i.e., the flood motion is a pure translation without attenuation or deformation.

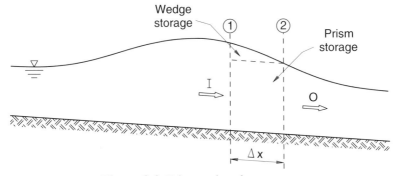

Figure 8-2. Prism and wedge storage.

Determination of $\hat{K}$ *and X*

The routing parameters $\hat{K}$ and X for a reach are determined from the inflow and outflow hydrographs of the past floods in the reach and Eqs. (8-8) and (8-11). From Eq. (8-11) $\hat{K}$ is the ratio of the channel storage to the weighted discharge $[XI + (1-X)O]$. Channel storage is determined by solving Eq. (8-8) for $\forall_2$:

$$\forall_2 = \forall_1 + \frac{\Delta t}{2}\left(I_1 + I_2 - O_1 - O_2\right) \qquad (8\text{-}14)$$

For various assumed values of X $(0 < X < 0.5)$, the weighted discharges $[XI + (1-X)O]$ are calculated. Each weighted flow is plotted against channel storage. The value of X that produces the narrowest loop (see Example 8-3) is chosen the representative value for the reach. The value of $\hat{K}$ is equal to the slope of the straight line through the narrowest loop.

The two routing parameters also can be obtained from the physical characteristics of the channel reach (Cunge, 1969) as explained in Section 8.4.

Example 8-2. The analysis of the past flood data for a certain reach of river yielded the values of the routing parameter as $\hat{K} = 1.2$ d and $X = 0.3$. Using these data, determine the outflow hydrograph resulting from the following inflow hydrograph to the reach of river.

Time (d)	0	1	2	3	4	5	6	7	8	9	10	11
Inflow (m³/s)	6	18	31	35	33	28	23	19	15	12	9	6

Selection of Δt: $\Delta t > 2\hat{K}X = 0.72$ d . It is convenient to choose $\Delta t = 1$ d.

With $\Delta t = 1$ d, $\hat{K} = 1.2$ d, and $X = 0.3$, the routing coefficients from Eq. (8-13) are: $C_0 = 0.104$, $C_1 = 0.641$, and $C_2 = 0.255$. Check the sum of the coefficients; it is 1, as it should be. The outflow hydrograph is obtained from Eq. (8-12). The routing calculations are shown in Table 8-3. The peak discharge decreases from 35 to 32.9 m³/s.

Table 8-3. Channel Routing

Time d	Inflow m³/s	$C_0 I_2$ m³/s	$C_1 I_1$ m³/s	$C_2 O_1$ m³/s	Outflow O m³/s
0	6	—	—	—	6
1	18	1.9	3.8	1.5	7.2
2	31	3.2	11.5	1.8	16.5
3	35	3.6	19.9	4.2	27.7
4	33	3.4	22.4	7.1	32.9
5	28	2.9	21.2	8.4	32.5
6	23	2.4	17.9	8.3	28.6
7	19	2.0	14.7	7.3	24.0
8	15	1.6	12.2	6.1	19.9
9	12	1.2	9.6	5.1	15.9
10	9	0.9	7.7	4.1	12.7
11	6	0.6	5.8	3.2	9.6

Example 8-3. Using the inflow and outflow hydrographs given in Example 8-2, determine $\hat{K}$ and X for the channel reach.

The procedure is summarized in Table 8-4. The channel storage in column 4 is obtained from Eq. (8-14). The weighted flows $[XI + (1 - X)O]$ in columns 5, 6, and 7 are calculated for different values of X ranging between 0 and 0.5. Each weighted flow is plotted against channel storage in the figure below. The value of X for which the storage versus weighted flow data plots closest to a straight line is taken as the correct value. The value of X from the figure is 0.3. The slope of the straight line in the figure for $X = 0.3$ yields the value of $\hat{K}$ equal to 1.2 d

.

Table 8-4. Calibration of Routing Parameters

1	2	3	4	5	6	7
				Weighted flow (m³/s)		
Time d	Inflow m³/s	Outflow m³/s	Storage (m³/s)-d	$X = 0.2$	$X = 0.3$	$X = 0.4$
0	6	6	0	–	–	–
1	18	7.2	5.4	9.4	10.4	11.5
2	31	16.5	18.1	19.4	20.9	22.3
3	35	27.7	29.0	29.2	29.9	30.6
4	33	32.9	32.7	32.9	32.9	32.9
5	28	32.5	30.5	31.6	31.2	30.7
6	23	28.6	25.5	27.5	26.9	26.4
7	19	24.0	20.2	23.0	22.5	22.0
8	15	19.9	15.3	18.9	18.4	17.9
9	12	15.9	10.9	15.1	14.7	14.3
10	9	12.7	7.1	12.0	11.6	11.2
11	6	9.6	3.5	8.9	8.5	8.2

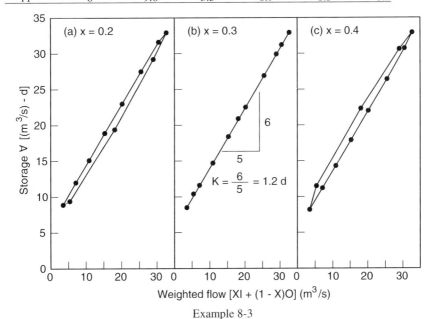

Example 8-3

PROBLEMS

8-7   A flood wave is moving through a reach of a river. When the flow at the upstream
       section of the river reach returned to a steady flow of 300 m^3/s, the flow at the
       downstream section was 600 m^3/s. How much time will it take for the flow at the
       downstream section to reduce to 450 m^3/s? The values of the routing parameters for
       the river reach are $\hat{K}$ = 2.0 days and X = 0.3.

8-8   Rework Example 8-2 with $\hat{K}$ = 0.5 day and X = 0.4.

8-9   Given the inflow and outflow hydrographs on the Iowa River, determine $\hat{K}$ and X
       for the river reach:

Time (h)	0	1	2	3	4	5	6	7	8
Inflow (m^3/s)	70	110	130	194	256	266	222	186	153
Outflow (m^3/s)	70	76	102	129	182	237	254	228	195

Time (h)	9	10	11	12	13	14	15	16	17
Inflow (m^3/s)	117	92	72	70	70	70	70	70	70
Outflow (m^3/s)	161	127	99	79	74	73	72	71	70

8-10   The following is a worked example illustrating the application of the Muskingum
       method. Determine the missing quantities that are indicated by asterisks.

$C_0 = 0.130;$ $C_1 = 0.304;$ $C_2 = *;$ $\Delta t = 24$ h; $\hat{K} = *;$ $X = *$

Time (days)	Inflow (m^3/s)	Outflow (m^3/s)
0	352	352
1	600	*

8-3 Simplification of the Momentum Equation

The momentum equation (7-2) can be written as

$$S_f = S_0 - \frac{\partial y}{\partial x} - \frac{V}{g}\frac{\partial V}{\partial x} - \frac{1}{g}\frac{\partial V}{\partial t} \qquad (8\text{-}15)$$

Through the order-of-magnitude analysis, it is possible to estimate the relative
importance of the various terms in Eq. (8-15). If some terms are small in
comparison to S_0, Eq. (8-15) can be simplified by omitting those terms. Consider a
typical flood in a river (see Figure 8-3) in which the unit discharge increased from
$q_A = 2$ m^3/s/m to $q_B = 20$ m^3/s/m in a day ($\Delta t = 24$ h). Let the bed slope of the river
be steep at about 0.001, i.e., 1.0 m/km and the Chézy coefficient be about 50. The
Chézy formula gives the flow depth and velocity at these two discharges as
$y_A = 1.2$ m, $y_B = 5.4$ m, $V_A = 1.7$ m/s, and $V_B = 3.7$ m/s. It is shown later that the

speed of a flood wave is about $1.5V$. The average speed of the flood wave under consideration is $c = 1.5 \times (1.7 + 3.7)/2 = 4.0$ m/s. The magnitudes of the last three terms in Eq. (8-15) for the assumed flood are

$$\frac{\partial y}{\partial x} \approx \frac{y_B - y_A}{\Delta x} \approx \frac{y_B - y_A}{c \, \Delta t} \approx \frac{5.4 - 1.2}{4.0 \times 24 \times 3600} \approx 0.012 \text{ m/km} \approx 1.2\% \text{ of } S_0$$

$$\frac{V}{g}\frac{\partial V}{\partial x} = \frac{\partial}{\partial x}\left(\frac{V^2}{2g}\right) \approx \frac{V_B^2 - V_A^2}{2g \, \Delta x} \approx \frac{3.7^2 - 1.7^2}{2 \times 9.81 \times 4.0 \times 24 \times 3600} \approx 0.0016 \text{ m/km}$$

$$\approx 0.16\% \text{ of } S_0$$

$$\frac{1}{g}\frac{\partial V}{\partial t} \approx \frac{V_B - V_A}{g \, \Delta t} \approx \frac{3.7 - 1.7}{9.81 \times 24 \times 3600} \approx 0.0024 \text{ m/km} \approx 0.24\% \text{ of } S_0$$

All three terms, the acceleration terms and the $\partial y/\partial x$ term, in Eq. (8-15) for the assumed flood are much smaller than S_0 and can be neglected. In a similar fashion it can be shown that relative magnitudes of the three terms for the same flood in a river with a gentle bed slope of 0.0001, i.e., 0.1 m/km, are

$$\frac{\partial y}{\partial x} = 56\% \text{ of } S_0 ; \quad \frac{V}{g}\frac{\partial V}{\partial x} = 0.7\% \text{ of } S_0 ; \quad \frac{1}{g}\frac{\partial V}{\partial t} = 1\% \text{ of } S_0$$

Only the acceleration terms can be neglected in Eq. (8-15) in the case of a flood in a river with a gentle bed slope.

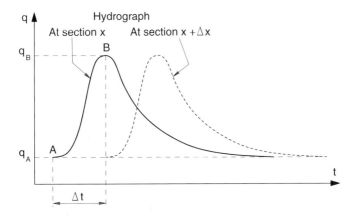

Figure 8-3. Translation of a typical flood hydrograph.

Methods of Flood Routing

Depending upon the number of terms kept in the momentum equation, there are three methods of flood routing in a channel: (i) kinematic-wave, (ii) diffusion-wave, and (iii) dynamic-wave methods. The momentum equations for the three methods are shown below:

$$S_f = S_0 - \frac{\partial y}{\partial x} - \frac{V}{g}\frac{\partial V}{\partial x} - \frac{1}{g}\frac{\partial V}{\partial t} \qquad (8\text{-}16)$$

Kinematic $\longrightarrow|$ $\quad|$ $\qquad\qquad\quad|$
Diffusion $\longrightarrow|$ $\qquad\qquad\qquad|$
Dynamic $\longrightarrow|$

8-4 Kinematic-Wave Method

The momentum equation for the kinematic-wave method reduces to [see Eq. (8-16)]

$$S_f = S_0 \qquad (8\text{-}17)$$

Substitution for S_f from Eq. (4-10) in terms of conveyance into Eq. (8-17) yields

$$Q = K\sqrt{S_0} \qquad (8\text{-}18)$$

According to Eq. (8-18), Q is a function of y (or A) alone; hence

$$\frac{\partial Q}{\partial t} = \frac{dQ}{dA}\frac{\partial A}{\partial t} \qquad (8\text{-}19)$$

The elimination of $\partial A/\partial t$ between the continuity equation (8-1) and Eq. (8-19) leads to the kinematic-wave equation, which can be written as

$$\frac{\partial Q}{\partial t} + c_k\frac{\partial Q}{\partial x} = 0 \qquad (8\text{-}20)$$

in which c_k is the speed of the kinematic wave and is given by

$$c_k = \frac{dQ}{dA} \qquad (8\text{-}21)$$

The speed of the kinematic wave is the same as that of the monoclinal wave given by Eq. (G-3). Note that c_k is positive because Q increases with A. Equation (8-20) can be written as

$$\frac{dQ}{dt} = \frac{\partial Q}{\partial t} + \frac{dx}{dt}\frac{\partial Q}{\partial x} = 0 \qquad (8\text{-}22)$$

if
$$\frac{dx}{dt} = c_k \qquad (8\text{-}23)$$

The integration of Eq. (8-22) yields

$$Q = \text{constant along Eq. (8-23)} \qquad (8\text{-}24)$$

Equation (8-23) defines the characteristics of the first-order partial differential equation (8-20). There is only one set of characteristics; the disturbance propagates only in one direction, i.e., the downstream direction. The value of Q remains constant along the characteristics, i.e., the wave is not damped. However, the slope of the characteristics may vary; the wave then will deform without attenuation. Such a wave is termed a kinematic wave by Lighthill and Whitham (1955).

Equation (8-21) is known as the Klietz-Seddon law and agrees satisfactorily with observed speeds of flood waves in rivers. It can be shown (Prob. 8-11) from Eq. (8-21) that the ratio β of the kinematic-wave velocity to the flow velocity is

$$\beta = \frac{c_k}{V} = \frac{A}{BK}\frac{dK}{dy} \qquad (8\text{-}25)$$

The value of β for three types of channel is tabulated below (Prob. 8-12):

	Ratio c_k/V	
Channel Type	Chézy Eq.	Manning Eq.
Wide rectangular	$1\frac{1}{2}$	$1\frac{2}{3}$
Wide parabolic	$1\frac{1}{3}$	$1\frac{4}{9}$
Triangular	$1\frac{1}{4}$	$1\frac{1}{3}$

Though both the Muskingum method and the kinematic-wave method are based on the continuity equation and the hypothesis of a single-valued stage-discharge relationship, the numerical solution of the former method gives rise to attenuation of the flood peak and the analytical solution of the latter method gives no attenuation. Therefore the flood wave attenuation in the Muskingum method is due to numerical diffusion of the finite-difference scheme used to reproduce the terms of the differential equation of continuity. It can be shown (Cunge, 1969) that a discretization of the temporal derivative in Eq. (8-20) by means of a weighted central difference and of the spatial derivative by means of nonweighted central difference (Figure 8-4) reduces Eq. (8-20) to

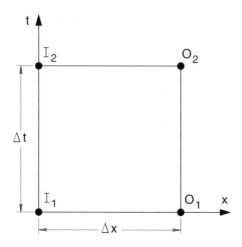

Figure 8-4. Rectangular mesh for space-time discretization.

$$\frac{X(I_2 - I_1) + (1 - X)(O_2 - O_1)}{\Delta t} + c_k \frac{(O_1 - I_1) + (O_2 - I_2)}{2\Delta x} = 0 \qquad (8\text{-}26)$$

where X is the weighting factor that lies between 0 and 0.5. Solving for O_2 and defining $\Delta x/c_k = \hat{K}$, Eq. (8-26) reduces to Eq. (8-13,) which, as shown earlier, yields wave diffusion. Because the analytical solution of Eq. (8-20) shows no attenuation of the flood peak, the kinematic-wave method is not suitable for flood routing in rivers with gentle slopes. However, this method has been applied successfully to overland flow, which is dealt with in Section 8-8.

It is instructive to understand the distinction between kinematic and dynamic waves. Kinematic waves travel only downstream at one value of speed given by Eq. (8-21) or Eq. (8-25), and dynamic waves can propagate both upstream and downstream at velocity $(V - c)$ given by Eq. (7-12) and $(V + c)$ given by Eq. (7-10), respectively. The speed of the kinematic wave in a rectangular channel with constant Chézy C is $c_k = \frac{3}{2}V$. The speed of the dynamic wave moving downstream is $c_d = (V + c) = (V + \sqrt{gy})$. The speed c_d is larger than c_k if $V/2 < \sqrt{gy}$, i.e. $F < 2$, a condition that is applicable to most rivers. The speed of the other dynamic wave is $c_d = (V - c) = (V - \sqrt{gy})$ and is less than c_k. Kinematic-wave motion requires that the terms representing the water-surface slope and local and convective acceleration are negligible, while these terms are important in dynamic-wave motion. Unless these terms are absolutely negligible, both types of wave motion are present in flood wave. The front of a disturbance advances as a positive dynamic wave at speed $c_d = V + c$; the center of the disturbance travels as a kinematic wave at speed c_k; and the rear of the disturbance propagates as a negative dynamic wave at speed $c_d = V - c$. The positive dynamic wave attenuates rapidly if $F < 2$, as described in Section 7-5, and makes little contribution to the passage of the flood wave. The negative dynamic wave propagates upstream if $F < 1$, as is the case in most rivers, and makes significant contribution only to flow with strong backwater conditions and flow reversals. In summary, the propagation of the flood wave in most cases can be represented adequately by the kinematic wave.

PROBLEMS

8-11   Prove Eq. (8-25).

8-12   Show that the ratio of the kinematic-wave velocity to the flow velocity for wide rectangular, parabolic and triangular channels is as shown in the table in Section 8-4. The cross section for these channels can be described by

$$\frac{B}{B_s} = \left(\frac{y}{y_s}\right)^i$$

and i is 0, 1/2, and 1 for rectangular, parabolic, and triangular channels, respectively. Note that for wide channels $P = B$.

8-13   Using the assumption that for a wide rectangular channel the wetted perimeter P is independent of flow area, show that the Manning equation can be written as $Q = \alpha A^\beta$ and the speed of the flood wave is given by $c = \beta V$.

8-5 Diffusion-Wave Method

The momentum equation (8-16) for the diffusion-wave method reduces to

$$S_f = S_0 - \frac{\partial y}{\partial x} \qquad (8\text{-}27)$$

Substitution for S_f from Eq. (4-10) in terms of conveyance K into Eq. (8-27) yields

$$\frac{Q^2}{K^2} = S_0 - \frac{\partial y}{\partial x} \qquad (8\text{-}28)$$

Differentiation of the continuity equation (8-1) with respect to x and Eq. (8-28) with respect to t and the use of $dA = B dy$ give

$$\frac{1}{B}\frac{\partial^2 Q}{\partial x^2} + \frac{\partial^2 y}{\partial x \partial t} = 0$$

$$\frac{2Q}{K^2}\frac{\partial Q}{\partial t} - \frac{2Q^2}{K^3}\frac{dK}{dA}\frac{\partial A}{\partial t} = -\frac{\partial^2 y}{\partial x \partial t}$$

Elimination of $\partial^2 y/\partial x \partial t$ between the above two equations, substitution for $\partial A/\partial t = -\partial Q/\partial x$ from the continuity equation (8-1), and rearrangement of the resulting equation yield

$$\frac{\partial Q}{\partial t} + c_f \frac{\partial Q}{\partial x} = \mu \frac{\partial^2 Q}{\partial x^2} \tag{8-29}$$

where the speed of the diffusion wave, c_f, is

$$c_f = \frac{Q}{K} \frac{dK}{dA} \tag{8-30}$$

and the diffusion coefficient, μ, is

$$\mu = \frac{K^2}{2BQ} \tag{8-31}$$

Equation (8-29) is a classical convection-diffusion (second-order parabolic partial differential) equation. The term on the right-hand side of Eq. (8-29) accounts for attenuation of flood wave. Equation (8-29) for constant μ and c_f has known analytical solutions. Hayami (1951) described the application of Eq. (8-29) to flood-wave problems.

From Eq. (8-30) the ratio β' of the diffusion-wave velocity to the flow velocity is

$$\beta' = \frac{c_f}{V} = \frac{A}{BK} \frac{dK}{dy} \tag{8-32}$$

Note that though Eq. (8-32) is identical to Eq. (8-25), the diffusion-wave velocity c_f is not identical to the kinematic-wave velocity c_k. The flow velocities in the two cases are determined from the different momentum equations, namely, Eqs. (8-17) and (8-27). However, the difference in the two velocities is negligible in most cases and c_f is assumed equal to c_k, i.e., $\beta = \beta'$.

The Muskingum-Cunge Method

Cunge (1969) showed that Eq. (8-13) is also a finite difference representation of the diffusion-wave equation (8-29) when

$$\mu = \left(\tfrac{1}{2} - X\right) c_f \Delta x \tag{8-33}$$

and

$$c_f = \frac{\Delta x}{\hat{K}} \tag{8-34}$$

Instead of the past flood data, Eqs. (8-33) and (8-34) can be used to determine the routing parameters $\hat{K}$ and X. From Eqs. (8-31) and (8-33) the routing parameter X is

$$X = \frac{1}{2}\left(1 - \frac{K^2}{\Delta x c_f BQ}\right) \tag{8-35}$$

and from Eq. (8-34) the routing parameter $\hat{K}$ is

$$\hat{K} = \frac{\Delta x}{c_f} \tag{8-36}$$

where c_f is equal to βV and is given by Eq. (8-25). In general, both $\hat{K}$ and X are functions of Q (or y). In most practical applications they are assumed as constant at some reference discharge Q_0 that is taken as either the average or the peak discharge. Equations (8-35) and (8-36) at the reference discharge can be written as

$$X = \frac{1}{2}\left(1 - \frac{Q_0}{\beta V_0 S_0 B_0 \Delta x}\right) \tag{8-37}$$

$$\hat{K} = \frac{\Delta x}{\beta V_0} \tag{8-38}$$

where the subscript 0 refers to the conditions at the reference discharge (see Example 8-4). It should be pointed out that the Muskinkum-Cunge method does not take into account any downstream influence such as backwater effect of river junctions.

Example 8-4. Determine the routing parameters of the Muskingum-Cunge method for the following flood and channel characteristics: Reference discharge $Q_0 = 48.0$ m^3/s; channel is rectangular in cross section of width 12 m and has a bottom slope of 0.0002; Manning $n = 0.012$; and reach length $\Delta x = 10$ km.

$$\frac{nQ_0}{b^{8/3}S_0} = \frac{0.012 \times 48}{12^{8/3}\sqrt{0.0002}} = 0.054$$

Upon entering Table D-1 with the above value and $m = 0$, one obtains

$$\frac{y_0}{b} = 0.198; \quad \text{whence} \quad y_0 = 0.198 \times 12 = 2.38 \text{ m}$$

$$V_0 = \frac{Q_0}{by_0} = \frac{48}{12 \times 2.38} = 1.68 \text{ m/s}$$

$$K = \frac{Q_0}{\sqrt{S_0}} = \frac{48}{\sqrt{0.0002}} = 3394.1 \text{ m}^3/\text{s}$$

$$\frac{dK}{dy} = \frac{d}{dy}\left(\frac{1}{n}AR^{2/3}\right) = \frac{d}{dy}\left[\frac{1}{n}\frac{(by)^{5/3}}{(b+2y)^{2/3}}\right] = \frac{b^{5/3}}{n}\left[\frac{(5/3)y^{2/3}}{(b+2y)^{2/3}} - \frac{(4/3)y^{5/3}}{(b+2y)^{5/3}}\right]$$

$$= \frac{12^{5/3}}{0.012}\left[\frac{5\times2.38^{2/3}}{3(12+2\times2.38)^{2/3}} - \frac{4\times2.38^{5/3}}{3(12+2\times2.38)^{5/3}}\right] = 2107.7 \text{ m}^2/\text{s}$$

From Eq. (8-25), $\quad \beta = \frac{A}{BK}\frac{dK}{dy} = \frac{12\times2.38}{12\times3394.1}\times2107.7 = 1.48$

The routing parameters from Eqs. (8-37) and (8-38) are

$$X = \frac{1}{2}\left(1 - \frac{Q_0}{\beta V_0 S_0 B_0 \Delta x}\right) = \frac{1}{2}\left(1 - \frac{48}{1.48\times1.68\times0.0002\times12\times10,000}\right) = 0.10$$

$$\hat{K} = \frac{\Delta x}{\beta V_0} = \frac{10\times1000}{1.57\times1.68} = 3791.3 \text{ s} = 1.05 \text{ h}$$

PROBLEMS

8-14 Calculate the outflow hydograph for the following inflow hydrograph and the channel data:

Time (h)	0	1	2	3	4	5	6	7	8	9
Inflow (m^3/s)	9	13	21	26	22	17	14	12	10	9

Bed slope = 0.0015; flow area at the peak discharge = 38 m^2; top width at the peak discharge = 15 m; Δx = 3 km; and β = 3/2.

8-15 A stretch of river 50 km long has a slope S_0 = 0.0003. Prior to the arrival of a flood wave, the river carries a steady flow of 1100 m^3/s with a mean surface width of 1.5 km. It is observed that that the flood wave requires 30 h to pass the upper end at which a maximum change in discharge of 300 m^3/s takes place. If it takes 24 h for the crest to reach the lower end, estimate the reduction of the peak flow.

8-16 Find the speed of the crest of a flood wave for a maximum flood discharge of 1100 m^3/s in a rectangular channel of 60 m in width. The bed slope of the channel is 0.0005 and the Manning n is 0.02.

8-17 Uniform, steady flow at depth of 2 m is occurring in a rectangular channel of width 15 m and longitudinal slope of 1:5000. The Manning coefficient for the channel is 0.012. If the Muskingum-Cunge method were to be used to compute the propagation of a flood wave, compute the coefficients C_0, C_1, and C_2 in Eq. (8-13) for Δx = 10 km and Δt = 1 h.

8-18 Determine the routing parameters of the Muskingum-Cunge method for the following flood and channel characteristics: peak discharge = 1000 m^3/s; channel bottom slope = 0.00087; water-surface width at the peak discharge = 100 m; flow area at the peak discharge = 400 m^2; reach length = 12.5 km; and parameter β = 1.6.

8-6 Dynamic-Wave Method

The complete momentum equation is used in the dynamic-wave method. This method is normally used where the backwater condition and the flow reversal exist. The complete momentum equation along with the continuity equation can be solved only by numerical methods. A number of numerical schemes are available for solving partial differential equations, each with specific advantages in terms of stability, convergence, accuracy, and efficiency. The numerical methods can be classified as the *method of characteristics, finite-difference method,* and *finite-element method.* In the method of characteristics each partial differential equation is transformed into two ordinary differential equations, which in turn are solved by the finite-difference method. In the finite-difference method, the partial differential equations are transformed into algebraic equations by replacing the partial derivatives with finite differences, and the resulting algebraic equations are then solved numerically. The finite-difference method is further classified into explicit and implicit methods. In the explicit method, the dependent variables in the finite-difference algebraic equations occur explicitly and can be determined at the end of each time step. In the implicit method, the dependent variables occur explicitly. In the finite-element method, the system is divided into a number of elements and the partial differential equations are integrated at the nodal points of the elements.

Presented herein is a simple method known as the *explicit finite-difference method.* This method is elected to illustrate some of the basic ideas of finite-difference methods. The main concept behind these methods is rewriting the partial differential equations in finite-difference form by using a fixed rectangular network in the *x-t* plane, such as shown in Figure 8-5. The method is based on the assumption that the values of the dependent variables V and y are known at the mesh points L and R. The governing equations are used to calculate the values of V and y at point P. Finite-difference approximations for the terms in Eqs. (7-1) and (7-2) can be written as

$$\left(\frac{\partial V}{\partial x}\right)_M = \frac{V_R - V_L}{2\Delta x} \tag{8-39a}$$

$$\left(\frac{\partial V}{\partial t}\right)_P = \frac{V_P - V_M}{\Delta t} \tag{8-39b}$$

$$\left(\frac{\partial y}{\partial x}\right)_M = \frac{y_R - y_L}{2\Delta x} \tag{8-39c}$$

$$\left(\frac{\partial y}{\partial t}\right)_P = \frac{y_P - y_M}{\Delta t} \tag{8-39d}$$

Substitution of the above equation in the continuity equation (7-1) results in

$$\frac{y_P - y_M}{\Delta t} + y_M \frac{V_R - V_L}{2\Delta x} + V_M \frac{y_P - y_M}{2\Delta x} = 0 \tag{8-40}$$

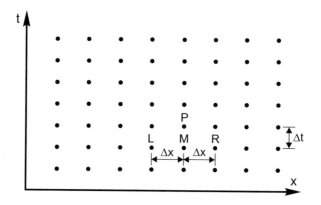

Figure 8-5. Finite-difference mesh.

In deriving Eq. (8-40) the cross section of the channel is assumed rectangular so that hydraulic depth D is equal to depth y. Substitution of Eq. (8-39) in the momentum equation (7-2) yields

$$\frac{V_P - V_M}{\Delta t} + V_M \frac{V_R - V_L}{2\Delta x} + g \frac{y_R - y_L}{2\Delta x} = g\left(S_0 - S_f\right)_M \qquad (8\text{-}41)$$

Equations (8-40) and (8-41) are algebraic equations in two unknown quantities, namely, V_P and y_P, and can be solved to determine these unknowns. By proceeding similarly, V and y can be determined at other mesh points

Different finite-difference methods, including the accuracy and stability of their solutions and implementation of initial and boundary conditions, can be found in books on computational hydraulics (Abott, 1979; Cunge et al., 1980).

8-7 Rating Curves

Instruments are set up on most rivers to record the variation of water level with time. These records must be converted into discharge-time records for flood routing. This conversion is accomplished by the "Jones formula", which is derived on the assumption that the acceleration terms in the momentum equation are negligible. The momentum equation is then given by Eq. (8-27), which can be written as

$$Q = K\sqrt{S_0 - \frac{\partial y}{\partial x}} \qquad (8\text{-}42)$$

Because the $\partial y/\partial x$ is positive on the falling stage of a flood and is negative on the rising stage, the discharge for a given depth is greater on the rising stage than the falling stage, which in turn results in a loop-rating curve as shown in Figure 8-6.

If Q_n is the discharge based on uniform flow, then

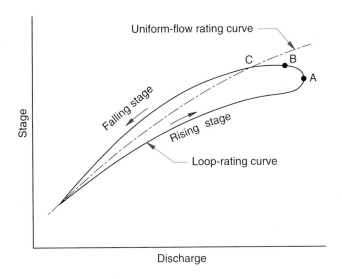

Figure 8-6. The loop-rating curve.

$$Q_n = K\sqrt{S_0} \qquad (8\text{-}43)$$

Elimination of K between Eqs. (8-42) and (8-43) gives

$$\frac{Q}{Q_n} = \sqrt{1 - \frac{1}{S_0}\frac{\partial y}{\partial x}} \qquad (8\text{-}44)$$

The term $\partial y/\partial x$ can be written in terms of $\partial y/\partial t$ on neglecting the subsidence of a flood wave. Elimination of Q between Eqs. (8-1) and (8-18) and the use of Eq. (8-21) give

$$\frac{\partial y}{\partial t} + c_k\frac{\partial y}{\partial x} = 0 \qquad (8\text{-}45)$$

Substitution for $\partial y/\partial x$ from Eq. (8-45) into Eq. (8-44) yields the Jones formula as

$$\frac{Q}{Q_n} = \sqrt{1 + \frac{1}{c_k S_0}\frac{\partial y}{\partial t}} \qquad (8\text{-}46)$$

Three points A, B, and C are identified on the loop-rating curve in Figure 8-6. The discharge is the maximum at point A, and the stage is the maximum at point B. The point C is at the intersection of the loop -rating curve with the uniform-flow rating curve along which $\partial y/\partial x$ is zero. At any instance $\partial y/\partial x$ is zero at the wave crest; point C therefore represents the wave crest. An observer on a bank sees first the point of maximum discharge, then the point of maximum stage, and then the wave crest.

PROBLEMS

8-19   Locate the relative positions of the three points A, B, and C shown in Figure 8-6 on curves representing (i) discharge vs. time, (ii) stage vs. time, and (iii) stage vs. distance along the channel.

8-20   Explain why for a given flow depth the discharge is greater on the rising stage of a flood than on the falling stage.

8-21   (a) Show that $dy/dt = \partial y/\partial t$ at point C in Figure 8-6. Note that $\partial y/\partial x$ is zero at point C.

(b) Using Eq. (8-27) and the Chézy's equation for the friction slope, derive an expression for $\partial Q/\partial x$ at point C in Figure 8-6 for flow in a rectangular channel.

(c) Using the continuity equation (8-25) for the speed of a flood wave and the relations derived above in (a) and (b), show that the rate of subsidence of the wave crest is given by

$$\frac{dy}{dx} = \frac{y_0}{3S_0}\frac{\partial^2 y}{\partial x^2}$$

where y_0 is the depth at the wave crest.

8-8 Overland Flow

Related to floods is the problem of surface runoff. The process of surface runoff from rainfall contains three elements: (a) overland flow as a thin sheet of water; (b) small stream flow with continuous overland flow as input; and (c) river flow without continuous lateral inflow. Traditionally, the runoff problem has been analyzed by the unit hydrograph theory, which is based on an assumption of linearity. Recently, the kinematic-wave theory has been used successfully to determine runoff hydrographs of overland flow (Henderson and Wooding, 1964; Wooding, 1965, 1966) from natural catchments. The catchment is divided in to a number of subcatchments, and each subcatchment is represented by a plane surface. The excess rainfall over the subcatchment flows down the plane to a small stream at the bottom in the form of a thin sheet of water, as shown in Figure 8-7. The overland flow can be analyzed by the equations of continuity and momentum. The continuity equation (1-9) can be written as

$$\frac{\partial y}{\partial t} + \frac{\partial q}{\partial t} = i_e \qquad (8-47)$$

where q is the unit discharge and i_e is the intensity of rainfall excess and has dimensions of LT^{-1}. Due to steep slopes in the overland flow region, all terms in

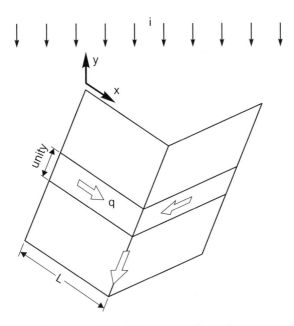

Figure 8-7. Rainfall over a subcatchment.

the momentum equation are small in comparison to the friction and slope terms. The momentum equation reduces to Eq. (8-17); therefore the discharge is a function of depth alone. Using the Chézy formula (2-4), the momentum equation (8-17) can be written as

$$q = C\sqrt{yS_0}$$

where C is the Chézy coefficient that is a function of the Reynolds number and relative roughness. The above equation can be written as

$$q = \alpha y^m \qquad (8\text{-}48)$$

where the coefficient α and the exponent m have different values for laminar and turbulent flows. For laminar flow $C = 4/R = 4v/Vy$, which yields $\alpha = gS_0/2v$ and $m = 3$. For turbulent flow C is given by the Manning equation (2-10), which yields $\alpha = K_n\sqrt{S_0}/n$ and $m = \frac{5}{3}$. Field data indicate that $m \approx 2$.

Equations (8-47) and (8-48) will be solved by the method of characteristics for the following initial and boundary conditions:

$$y = 0\begin{cases} 0 \leq x \leq L, & t=0 \\ x=0, & t>0 \end{cases} \qquad (8\text{-}49)$$

where x and L are defined in Figure 8-7. Elimination of q between Eqs. (8-47) and (8-48) yields

$$\frac{\partial y}{\partial t} + m\alpha y^{m-1}\frac{\partial y}{\partial x} = i_e \tag{8-50}$$

The above equation can be written as

$$\frac{dy}{dt} = i_e \tag{8-51}$$

where

$$\frac{dy}{dt} = \frac{\partial y}{\partial t} + \frac{dx}{dt}\frac{\partial y}{\partial x}$$

and

$$\frac{dx}{dt} = c = m\alpha y^{m-1} \tag{8-52}$$

The partial differential equation (8-50) reduces to the ordinary differential equation (8-51), which is valid only along the path given by Eq. (8-52). The path is known as a characteristic. The parameter c in Eq. (8-52) is the celerity of a small disturbance along the characteristics. By integrating Eq. (8-52), the characteristics are derived as

$$x - x_0 = m\alpha \int_{t_0}^{t} y^{m-1}\,dt \tag{8-53}$$

where y is obtained by integrating Eq. (8-51) as

$$y - y_0 = \int_{t_0}^{t} i_e\,dt \tag{8-54}$$

and y_0 is the depth at point (x_0, t_0) along the characteristics. Equation (8-54) will be solved for constant rainfall excess such that

$$i_e(t) = \begin{cases} i_* = \text{constant}, & 0 \le t \le t_r\,, \ x \ge 0 \\ 0 & , \quad t < 0, t > t_r\,, \ x < 0 \end{cases} \tag{8-55}$$

where t_r is the duration of the rainfall excess. Substitution for i_e from Eq. (8-55) into Eq. (8-54) yields

$$y - y_0 = i_*\left(t - t_0\right), \qquad t \le t_r \tag{8-56}$$

Substitution for y from Eq. (8-56) into Eq. (8-53) gives the following equation for the characteristics:

$$x - x_0 = m\alpha \int_{t_0}^{t} \left[i_*\left(t - t_0\right) + y_0\right]^{m-1} dt \tag{8-57}$$

Using Eq. (8-57), three characteristics are drawn in Figure 8-8a. These characteristics are termed the j-characteristic, the limiting characteristic, and the k-characteristic. The j-characteristic CD initiates from point C$(x_j,0)$ on the x-axis where $t_0 = 0$ and $y_0 = 0$. From Eq. (8-57) the j-characteristic is given by

$$x - x_j = \alpha\, i_*^{m-1} t^m \tag{8-58}$$

The k-characteristic EF initiates from point E$(0,t_k)$ on the t-axis where $x_0 = 0$ and $y_0 = 0$. The k-characteristic from Eq. (8-57) is given by

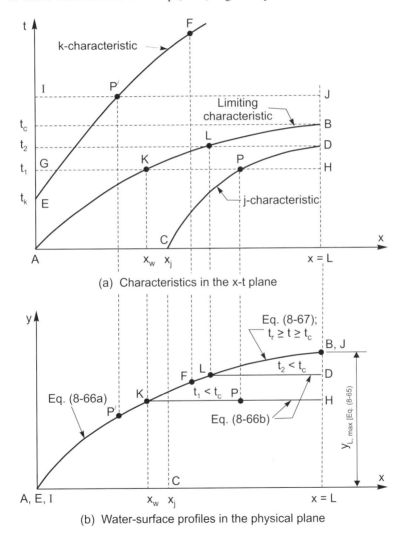

(a) Characteristics in the x-t plane

(b) Water-surface profiles in the physical plane

Figure 8-8. Characteristic diagram and water-surface profile.

$$x = \alpha \, i_*^{m-1} \left(t - t_k\right)^m \tag{8-59}$$

The limiting characteristic AB initiates from the origin where $x_0 = 0$, $t_0 = 0$, and $y_0 = 0$; it is given by

$$x = \alpha \, i_*^{m-1} t^m \tag{8-60}$$

The limiting characteristic reaches the downstream end of the subcatchment $(x = L)$ at $t = t_c$. By substituting $x = L$ in Eq. (8-60), an expression for t_c can be obtained as

$$t_c = \left(\frac{L \, i_*^{1-m}}{\alpha}\right)^{1/m} \tag{8-61}$$

The time t_c is referred to as the *time of concentration* and is equal to the time taken by a small disturbance to travel along the limiting characteristic from $x = 0$ to $x = L$. The flow condition at any point in the x-t plane can be determined as follows. Using Eq. (8-57) draw the characteristic through the point under consideration, and using Eq. (8-54) or (8-56) determine the depth at the point. The discharge at the point is determined from Eq. (8-48). The discharge hydrograph at the subcatchmant outfall $(x = L)$ is derived for a range of the duration of the rainfall excess.

Case A: $t_r = \infty$. First consider a storm of an infinite duration of rainfall excess. The region in the x-t plane in Figure 8-8a consists of two subregions: one below the limiting characteristic where $t_0 = 0$ and $y_0 = 0$ and the other above the limiting characteristic where $x_0 = 0$ and $y_0 = 0$. Consider two points in the x-t plane: point $P(x,t)$ below the limiting characteristic and point $P'(x,t)$ above the limiting characteristic. Draw the j-characteristic CD through point P and the k-characteristic EF through point P'. The j-characteristic CD initiates from point C where $y_0 = 0$ and $t_0 = 0$. Depth at point P on the j-characteristics is given from Eq. (8-56) as

$$y = i_* t \tag{8-62}$$

The k-characteristic EF initiates from point E where $y_0 = 0$ and $t_0 = t_k$. Depth at point P' on the k-characteristic from Eq. (8-56) is

$$y = i_* \left(t - t_k\right) \tag{8-63}$$

From Eq. (8-59) for the k-characteristic

$$\left(t - t_k\right) = \left(\frac{x \, i_*^{1-m}}{\alpha}\right)^{1/m} \tag{8-64}$$

Substitution for $(t - t_k)$ from Eq. (8-64) into Eq. (8-63) yields the depth at point P' as

$$y = i_* \left(\frac{xi_*^{1-m}}{\alpha} \right)^{1/m} = \left(\frac{xi_*}{\alpha} \right)^{1/m} \tag{8-65}$$

Examine the variation of y with x at two different times: one at time $t \le t_c$ along the horizontal line GH through point P and the other at time $t > t_c$ along the horizontal line IJ through point P'. Line GH intersects the limiting characteristic at K where $x = x_w$, which can be determined from Eq. (8-60). The points along GK ($x \le x_w$) lie above the limiting characteristic, and the depth there is given by Eq. (8-65). Note that the depth along GK increases as x increases from 0 to x_w. The points along KH ($x \ge x_w$) lie below the limiting characteristic, and the depth there is given by Eq. (8-62). Note that the depth along KH is constant as t is constant along KH. Depth at $t \le t_c$ is therefore given by

$$y = \begin{cases} \left(\dfrac{xi_*}{\alpha} \right)^{1/m}, & x_w \ge x \ge 0, \quad t \le t_c \quad \text{(a)} \\ i_* t, & x \ge x_w, \quad t \le t_c \quad \text{(b)} \end{cases} \tag{8-66}$$

Characteristic EF intersects Line IJ at point P' where the depth is given by Eq. (8-65). Depth at $t \ge t_c$ is therefore given by

$$y = \left(\frac{xi_*}{\alpha} \right)^{1/m}, \qquad L \ge x \ge 0, \quad t \ge t_c \tag{8-67}$$

Plots of Eqs. (8-66) and (8-67) are shown in Figure 8-8b. The curve AKH represents the water-surface profile at $t < t_c$ during the storm. Curve AB is the limiting, steady-state water-surface profile which is reached at $t = t_c$. The points in the x-y plane corresponding to the points in the x-t plane can easily be located as they are represented by the same letters in both planes. From Eq. (8-66b) depth at $x = x_w = L$ increases linearly with time from 0 to $i_* t_c$. The maximum depth at $x = L$ is $i_* t_c$, which upon substituting for t_c from Eq. (8-61) becomes

$$y_{L,max} = i_* t_c = \left(\frac{Li_*}{\alpha} \right)^{1/m} \tag{8-68}$$

where $y_{L,max}$ is the maximum depth at $x = L$. From Eq. (8-48) the discharge hydrograph for the subcatchment at $x = L$ is given by

$$q_L = \alpha y_L^m, \qquad y_L = \begin{cases} i_* t, & t \le t_c \quad \text{(a)} \\ \left(\dfrac{Li_*}{\alpha} \right)^{1/m}, & t \ge t_c \quad \text{(b)} \end{cases} \tag{8-69}$$

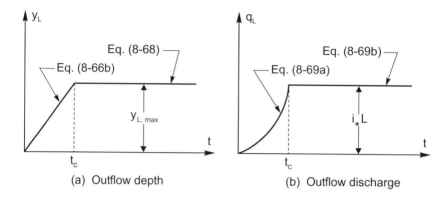

Figure 8-9. Outflow depth and discharge for $t_r = \infty$.

The variations with time of depth and discharge at $x = L$ are shown in Figure 8-9. The depth inreases linearly for $0 \le t \le t_c$ and thereafter remains constant for $t \ge t_c$. Similarly the discharge increases until $t = t_c$ and later on stays constant. The maximum value of the outflow must equal the inflow, i.e., $q_{L,\max} = i_* L$.

Case B: $t_c \le t_r < \infty$. The solution given above is valid up to $t = t_r$ when the rainfall excess stops. It is expected that depth at a point should decrease after the storm. Examine the flow condition at point $P(x,t; t > t_r)$ in the x-t plane in Figure 8-10a. Draw the characteristic CP through point P. Let the point on characteristic CP at $t = t_r$ be $D(x_{wr}, t_r)$. From Eq. (8-67) the location of point D along the x-axis is given by

$$x_{wr} = \frac{\alpha y^m}{i_*} \tag{8-70}$$

After the rainfall excess stops, the variation of depth with time along a characteristic from Eq. (8-51) is given as

$$\frac{dy}{dt} = 0, \qquad t > t_r \tag{8-71}$$

According to Eq. (8-71) depth, and hence discharge, along a characteristic remains constant. In other words the depth at point P is the same as at point D. Point P is Δx distance downstream of point D. The distance Δx is determined as follows. Because depth is constant along DP, the slope of the characteristic from Eq. (8-52) is constant, i.e.,

$$\frac{dx}{dt} = m\alpha y^{m-1} = \text{constant}$$

The characteristic is a straight line beyond point D. From the slope of the straight line the distance Δx is

$$\Delta x = m\alpha y^{m-1}\left(t - t_r\right) \tag{8-72}$$

Using Eqs. (8-70) and (8-72), the locus of constant depth for $t > t_r$ is given by

$$x = x_{wr} + \Delta x = \frac{\alpha y^m}{i_*} + m\alpha y^{m-1}\left(t - t_r\right) = \alpha y^{m-1}\left[yi_*^{-1} + m\left(t - t_r\right)\right] \tag{8-73}$$

The water-surface profile at $t = t_r \geq t_c$ is given by Eq. (8-67) and is shown as curve AB in Figure 8-10b. The water-surface profile at $t > t_r$ is given by Eq. (8-73) and is

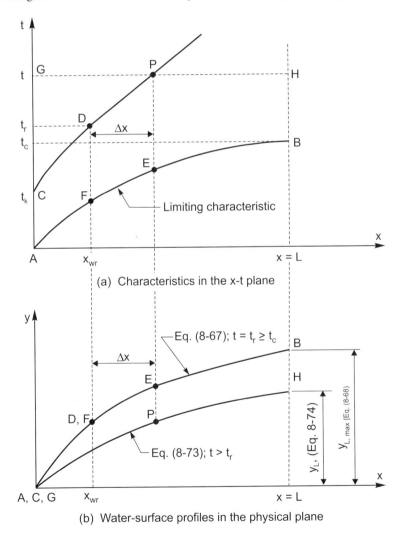

(a) Characteristics in the x-t plane

(b) Water-surface profiles in the physical plane

Figure 8-10. Characteristic diagram and water-surface profile.

shown as curve AH in Figure 8-10b. Note that point D in duration $(t - t_r)$ moves to point P. For $t > t_r$ the depth at $x = L$ from Eq. (8-73) is given by the implicit relation

$$L = \alpha y_L^{m-1} \left[y_L i_*^{-1} + m(t - t_r) \right] \qquad (8-74)$$

It can be seen from Eq. (8-74) that as y_L approaches zero, t approaches to infinity as long as $m > 1$. From Eq. (8-48) the discharge hydrograph at $x = L$ is given by

$$q_L = \alpha y_L^m \qquad y_L = \begin{cases} i_* t \ , & t_r \geq t_c \geq t \geq 0 & (a) \\ \left(\dfrac{L i_*}{\alpha} \right)^{1/m} , & t_r \geq t \geq t_c & (b) \\ \text{Eq.}(8-74), & t > t_r & (c) \end{cases} \qquad (8-75)$$

The variations with time of depth and discharge at $x = L$ are shown in Figure 8-11. The depth increases linearly from 0 to $y_{L,\max}$ between $t = 0$ and $t = t_c$, stays constant at $y_{L,\max}$ between $t = t_c$ and $t = t_r$, and decreases asymptotically to zero for $t > t_r$. Similarly the discharge first increases, remains constant at the maximum value, and then decreases with increasing time.

   Case C: $t_r < t_c$. The rainfall excess ceases before the initial disturbance has reached the subcatchment outfall at $x = L$. The water-surface profile at $t = t_r < t_c$ is given by curve ACD in Figure 8-12b (or curve AKH in Figure 8-8b). From Eq. (8-66b) the end depth $y_{Lr} = i_* t_r$. At $t = t_r$ the initial disturbance has reached point C along the limiting characteristic AB. As shown earlier, the characteristics beyond t_r are straight lines, such as characteristic CE passing through point C, because the rainfall excess is zero for $t > t_r$. Examine the variation of depth with x at time $t > t_r$, i.e., along line FG in Figure 8-12a. The line FG intersects the characteristic AE at point L. Consider two points P and P′ on line FG. Draw characteristic HI through point P′ and characteristic JK through point P.

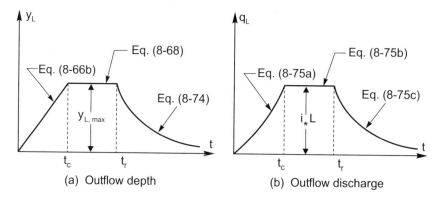

(a) Outflow depth   (b) Outflow discharge

Figure 8-11. Outflow depth and discharge for $t_c < t_r < \infty$.

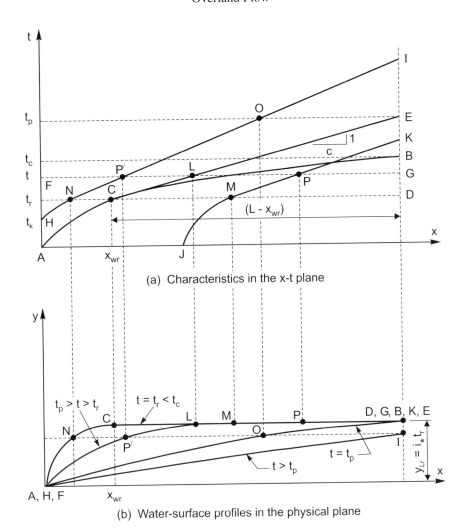

(a) Characteristics in the x-t plane

(b) Water-surface profiles in the physical plane

Figure 8-12. Characteristic diagram and water-surface profiles.

Characteristics HI and JK intersect the horizontal line $t = t_r$ at points N and M, respectively. Characteristics HI between points N and I and characteristic JK between points M and K are straight lines. The depth along the straight portion of the characteristics remains constant. It means that depths at points N, P′, O, and I, points C, L, and E, and points M, P, and K are the same. During $(t - t_r)$ point N moves to point P′, point C to point L, and point M to point P. The water-surface profile at $t > t_r$ is given by curve AP′LG. The end depth y_{Lr} remains constant until points C and E coincide at $t = t_p$. From the geometry of characteristic CE in Figure 8-12a, the time t_p is given by

$$t_p = t_r + \frac{L - x_{wr}}{c} = t_r + \frac{L - \alpha y_{Lr}^m / i_*}{\alpha m y_{Lr}^{m-1}} = t_r + \frac{t_c^* - t_r}{m} \tag{8-76}$$

in which

$$t_c^* = \frac{L}{\alpha y_{Lr}^{m-1}} \tag{8-77}$$

For the time $t > t_p$, the end depth is given by Eq. (8-74). From Eq. (8-48) the discharge hydrograph at $x = L$ is given by

$$q_L = \alpha y_L^m, \qquad y_L = \begin{cases} i_* t, & 0 < t \le t_r < t_c & \text{(a)} \\ i_* t_r, & t_c > t_r < t \le t_p & \text{(b)} \\ \text{Eq.}(8-74), & t > t_p & \text{(c)} \end{cases} \tag{8-78}$$

The variations with time of depth and discharge at $x = L$ are shown in Figure 8-13.

The continuity and momentum equations for flow condition in small streams are similar to Eqs. (8-47) and (8-48) in form and can be solved by the method of characteristics, as explained above. Details of the analysis are given by Wooding (1965, 1966) and Eagleson (1970).

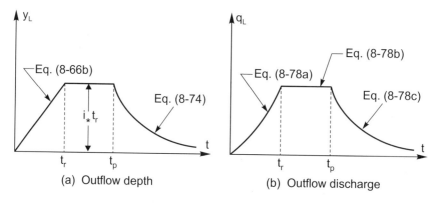

Figure 8-13. Outflow depth and discharge for $t_r < t_c$.

PROBLEMS

8-22   Rainfall excess over a uniform, plane catchment of length L is given by

$$i_e = \begin{cases} t, & t > 0 \\ 0, & t \le 0 \end{cases}$$

Show that for a depth-discharge relation $q = \alpha y^2$, the time of concentration is given by

$$t_c = (3L/\alpha)^{1/3}$$

8-23 Consider a uniform, plane subcatchment with the following parameters: $m = 5/3$, $\alpha = 4.5$ s^{-1}, and $L = 100$ m. The subcatchment is subjected uniformly to the following rainfall, all of which runs off:

Time t (min)	Rainfall i_e (cm/h)
$0 \le t \le 5$	4
$5 < t \le 10$	9
$t > 10$	0

Calculate and plot the outflow hydrograph. Assume that two storms can be superposed linearly, each beginning with a dry bed.

8-24 Show that t_p in Eq. (8-76) is equal to $5t_c/3$ for $t_r = t_c/3$ and $m = 2$.

8-25 A basin is subjected uniformly to a precipitation, all of which runs off. For the following data determine the outflow discharge 20 minutes after the rainfall stops: $i_* = 5$ cm/h, $t_r = 15$ min, $L = 30$ m, $m = 2$, and $\alpha = 5$ s^{-1}.

CHAPTER 9

ARTIFICIAL CHANNEL CONTROLS

An artificial channel control (ACC), as defined earlier in Chapter 3, is a man-made feature in channels, such as a weir, spillway, gate, or venturi flume. These structures produce a "large enough" disturbance to "choke" the flow; consequently the flow upstream and downstream of such structures is subcritical and supercritical, respectively. Relations between discharge and depth in the neighborhood of the artificial channel controls are presented in this chapter. These relations serve as a boundary condition in analyzing flow profiles and also are used to determine discharge from a known depth. The ACCs in the latter capacity serve as a flow-measuring device. The flow in the vicinity of ACCs is rapidly varied flow wherein the pressure distribution is nonhydrostatic and the one-dimensional governing equation is not valid. Recently some problems of rapidly varied flow have been solved numerically by using two-dimensional governing equations (Chaudhry, 1993). Traditionally rapidly varied flow has been studied by conducting physical model studies in laboratories, and empirical relations between depth and discharge have been obtained.

9-1 Weirs, Sills, and Overfalls

Sharp-Crested Weirs

A weir may be defined as any regular obstruction built across a channel over which the flow takes place. A weir may be classified according to

    (a) shape of the opening: rectangular, triangular, trapezoidal, etc.;
    (b) shape of the edge: sharp crested, broad crested;
    (c) discharge condition: free or submerged; and
    (d) ends condition: contracted or suppressed.

A sharp-crested weir consists of a vertical plate mounted perpendicular to the flow direction. The top of the plate has a sharp, beveled edge that clearly demarcates the line of separation of the nappe from the plate, as shown in Figure 9-1. The water surface curves rapidly in the vicinity of the weir. The lower surface of the nappe initially goes up above the crest before plunging down. The pressure distribution within the nappe approaches zero well past the weir in a section at point E.

The true section of minimum energy is that at which the sloping energy line reaches its minimum elevation above the lower boundary of the flow. The true critical section lies at point D of the free nappe in Figure 9-1 at which the lower surface has attained its maximum elevation. However, neither is the discharge a constant function of this true critical depth, since it may vary with change in the geometry of the weir, nor is it yet possible to determine analytically the form of

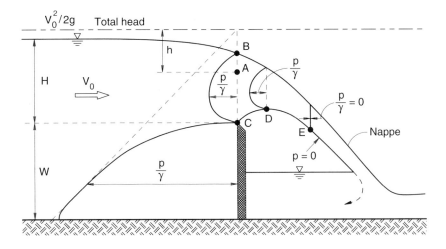

Figure 9-1. Flow over a sharp-crested weir.

this function.

The shape of the complex flow field, and consequently the relationship between the discharge and head over the weir, cannot be analyzed analytically. The derivation of a formula for weirs requires major simplifications of the problem. Such a formula will include an experimental coefficient. The derivation of the discharge relation is based on the following simplified assumptions: (i) the velocity distribution of the approach flow is uniform; (ii) the pressure under the nappe is atmospheric; (iii) the streamlines are horizontal over the crest so that the pressure within the nappe is zero; and (iv) the effect of viscosity is negligible. The relation is first derived for a rectangular weir that occupies the full width of the channel. Such a weir, called a suppressed weir, does not have lateral contraction effects and the lengths of the nappe and weir are equal. On the basis of these assumptions the velocity at any point such as A is $\sqrt{2gh}$, where h is the vertical distance of point A below the energy-grade line. The discharge q per unit length of the crest is

$$q = \int_{V_0^2/2g}^{H+V_0^2/2g} \sqrt{2gh}\,dh = \frac{2}{3}\sqrt{2g}\left[\left(H+\frac{V_0^2}{2g}\right)^{3/2} - \left(\frac{V_0^2}{2g}\right)^{3/2}\right] \tag{9-1}$$

where V_0 is the approach flow velocity and H is the head over the weir of height W. The actual discharge over the weir will be different from that given by Eq. (9-1) due to the simplified assumptions made in its derivation. The actual discharge is obtained by multiplying the theoretical discharge by a contraction coefficient C_c. The effects of the contraction coefficient and the approach velocity can be integrated in a single coefficient, called discharge coefficient C_d, leading to the final expression for the discharge as

$$q = \frac{2}{3} C_d \sqrt{2g} H^{3/2} \tag{9-2}$$

where C_d is the coefficient of discharge for the weir and it accounts for the simplifications and the velocity of approach in Eq. (9-1). The coefficient of discharge is determined experimentally. It can be shown by means of dimensional analysis that the coefficient of discharge is a function of H/W and Reynolds and Weber numbers, which respectively accounts for viscous and surface tension effects. The effect on C_d of Reynolds and Weber numbers is usually negligible, except at very low heads when viscosity and surface tension effects may be large. Rehbock (1929) proposed the following expression for the coefficient of discharge:

$$C_d = 0.611 + 0.075 \frac{H}{W} + \frac{0.36}{H\sqrt{\rho g/\sigma} - 1} \tag{9-3}$$

where σ is the surface tension of liquid. Equation (9-3) clearly indicates the influence of H/W on C_d. Also for small values of H, the last term in Eq. (9-3) becomes significant. With increasing H, C_d first decreases, reaches a minimum, and then increases. The effect of surface tension in Eq. (9-3) can be neglected for H greater than the head H^* corresponding to the minimum value of C_d. Differentiating C_d with H and equating it to zero, one obtains an expression for H^* as

$$H^* = \sqrt{\frac{\sigma}{\rho g}} + 2.19 \left(\frac{W^2 \sigma}{\rho g} \right)^{1/4} \tag{9-4}$$

Thus Eq. (9-3) for $H > H^*$ reduces to

$$C_d = 0.611 + 0.075 \frac{H}{W} \tag{9-5}$$

It should be pointed out that the above equation for C_d is applicable only if the pressure under the nappe is atmospheric. In suppressed weirs the air under the nappe has no contact with the atmosphere. The trapped air, unless ventilated, is carried away by the flowing water; thus creating a negative pressure under the nappe. The nappe is deflected towards the weir, and the discharge over the weir is more than given by the above-mentioned formulas. In addition the tailwater level should not drown the portion of the nappe where the pressure within the nappe is above zero, i.e., it should be below point E in Figure 9-1; such a weir is termed free-flow weir. If the tailwater level rises above point E, the weir is called a submerged weir. Villemonte (1947) suggested the following equation for discharge over a submerged weir:

$$\frac{q}{q_f} = \left(1 - \left(\frac{H_2}{H_1}\right)^n\right)^{0.385} \tag{9-6}$$

where H_1 = head over the weir crest; H_2 = downstream water level above the weir crest; q_f = free-flow discharge for head H_1; and n is an exponent that is equal to $\frac{3}{2}$ for rectangular weirs. Abou-Seida and Quraishi (1976) proposed the following relation:

$$\frac{q}{q_f} = \left(1 + \frac{H_2}{2H_1}\right)\sqrt{1 - \frac{H_2}{H_1}} \tag{9-7}$$

Often a weir does not span the full width of the channel, as shown in Figure 9-2. Such weirs are termed contracted weirs. The flow issuing from contracted weirs undergoes contractions at the sides so that the effective width of the weir is less than B. Experiments show that the reduction in width per end contraction is about $0.1H$ when $B/H > 3$. The discharge equation for contracted weirs with two end contractions can be written as

$$Q = \frac{2}{3}C_d\sqrt{2g}\left(B - 0.2H\right)H^{3/2} \tag{9-8}$$

Weirs of various shapes have been used in practice for meeting specific requirements. The general form of discharge equation for a weir can be expressed as $Q = KH^n$, where K and n are coefficients. The discharge over a triangular weir (Prob. 9-6) is given as

$$Q = \frac{8}{15}C_d\sqrt{2g}\tan\frac{\theta}{2}H^{5/2} \tag{9-9}$$

where θ is the angle between the sides of the triangle. For a 90° triangular weir C_d

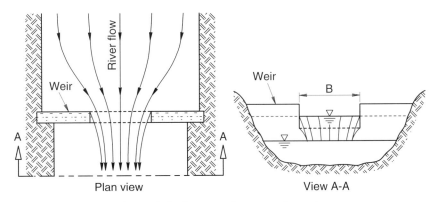

Figure 9-2. Contracted rectangular weir.

is about 0.585 and Eq. (9-9) becomes

$$Q = 1.38H^{5/2}$$

(9-10)

where Q is in cubic meters per second and H is in meters.

Sills

Clearly Eq. (9-5) cannot be valid for small values of W since C_d tends to infinity as W tends to zero. Equation (9-5) is valid for $H/W < 5$ (see Figure 9-3). For large values of $H/W > 20$, the weir is termed a "sill." The flow depth upstream of a sill is equal to the critical depth y_c, as in the case of a free overfall. The critical-depth relationship for rectangular cross sections is

$$q = \sqrt{gy_c^3} = \sqrt{g(H+W)^3}$$

(9-11)

The expression for C_d for sills can be derived from Eqs. (9-2) and (9-11) as

$$C_d = 1.06\left(1 + \frac{W}{H}\right)^{3/2}$$

(9-12)

By fitting a curve to the experimental data of Kandaswamy and Rouse (1957), Swamee (1988) obtained the following equation that is valid for any value of H/W:

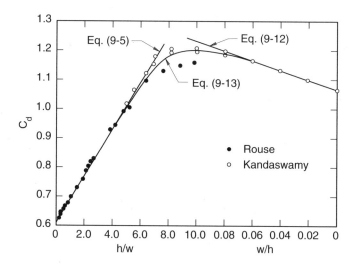

Figure 9-3. Discharge coefficient for the entire weir-sill range.

$$C_d = 1.06\left[\left(\frac{14.14W}{8.15W + H}\right)^{10} + \left(\frac{H}{H + W}\right)^{15}\right]^{-0.1} \qquad (9\text{-}13)$$

A comparison of Eq. (9-13) to the experimental data and Eqs. (9-5) and (9-12) is shown in Figure 9-3. Based on the good agreement among them, Eq. (9-13) can be used for the full range of H/W.

Broad-Crested Weirs

It was shown in Section 3-2 that if the height of a hump in a channel floor exceeds a critical value Δz_c and the hump is long enough for parallel flow to occur over it, the flow over the hump is critical. The hump acts as a critical-depth control. Such a rectangular or trapezoidal hump is called a broad-crested weir (Figure 9-4b). An expression for discharge over a broad-crested weir can be derived assuming critical depth over the crest and neglecting the energy loss between the upstream section where the head over the crest is measured and the section of the critical depth. The unit discharge is given by

$$q = \sqrt{g y_c^3} = \sqrt{g\left(\tfrac{2}{3}E\right)^3} = \sqrt{\tfrac{8}{27}g}\, E^{3/2} \qquad (9\text{-}14)$$

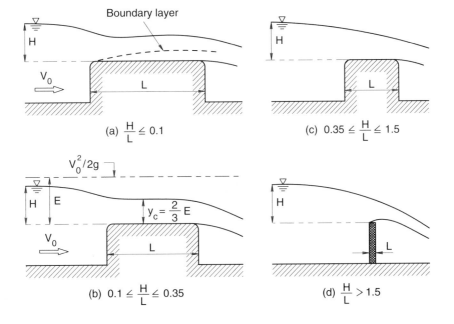

Figure 9-4. Flow profiles over broad-crested weirs of different lengths.

where E is the height of the upstream total energy line over the weir crest and is equal to $H + V_0^2/2g$. Introducing a coefficient of discharge C_d to account for the boundary layer development over the surface of the weir, the total discharge is given by

$$Q = C_d B \sqrt{\tfrac{8}{27} g} \, E^{3/2} \tag{9-15}$$

A theoretical expression for the coefficient of discharge (Rouse, 1950) is

$$C_d = \left(1 - \frac{2\delta^*}{B}\right)\left(1 - \frac{\delta^*}{E}\right)^{3/2} \tag{9-16}$$

where δ^*, the displacement thickness of the boundary layer, depends on the crest length L and the Reynolds number $R_L = VL/v$, where V is the flow velocity at the crest and v is the kinematic viscosity of liquid. The parameter V is equal to $\sqrt{gy_c} = \sqrt{2gE/3}$. The displacement thickness can be estimated from

$$\frac{\delta^*}{L} = \begin{cases} 1.73 R_L^{-0.5} & \text{if } R_L < 3 \times 10^5 \\ 0.037 R_L^{-0.2} & \text{if } R_L > 3 \times 10^5 \end{cases} \tag{9-17}$$

Equation (9-15) is not applicable to a short weir. There is no region of parallel flow over the short crest weir and the critical-depth relation used in Eq. (9-14) cannot be used. Based on the value of H/L, four types of finite-crest-width weirs can be identified (Govinda Rao and Muralidhar, 1963):

1. Long-crested weir (Figure 9-4a); $H/L < 0.1$: The critical flow control section is near the downstream end of the weir. The flow over the weir crest is subcritical and the value of C_d depends on the resistance of the weir surface. This type of weir is of limited use as a dependable device for flow measurements.
2. Broad-crested weir (Figure 9-4b); $0.1 < H/L < 0.35$: The region of parallel flow occurs near the middle section of the crest. The variation in the coefficient of discharge with H/L is small.
3. Narrow-crested weir (Figure 9-4c); $0.35 < H/L < 1.5$: The streamlines are curved over the entire crest; there is no region of parallel flow over the crest.
4. Sharp-crested weir (Figure 9-4d); $1.5 < H/L$: The flow separates at the upstream end and does not reattach to the crest. The weir behaves as a sharp-crested weir.

Swamee (1988) proposed the following equation for the coefficient of discharge for finite-crest weirs:

$$C_d = 0.5 + 0.1 \left[\frac{(H/L)^5 + 1500(H/L)^{13}}{1 + 1000(H/L)^3}\right]^{0.1} \tag{9-18}$$

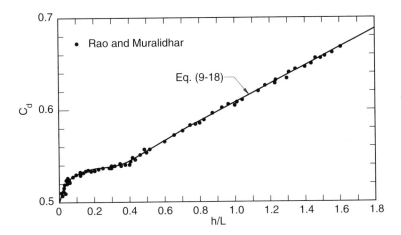

Figure 9-5. Discharge coefficient for finite-crest weirs.

Equation (9-18) along with the experimental data of Govinda Rao and Muralidhar (1963) is presented in Figure 9-5. It can be seen from Eq. (9-18) that C_d increases with increasing H/L. It implies that for a given discharge the head over a sharp-crested weir is less than that over a broad-crested weir. In other words less energy is needed to pass a flow over sharp-crested weirs than over broad-crested weirs (Prob. 9-1).

It should be mentioned that if the upstream end of the broad-crested weir is not rounded, flow will separate from the boundary and reattach to the boundary further downstream. The control section will coincide with the maximum elevation of the separation surface, where the specific energy attains its true minimum value.

Overfalls

A free overfall is a special case of sills wherein $W = 0$. As explained earlier, the freefall is a CDC if the flow upstream is subcritical and it is an NDC if the flow upstream is supercritical. One would expect the depth at the brink to be equal to the critical depth in subcritical flow and the normal depth in supercritical flow. Experimental data show that the brink depth is less than the respective critical and normal depths. The reason for it is the nonhydrostatic pressure distribution at the brink, as shown in Figure 9-6. In the case of complete aeration under the nappe, the pressure is zero at both points B and C, but not within the nappe at the brink. The internal pressure is much smaller than the hydrostatic pressure. From the energy equation the decrease in pressure implies an increase in velocity, and the continuity equation in turn requires an accompanying decrease in depth.

A short distance upstream from the brink in section 1 the pressure distribution is hydrostatic since the flow is almost parallel. The depth y_1 in section 1 is equal to y_c in subcritical flow and y_n in supercritical flow. The brink depth y_b is less than y_c

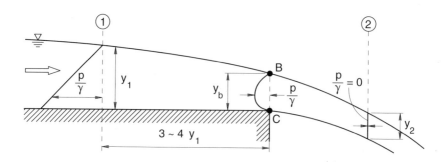

Figure 9-6. The free overfall.

(or y_n). The lower limit of y_b can be determined by applying the momentum and continuity equations between section 1 and section 2 where the pressure within the nappe is zero. It can be shown (Prob. 9-2) that

$$\frac{y_2}{y_1} = \frac{2F_1^{\,2}}{1+2F_1^{\,2}} \tag{9-19}$$

If $y_1 = y_c$, i.e., $F_1 = 1$, Eq. (9-19) becomes

$$\frac{y_2}{y_c} = \frac{2}{3} \tag{9-20}$$

Since there is some residual pressure at the brink, the brink depth y_b must be greater than y_2. Experimental data of Rouse (1936) on the end depth for subcritical flow in a rectangular channel with sidewall on both sides continuing downstream of the free nappe yielded $y_b = 0.715 \, y_c$. Since the location of the critical depth is not fixed and it changes with discharge and channel slope and roughness, the end depth is a more suitable choice for measuring discharge. Substitution of the relation between y_c and y_b in Eq. (9-6) yields

$$q = \sqrt{g(y_b/0.715)^3} = 1.65\sqrt{g}\, y_b^{3/2} \tag{9-21}$$

Experimental data on the brink depth for various channel shapes are available in the literature (Diskin, 1961; Rajaratnam and Muralidhar, 1964a, 1964b).

Example 9-1. Find the height of a sharp-crested free-flow rectangular weir to pass a discharge of 0.4 m³/s with an upstream depth of 1.0 m. The width of the weir is 2.0 m and is equal to the width of the channel.

Given $H+W = 1.0$ or $H = 1 - W$ and $q = 0.4/2 = 0.2$ m³/s/m. Substituting for H in Eqs. (9-2) and (9-5), one obtains

$$0.2 = \frac{2}{3}\left(0.611 + \frac{0.075(1-W)}{W}\right)\sqrt{2\times9.81}(1-W)^{3/2}$$

Solution of the above equation by the trial-and-error procedure gives $W = 0.77$ m.

PROBLEMS

9-1 Specific energy E of flow over a sharp-crested weir in a section is given by

$$E = \frac{1}{Q}\int_A\left(\frac{p}{\gamma} + z + \frac{v^2}{2g}\right)v\,dA$$

wherein p, z, and v respectively represent the pressure, elevation, and velocity in an elementary section dA and Q is discharge. In order to obtain E as a function of head H over the weir crest, derive an expression for the pressure distribution within the section assuming (i) streamlines are concentric and (ii) $vr = $ constant, where r is the radius of curvature of a streamline. Find the minimum value of E for $r_c = 5H$, where r_c is the radius of curvature of the lower surface of the nappe. Compare the minimum value of E with that for a flow over a broad-crested weir.

9-2 Prove Eq. (9-19) by applying the momentum and continuity equations between sections 1 and 2 in Figure 9-6.

9-3 The vertical distance between the upper and lower nappe profiles of a jet at a section downstream from a free overfall, where the pressure throughout the jet is atmospheric, is equal to $\frac{2}{3}y_c$ [see Eq. (9-20)]. Determine whether the corresponding distance in a jet downstream from a sill of height $W/H < 15$ is larger or smaller than $\frac{2}{3}y_c$.

9-4 Depth upstream of a control structure in a 10-m-wide rectangular channel is measured to be 1.2 m. Determine the channel discharge if the control structure is

a. a sharp-crested suppressed weir with depth measured above the weir crest.
b. a free overfall with depth measured at the brink.
c. a broad-crested weir with depth measured in the middle of the weir in the zone of parallel flow.
State the assumptions made, if any, in each case.

9-5 Calculate the height of a 1.5-m-long suppressed weir to pass a flow of 0.8 m³/s while maintaining an upstream flow depth of 2.0 m.

9-6 Derive Eq. (9-9) for the discharge over a triangular weir.

9-7 It is desired to measure discharge ranging from 2 to 10 liters per second with a relative accuracy of 0.5 %. The discharge is determined by measuring the head over a sharp-crested weir. The head can be measured to the nearest 1 mm. What should be the maximum width of the weir to satisfy these conditions.

9-8   Does the discharge coefficient for a broad-crested weir shown in Figure 9-4a
       increase or decrease with L and why?

9-9   Why is the pressure distribution at the brink in Figure 9-6 not hydrostatic?

9-10   Determine discharge over a high, broad-crested weir that is 12 m long if the head on
       the weir is 0.5 m.

9-2 Ogee-Crest Spillway

Both sharp-crested and broad-crested weirs have deficiencies that make them
unsuitable as high head control structures. The former is limited in its structural
strength and the latter has relatively small discharge coefficient and zones of
negative pressure. These limitations can be removed by shaping the crest to
conform to the profile of the lower nappe surface from a sharp-crested weir, as
shown in Figure 9-7. The shape of such a profile depends on the head over the
crest, referred to as the design head, the inclination of the upstream face of the
overflow section, and the velocity of approach. Crest shapes have been studied
extensively by U. S. Bureau of Reclamation and the results are published in a
report (USBR, 1948, 1977). At the design head the pressure over the crest is
atmospheric and the coefficient of discharge for the ogee crest is nearly the same
as that for the sharp-crested weir. The head over the ogee crest is measured above
the high point of the nappe, i.e., point D in Figure 9-7. For a high spillway,
experiments show that the vertical distance between point D and point C (the weir
crest) is $0.11H$. The design head H_D is then $H_D = 0.89H$. The ratio of the discharge
coefficient for the ogee crest to that of the sharp-crested weir is about
$(1/0.89)^{3/2} = 1.19$.

It should be pointed out that the pressure on the crest is atmospheric only at
the design head. If the head is less than the design head, the nappe tries to deflect
towards the weir; consequently the pressure on the crest is above atmospheric and
the discharge coefficient decreases. On the other hand for heads higher than the

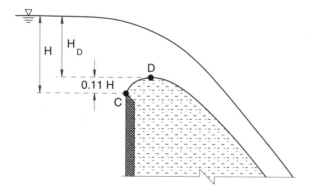

Figure 9-7. The ogee-crest spillway.

design head the pressure on the crest is negative and the discharge coefficient increases. This feature is advantageously exploited by selecting the design head that is much smaller than the maximum head. However, a large underpressure can lead to flow separation and cavitation.

PROBLEMS

9-11 Why does the discharge coefficient for a spillway increase with increasing head?

9-12 Determine the head over a high spillway for a unit discharge of 30 m³/s/m.

9-3 Underflow Gates

Gates are used for controlling discharge in canals, in outlets, and on spillways. Some common underflow gates are sluice and tainter gates, as shown in Figure 9-8. Flow under a gate can be analyzed using the energy equation between sections 1 and 2 in Figure 9-8:

$$y_1 + \frac{q^2}{2gy_1^2} = y_2 + \frac{q^2}{2gy_2^2} = C_c w + \frac{q^2}{2g(C_c w)^2} \tag{9-22}$$

where C_c is the contraction coefficient of the jet and w is the gate opening. Equation (9-22) can be written as

$$q = C_c w \sqrt{\frac{y_1}{y_1 + C_c w}} \sqrt{2gy_1} = C_d w \sqrt{2gy_1} \tag{9-23}$$

where

$$C_d = C_c \sqrt{\frac{1}{1 + C_c w / y_1}} \tag{9-24}$$

in which C_d is a function of the gate opening and the gate geometry.

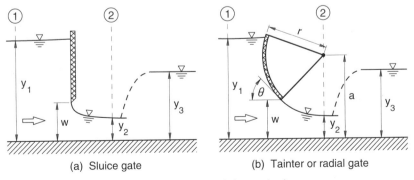

(a) Sluice gate   (b) Tainter or radial gate

Figure 9-8. Flow under a sluice and tainter gate.

Equation (9-23) is derived for free outflow. For submerged outflow, Eqs. (3-56) and (3-57) are applicable. However, Eq. (9-23) can also be used for submerged outflow; C_d in that case is a function of downstream depth also. The C_d for the vertical sluice gate as determined experimentally by Henry (1950) is shown in Figure 9-9. The dotted line for y_3/w in Figure 9-9 shows the computational results based on Eqs. (3-43) and (3-44) with $C_c = 0.6$. The experimental results of Toch (1955) on the discharge coefficient for the radial gate are presented in Figure 9-10. For free outflow, Toch found that C_c in Eq. (9-24) can be approximated by

$$C_c = 1 - 0.75\frac{\theta^0}{90} + 0.36\left(\frac{\theta^0}{90}\right)^2, \quad \theta \le 90^0 \tag{9-25}$$

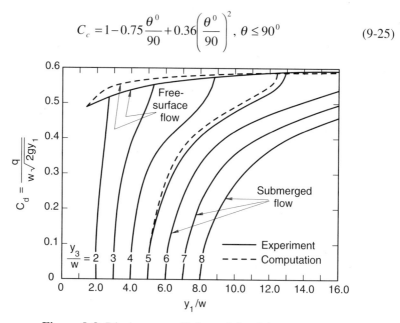

Figure 9-9. Discharge coefficient of the sluice gate.

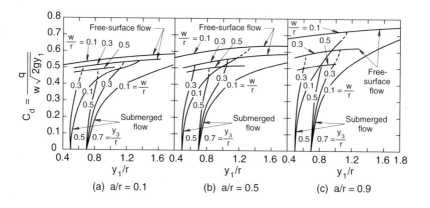

Figure 9-10. Discharge coefficient of the radial gate.

The model results presented in Figures 9-9 and 9-10 must be corrected for scale effects, because the model Reynolds number is too small in most cases. An example for correcting the model data is presented by Naudascher (1984).

PROBLEMS

9-13 A tainter gate (Figure 9-8) 15 m long and 6 m in radius is installed on top of a high spillway. The trunion height $a = 3$ m. Determine the rating curve of the gate opening ranging from 1 to 2 m.

9-14 Find the discharge through a free-flowing sluice gate having a gate opening of 0.5 m under an upstream head of 3.0 m. What would be the discharge if the gate is submerged and the tailwater depth is 2.3 m?

9-15 Discuss the effect of w/r on the discharge coefficient of the radial gate shown in Figure 9-10.

9-16 A rectangular channel carries a discharge of 2.9 $m^3/s/m$. Find the opening of a sluice gate to pass the flow with an upstream depth of 4.5 m. Assume a free flow under the gate.

9-4 Venturi Flumes

A venturi flume is a *critical-flow flume* wherein the critical depth is created by a contraction in width of the channel. Thus the contracted section serves as a control. Venturi flumes have two advantages over weirs where the critical depth is created by a vertical constriction. First, the head loss is smaller in flumes than in weirs. Second, there is no dead zone in flumes where sediment and debris can accumulate; such a dead zone exists upstream of the weirs. One of the most common venturi flumes is the *Parshall flume* shown in Figure 9-11. It consists of three sections: a converging section, a throat section, and a diverging section. The critical depth develops in the throat section. The flow upstream and downstream of the throat is subcritical and supercritical, respectively. A hydraulic jump forms in the diverging section. The flow in the flume is a "free flow" if the ratio of the downstream head H_s measured in the downstream stilling well to the upstream head H_a measured in the upstream stilling well, H_s/H_a, is less than about 0.70. The discharge for the free-flow condition is given by

$$Q = KH_a^n \qquad (9-26)$$

where K and n are functions of throat width W. The discharge equation in English units for Parshall flumes with widths ranging from 1 to 8 ft for the free-flow condition is

$$Q = 4WH_a^{1.522W^{0.026}} \qquad (9-27)$$

where Q is in cubic feet per second and H_a and W are in feet. Discharge equations

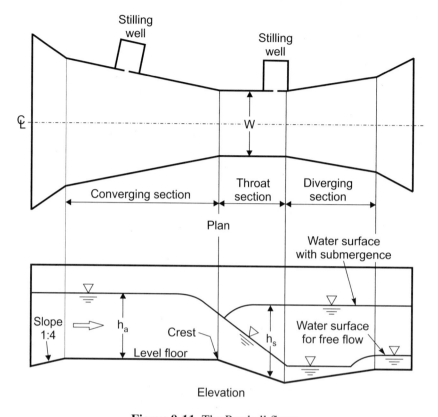

Figure 9-11. The Parshall flume.

for outside the range of W noted above and for submerged flow conditions and the dimensions of the Parshall flume for various throat widths can be found in Parshall (1926) and Chow (1959). Equation (9-26) can be written in a form that is dimensionally homogeneous as

$$Q = C_d \sqrt{2g} W H_a^{3/2} \qquad (9-28)$$

where the discharge coefficient C_d is a function of H_a/W. This function can be found in Roberson et al. (1988).

PROBLEMS

9-17   Determine the discharge through a Parshall flume if $W = 2$ m and $H_a = 0.6$ m. The flume runs in a free-flow condition.

9-18   Find the width of a Parshall flume to measure 0.5 m³/s in a channel having a depth of flow equal to 0.6 m.

CHAPTER 10

SPECIAL TOPICS

A number of special topics, including problems that cannot be analyzed with the one-dimensional differential equations, are discussed in this chapter. The pressure distribution in some problems is either nonhydrostatic or such flows are nonuniform in the lateral direction.

10-1 Contractions and Expansions

Transitions such as contractions and expansions are often provided in canals for various practical purposes. Though the flow behavior in a channel contraction and expansion was analyzed earlier in Chapter 3, the analysis did not include the energy loss and the possibility of formation of oblique waves in the transition. The energy loss in the transition depends on the geometry of the transition, which in turn is dictated by the economic consideration. The design of a transition in subcritical flow is governed by the energy loss in the transition and the cost of the transition. The oblique waves are significant only in supercritical flow.

Subcritical Flow

The energy loss in a transition is composed of the friction loss and the form loss. The friction loss can be estimated with the Manning equation. The form loss depends on the shape of the transition and is expressed as

$$\Delta H = C_L \Delta h_v \qquad (10\text{-}1)$$

where ΔH = form loss, C_L = loss coefficient, and Δh_v = difference in velocity head across the transition. The loss coefficients are usually determined experimentally. It is well known that the energy loss is generally smaller in contractions than expansions; therefore the loss coefficient is larger for expansions than contractions. The three typical transitions for subcritical flow, as shown in Figure 10-1, are (a) the cylinder-quadrant transition, (b) the wedge-type transition, and (c) the warped transition. The cylinder-quadrant transition is the least expensive to construct, while the warped transition is the costliest. On the other hand the loss coefficient is the smallest for the warped transition and is the biggest for the cylinder-quadrant transition. Approximate values of the loss coefficient are given in Table 10-1. The water-surface profile in the transition can be determined with the energy equation, which between sections i and $i+1$ can be written as

$$E_{i+1} = E_i + \Delta z_{i,i+1} - \Delta H_{i,i+1} \qquad (10\text{-}2)$$

where $\Delta z_{i,i+1}$ = change in elevation of channel bed between sections i and $i+1$ and

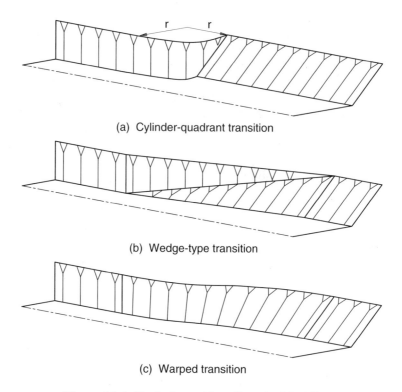

(a) Cylinder-quadrant transition

(b) Wedge-type transition

(c) Warped transition

Figure 10-1. Typical transitions for subcritical flow.

Table 10-1. Loss Coefficients for Transitions

Type of Transition	Contraction	Expansion
Warped	0.10	0.20
Wedge	0.15	0.25
Cylinder-quadrant	0.20	0.30

$\Delta H_{i,i+1}$ = head loss between sections i and $i+1$. The total head loss in the transition is distributed uniformly along the length of the transition. The various steps in designing a transition are as follows:

(a) Choose the type of transition to be used.
(b) Determine the length of the transition. In accordance with recommendations by Hinds (1928), a length of the transition is chosen so that the line connecting the water surface at the channel sides between the

entrance and exit sections of the transition forms an angle not larger than 12.5°.

(c) Efficiency of energy conversion requires that the flow profile be continuous and as smooth as possible. An arbitrary smooth water-surface profile is sketched between the known end points of the profile.

(d) The loss of head between entrance and exit sections is assumed to be the sum of the friction loss and form loss. The friction loss is assumed equal to $(S_{f1} + S_{f2})L/2$, where S_{f1} and S_{f2} are the friction slopes in the two channels and L is the length of the transition. The form loss is determined from Eq. (10-1). From the known total head at one end of the transition compute the total head at the other end of the transition.

(e) Select a number of equally spaced sections along the transition where the flow profile and the dimensions of the transition will be computed. The total head loss is distributed uniformly along the length of the transition. Sketch the energy grade line.

(f) Scale off the velocity head at the various sections from the energy grade line and the assumed water-surface profile. Compute the velocity and the required cross-sectional area in the various sections.

(g) The cross-sectional area is a function of depth and the shape of the cross section. All other dimensions of the cross sections except one are fixed by the geometry of the transition. The remaining dimension is determined so that the cross-sectional area is equal to the computed value. For example, the side slope and the channel width are assumed and the bed elevation is computed.

(h) Plot the bed profile; it should be smooth and continuous so that no flow separation occurs. If the bed profile has sharp breaks and curves, modify the assumed water-surface profile and repeat the calculations.

Vittal and Chiranjeevi (1983) examined methods for designing channel transitions for subcritical flow and suggested a direct method for designing transitions.

Supercritical Flow

It was shown earlier that a disturbance, unless it is "large," cannot travel upstream in supercritical flow. But such a disturbance travels across the flow as it is carried downstream; consequently the boundary of the disturbed zone in supercritical flow, called the wavefront, is oblique. The formation of oblique wavefronts produced by the deflection of vertical channel walls and the flow conditions in the disturbed zones are analyzed first. The analysis is then used in designing contractions and expansions in supercritical flow.

Sudden Inward Deflection in the Boundary. The oblique wavefront produced by a vertical channel wall deflected towards the flow through a finite angle θ is shown in Figure 10-2. The wavefront makes an angle β from the initial flow direction. The flow downstream from the section where the wall is deflected is nonuniform in the lateral direction. The depth is higher downstream than

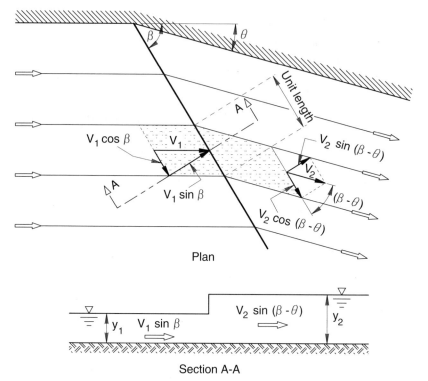

Figure 10-2. Oblique wavefront in supercritical flow.

upstream of the wavefront; such a deflection in the wall is termed positive deflection. Evaluation of the effect on flow of the wall deflection requires determination of angle β of the wavefront in addition to depth and velocity downstream of the wavefront. The solution for the three unknown quantities requires three equations that are the continuity equation and the momentum equations along and normal to the wavefront. The gravity forces and the boundary resistance can be neglected in the momentum equations. The three equations for a control volume shown as the speckled region in Figure 10-2 can be written as follows:

Continuity equation:
$$y_1 V_1 \sin \beta = y_2 V_2 \sin(\beta - \theta)$$
(10-3)

i.e., the unit discharges normal to the wavefront are equal.

Momentum equation normal to the wavefront:

$$\frac{y_1^2}{2} + \frac{y_1 V_1^2 \sin^2 \beta}{g} = \frac{y_2^2}{2} + \frac{y_2 V_2^2 \sin^2(\beta - \theta)}{g}$$
(10-4)

Momentum equation along the wavefront:

$$V_1 \cos \beta = V_2 \cos(\beta - \theta) \qquad (10\text{-}5)$$

Equation (10-5) is based on the fact that there is no net force on the control volume along the wavefront, and therefore the net momentum flux along the wavefront is zero, i.e., the velocity components along the wavefront are equal. Substitution for $V_2 \sin(\beta - \theta)$ from Eq. (10-3) into Eq. (10-4) and some manipulation of the terms in the resulting equation yield

$$\sin \beta = \frac{1}{F_1} \sqrt{\frac{1}{2} \frac{y_2}{y_1} \left(\frac{y_2}{y_1} + 1 \right)} \qquad (10\text{-}6)$$

Equation (10-6) can be rearranged as

$$\frac{y_2}{y_1} = \frac{1}{2} \left(\sqrt{1 + 8F_1^2 \sin^2 \beta} - 1 \right) \qquad (10\text{-}7)$$

Note the similarity between Eq. (10-7) and Eq. (3-17) for the hydraulic jump. Elimination of V_2 / V_1 between Eqs. (10-3) and (10-5) gives

$$\frac{y_2}{y_1} = \frac{\tan \beta}{\tan(\beta - \theta)} \qquad (10\text{-}8)$$

Elimination of y_2/y_1 between Eqs. (10-7) and (10-8) yields

$$\frac{\tan \beta}{\tan(\beta - \theta)} = \frac{1}{2} \left(\sqrt{1 + 8F_1^2 \sin^2 \beta} - 1 \right) \qquad (10\text{-}9)$$

For a given F_1 and θ, Eq. (10-9) can be solved for β using the trial-and-error method. A graphical solution of Eq. (10-9) is presented in Figure 10-3 that can be used to find out β. The figure also includes a dividing line between subcritical and supercritical flow, where $F_2 = 1$. For a given value of θ, two values of β are obtained. The smaller value is of practical interest for which $F_2 > 1$. The value of y_2/y_1 then can be determined from Eq. (10-7). Equation (10-5) can be written in terms of F_1 and F_2 as

$$\frac{F_2}{F_1} = \sqrt{\frac{y_1}{y_2}} \frac{\cos \beta}{\cos(\beta - \theta)} \qquad (10\text{-}10)$$

The value of F_2 can be obtained from Eq. (10-10). As an example for $F_1 = 8$ and $\theta = 10°$, the value of β from Figure 10-3 is $\beta = 16°$, and Eqs. (10-7) and (10-10) yield $y_2/y_1 = 2.73$ and $F_2 = 4.68$.

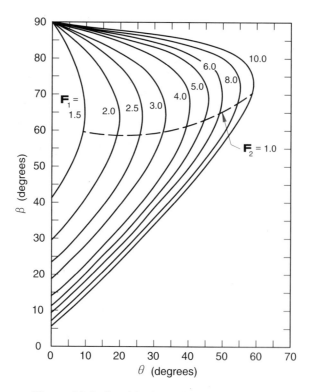

Figure 10-3. Graphical solutions of Eq. (10-9).

Note that Figure 10-3 is not applicable to negative deflection angles, because in such cases no steep wavefronts are formed. The analyses developed in the following sections are applicable to negative deflection angles.

Example 10-1. A wide rectangular channel carries water at a depth of 40 cm and a velocity of 10 m/s. One of the sidewalls of the channel is deflected inward at an angle of 15°. Determine the angle of the oblique wave formed due to wall deflection and the flow conditions downstream of the oblique jump.

To obtain the angle of the oblique wave from Figure 10-3, one requires the value of F_1:

$$F_1 = \frac{10}{\sqrt{9.81 \times 0.40}} = 5.05$$

From Figure 10-3, $\beta = 25°$

From Eq. (10-8), $\dfrac{y_2}{y_1} = \dfrac{\tan \beta}{\tan(\beta - \theta)} = \dfrac{\tan 25°}{\tan 10°} = 2.64$

$$y_2 = 2.64 \times 0.40 = 1.06 \text{ m}$$

From Eq. (10-5), $$V_2 = V_1 \frac{\cos \beta}{\cos(\beta - \theta)} = 10 \times \frac{\cos 25°}{\cos 10°} = 9.20 \text{ m/s}$$

Gradual Change in the Boundary. Consider supercritical flow along a concave and convex boundary shown in Figure 10-4. For visualizing the surface configuration the curved sections of channel walls are replaced by a sequence of short chords, each one deflected relative to the preceding one by a small angle θ, say 4°, as indicated in Figure 10-4.

For the concave wall curving toward the stream having the flow characteristic F_1, the first wavefront crosses the flow at angle β_1, representing an increase in depth. As the flow passes this line, the streamlines are deflected through the angle $\theta = 4°$ towards the positive wavefront—regardless of, but with a longitudinal delay proportional to, the lateral distance from the wall. At the beginning of the wall curvature, therefore, only the filament of fluid adjacent to the wall is deflected through 4°. The lines along the concave wall are called positive disturbance lines, since each causes a rise in depth corresponding to the assumed deflection $\theta = 4°$, so that the flow between them is characterized by decreasing values of F from F_1 to F_{+5} and by decreasing velocities. The wave angles β hence increase, and this fact combined with the sense of flow direction produce general convergence of the wavefronts. The surface gradients increase with distance from the wall, and eventually steep fronts will develop.

Negative wavefronts are caused by convex walls wherein streamlines are deflected away from the wavefronts, and in the example of Figure 10-4 each again represents a surface depression corresponding to $\theta = -4°$. As depth y and angle β of the wavefront decrease, velocity V and Froude number F increase. Since wavefront angles are measured relative to the direction of the preceding flow, the disturbance line must diverge. It is not possible, therefore, for steep wavefronts to form as the result of convex walls. As these lines cross the positive lines from the opposite concave wall, the effect of the latter is canceled to the extent of the deflection carried by the negative lines. A detailed study of surface contours by means of the so-called method of characteristics is possible.

The change in flow conditions along a curved wall is gradual, unless the wavefronts intersect. Equations derived above can be used for the establishment of the general differential equations if infinitesimal wave heights are assumed. Let $y_1 = y$, $y_2 = y + dy$, and $dy/y \ll 1$. Equations (10-6) and (10-8) reduce to

$$\sin \beta = \frac{1}{F} = \frac{\sqrt{gy}}{V} \qquad (10\text{-}11)$$

and

$$\frac{dy}{y} = \frac{\tan \theta (1 - \tan^2 \beta)}{\tan \beta - \tan \theta} \qquad (10\text{-}12)$$

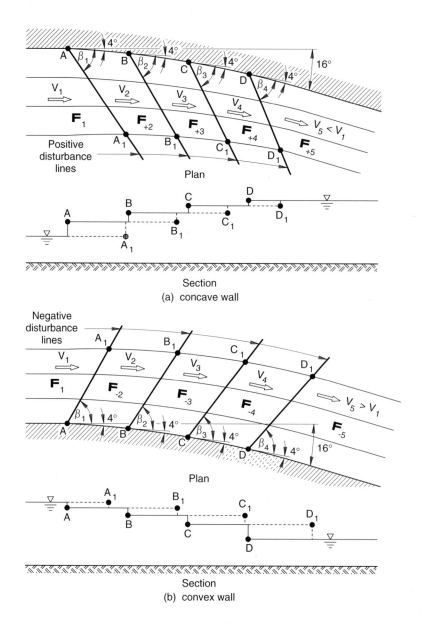

Figure 10-4. Oblique wavefronts due to flow along concave and convex walls.

For small θ, $\tan \theta = \theta$. Also $\tan \beta \gg \tan \theta$. With these approximations, Eq. (10-12) becomes

$$\frac{dy}{d\theta} = \frac{y \sec^2 \beta}{\tan \beta} = \frac{y}{\sin \beta \cos\beta} \qquad (10\text{-}13)$$

Elimination of y between Eqs. (10-11) and (10-13) gives

$$\frac{dy}{d\theta} = \frac{V^2}{g} \tan \beta \qquad (10\text{-}14)$$

Equation (10-14) can be integrated on the assumption that the specific energy E remains constant. Such an assumption is reasonable for gradual wall deflections. One can write for $\tan \beta$ from Eq. (10-11) as

$$\tan \beta = \sqrt{\frac{gy}{V^2 - gy}} \qquad (10\text{-}15)$$

From the definition of the specific energy, one can write for V^2 as

$$V^2 = 2g(E - y) \qquad (10\text{-}16)$$

Elimination of V^2 from Eqs. (10-15) and (10-16) gives

$$\tan \beta = \sqrt{\frac{y}{2E - 3y}} \qquad (10\text{-}17)$$

Substitution for V^2 from Eq. (10-16) and for $\tan \beta$ from Eq. (10-17) into Eq. (10-14) yields

$$\frac{dy}{d\theta} = 2(E - y)\sqrt{\frac{y}{2E - 3y}} \qquad (10\text{-}18)$$

The solution of Eq. (10-18) is

$$\theta = \sqrt{3} \tan^{-1}\sqrt{\frac{3y}{2E - 3y}} - \tan^{-1}\frac{1}{\sqrt{3}}\sqrt{\frac{3y}{2E - 3y}} - \theta_1 \qquad (10\text{-}19)$$

where θ_1 is a constant of integration. Because

$$E = y + \frac{V^2}{2g} = \frac{y}{2}\left(2 + F^2\right) \qquad (10\text{-}20)$$

Eq. (10-19) can be written as

$$\theta = \sqrt{3}\,\tan^{-1}\frac{\sqrt{3}}{\sqrt{F^2-1}} \;-\; \tan^{-1}\frac{1}{\sqrt{F^2-1}} \;-\; \theta_1 \tag{10-21}$$

The constant of integration is determined from the condition that $F = F_1$ for $\theta = 0$. A plot of Eq. (10-21) is presented in Figure 10-5. Equation (10-20) for constant E can be written in terms of y_2/y_1 as

$$\frac{y_2}{y_1} = \frac{F_1^2+2}{F_2^2+2} \tag{10-22}$$

As an example for $F_1 = 6.5$ and $\theta = 0°$, θ_1 from Figure 10-5 is equal to 17°. If the gradual wall deflection is +9°, then $\theta + \theta_1 = 9° + 17° = 26°$, and F_2 from Figure 10-5 is 4.2. The value of y_2/y_1 obtained from Eq. (10-22) is 2.25.

Sudden Outward Deflection in the Boundary. The flow in a channel in which the channel wall is turned outward from the flow at an angle θ is shown in

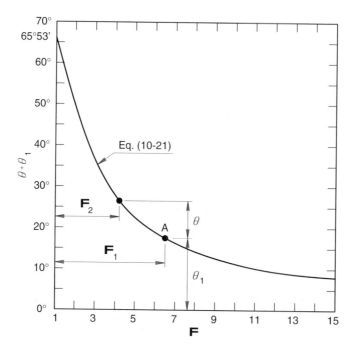

Figure 10-5. Graphical solution of Eq. (10-21).

Figure 10-6. The corner (point A) can be considered to be a short convex curve joining the two straight boundaries. A number of wavefronts diverge from the convex boundary. In the limit all wavefronts originate from point A. Depth of flow decreases in the fan-shaped region of the wavefronts delineated by wave angles β_1 and β_2. As the angle between wavefronts approaches zero, the flow direction and Froude number immediately upstream of wavefront AC are equal to the final flow direction and Froude number parallel to the downstream boundary. The angles β_1 and β_2 are given by Eq. (10-11) as

$$\beta_1 = \sin^{-1} F_1 \quad \text{and} \quad \beta_2 = \sin^{-1} F_2$$

where F_1 = initial Froude number and F_2 = final Froude number. The relationship among F_1, F_2, and θ is governed by Eq. (10-21) and between y_1 and y_2 by Eq. (10-22). Note that the angle θ is negative.

Example 10-2. Determine V_2 and y_2 in Figure 10-6 for V_1 = 9 m/s, y_1 = 0.5 m, and θ = 10°:

$$F_1 = \frac{V_1}{\sqrt{gy_1}} = \frac{9}{\sqrt{9.81 \times 0.5}} = 4.06$$

For F_1 = 4.06 and θ = 0°, θ_1 from Figure 10-5 is 28°. Now for $\theta + \theta_1 = -10° + 28° = 18°$, F_2 from Figure 10-5 is 6.4.

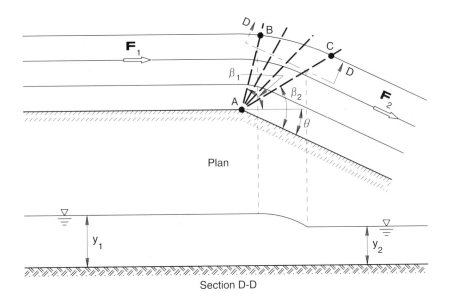

Plan

Section D-D

Figure 10-6. Flow past a corner.

From Eq. (10-22),
$$\frac{y_2}{y_1} = \frac{F_1^2 + 2}{F_2^2 + 2} = \frac{4.06^2 + 2}{6.4^2 + 2} = 0.43$$

$$y_2 = 0.43 \times 0.5 = 0.22 \text{ m}$$

$$V_2 = F_2 \sqrt{gy_2} = 6.4\sqrt{9.81 \times 0.22} = 9.40 \text{ m/s}$$

Reflection of a Positive Wavefront. Consider the flow through a rectangular channel of which one of the walls is being deflected by an angle θ, as shown in Figure 10-7. An oblique positive wavefront develops at the corner A. Angle β_1 of the wavefront AB from the initial flow direction is given by Eq. (10-9) or Figure 10-3. The flow downstream of the wavefront AB is parallel to the inclined wall AC and the Froude number is F_2. Note that F_2 is less than F_1. The straight wall BD deflects the oncoming flow through an angle θ, resulting in another positive wavefront BC that makes an angle ($\beta_2 - \theta$) with BD. The flow downstream of BC is parallel to the wall BD and the Froude number is F_3, which is less than F_2. The positive wavefront AB is considered reflected by the wall as positive wavefront BC. In similar fashion the positive wavefront BC is reflected as positive wavefront CD, which in turn is reflected again as positive wavefront DE, and so on. After each reflection the value of the Froude number decreases and the value of angle β increases.

Example 10-3. Find the flow conditions at a point on the wall immediately downstream of point B in Figure 10-7 if the flow velocity and depth of the approach flow are 10 m/s and 0.5 m, respectively. The wall deflection is 10°:

$$F_1 = \frac{V_1}{\sqrt{gy_1}} = \frac{10}{\sqrt{9.81 \times 0.5}} = 4.52$$

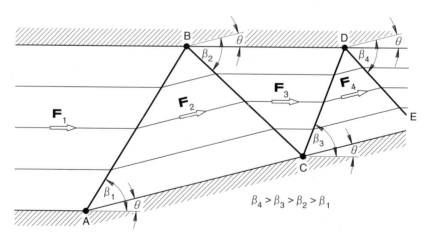

Figure 10-7. Reflections of a positive wavefront.

For $F_1 = 4.52$ and $\theta = 10°$, the value of β_1 from Figure 10-3 is $21.5°$. From Eq. (10-7)

$$\frac{y_2}{y_1} = \frac{1}{2}\left(\sqrt{1+8F_1^2 \sin^2 \beta_1} - 1\right) = \frac{1}{2}\left(\sqrt{1+8\times4.52^2 \sin^2 21.5} - 1\right) = 1.90$$

From Eq. (10-10), $F_2 = F_1\sqrt{\dfrac{y_1}{y_2}\dfrac{\cos\beta_1}{\cos(\beta_1-\theta)}} = 4.52\sqrt{\dfrac{1}{1.90}\dfrac{\cos 21.5°}{\cos 11.5°}} = 3.11$

For $F_2 = 3.11$ and $\theta = 10°$, the value of β_2 from Figure 10-3 is $33°$. From Eq. (10-7)

$$\frac{y_3}{y_2} = \frac{1}{2}\left(\sqrt{1+8F_2^2 \sin^2 \beta_2} - 1\right) = \frac{1}{2}\left(\sqrt{1+8\times3.11^2 \sin^2 33} - 1\right) = 1.95$$

From Eq. (10-10), $F_3 = F_2\sqrt{\dfrac{y_2}{y_3}\dfrac{\cos\beta_2}{\cos(\beta_2-\theta)}} = 3.11\sqrt{\dfrac{1}{1.95}\dfrac{\cos 33°}{\cos 23°}} = 2.03$

$$y_3 = \frac{y_3}{y_2}\frac{y_2}{y_1}y_1 = 1.95\times1.90\times0.5 = 1.85 \text{ m}$$

$$V_3 = F_3\sqrt{gy_3} = 2.03\sqrt{9.81\times1.85} = 8.65 \text{ m/s}$$

Interaction of Two Positive Wavefronts. Consider the flow through a rectangular channel of which one of the walls is being deflected by an angle θ_1 and the other wall by an angle θ_2, as shown in Figure 10-8. Let $\theta_1 > \theta_2$. Positive wavefronts AC and BC emanate from corners A and B and meet at point C. The flows immediately downstream of wavefronts AC and BC are parallel to the wall AD at an angle θ_1 with the horizontal and BE at an angle θ_2 with the horizontal, respectively. The Froude number is F_2 downstream of AC and F_3 downstream of BC. The Froude numbers F_2 and F_3 and wavefront angles β_1 and β_2 can be determined from Eqs. (10-9), (10-7), and (10-10). The flow downstream of wavefront AC sees the flow downstream of the wavefront BC as a positive deflection resulting in formation of a positive wavefront CD. Similarly the flow downstream of the wavefront BC sees the flow downstream of the wavefront AC as a positive deflection resulting in formation of a positive wavefront CE. The flow downstream of wavefronts CD and DE has the same Froude number F_4 and direction at an angle θ_3 with the horizontal. The angle θ_3 and the Froude number F_4 are determined by the trial-and-error procedure. For an assumed value of θ_3, the deflection angles for the two flows are $(\theta_1 + \theta_3)$ and $(\theta_2 - \theta_3)$. The two values of F_4 are determined for the two flows from Eqs. (10-9), (10-7), and (10-10): one at the Froude number F_2 and angle $\theta = (\theta_1 + \theta_3)$; and the other at the Froude number F_3 and angle $\theta = (\theta_2 - \theta_3)$. If the two values of F_4 are the same, the assumed θ_3 is correct. Otherwise repeat the procedure for another value of θ_3. Note that for $\theta_1 = \theta_2$, $\theta_3 = 0$ and $\beta_3 = \beta_4$.

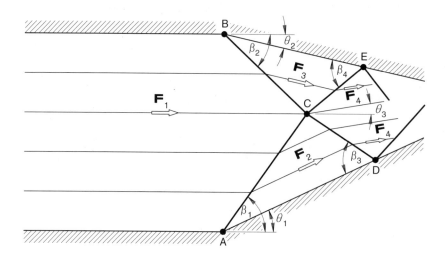

Figure 10-8. Interaction of two positive wavefronts.

Interaction of Positive and Negative Wavefronts. Consider the flow with a positive wall deflection followed by a negative wall deflection, as shown in Figure 10-9. A positive wavefront develops at point A and a number of wavefronts originate from point B. The positive wavefront is deflected as it intersects the negative wavefronts and the flow depths at the intersections are obtained by superposition.

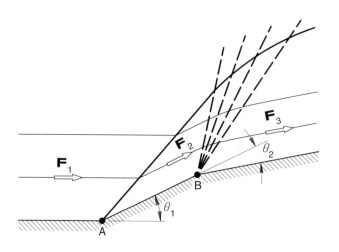

Figure 10-9. Interaction of positive and negative wavefronts.

Contractions. A few feasible transition for supercritical flow, as shown in Figure 10-10 are (a) Straight-wall transition, (b) convex-wall transition, and (c) S-shaped-wall transition. It can be easily seen that for the same length of the transition the angles θ, θ', and θ'' in Figure 10-10 follow the relation

$$\theta'' > \theta' > \theta$$

Because the magnitude of the disturbance in supercritical flow increases with increasing deflection angle, a straight-wall transition that produces the least increase in depth is the optimal shape for contractions in supercritical flow. The wave patterns for two values of deflection angle θ of the straight-wall transition are shown in Figure 10-11. The positive wavefront that emanates from A (A') in Figure 10-11a, after being reflected at B, arrives exactly at the intersection C (C'). The streamlines downstream of wavefronts BC and BC' are parallel to the sidewalls of the channel downstream from the contraction, and there are no further

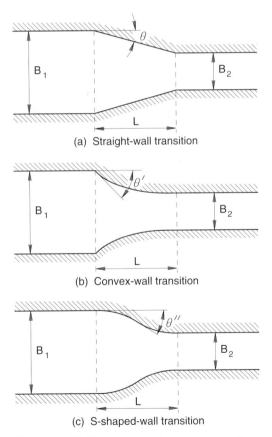

(a) Straight-wall transition

(b) Convex-wall transition

(c) S-shaped-wall transition

Figure 10-10. Rectangular channel contractions.

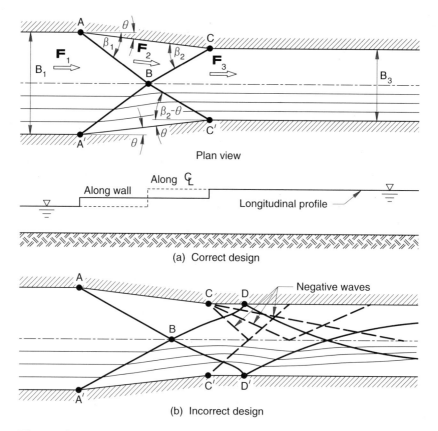

Figure 10-11. Wave pattern in a straight-wall contraction: (a) correct design, (b) incorrect design.

reflections. On the other hand the wavefronts in Figure 10-11b, after being reflected at B, arrive at other points such as D (D') and lead to a diamond-shaped pattern of standing waves.

For given values of F_1, B_1, and B_3, the deflection angle θ of the straight-wall transition can be determined by the trial-and-error procedure. For an assumed value of θ, determine β_1, y_2/y_1, F_2, β_2, y_3/y_2, and F_3 with Eqs. (10-9), (10.7) or (10-8), and (10-10). From the geometry in Figure 10-11a, the length of the transition is given by

$$L = \frac{B_1 - B_3}{2\tan\theta} \tag{10-23}$$

and

$$L = L_1 + L_2 = \frac{B_1}{2}\tan\beta_1 + \frac{B_3}{2}\tan(\beta_2 - \theta) \tag{10-24}$$

If the two values of L from Eqs. (10-23) and (10-24) are the same, the assumed value of θ is correct; otherwise try another value of θ. The use of the plot in Figure 10-3 may introduce some errors. The final solution should satisfy the continuity equation

$$B_1 y_1 V_1 = B_3 y_3 V_3$$

or

$$F_3 = F_1 \left(\frac{B_1}{B_3} \right) \left(\frac{y_3}{y_1} \right)^{3/2}$$

Example 10-4. A rectangular channel is 5.0 m in width and carries a flow at a Froude number of 6. The width of the channel is to be reduced to 2.5 m. Design a straight-wall contraction for the channel.

Calculations are done in tabular form for various assumed values of θ until two values of L from Eqs. (10-23) and (10-24) are same. Given $B_1 = 5.0$ m; $B_3 = 2.5$ m; and $F_1 = 7$.

θ deg.	L m	β_1 deg.	y_2/y_1	F_2	β_2 deg.	y_3/y_2	F_3	L_1+L_2 m
6.0	11.89	15.0	1.75	4.44	18.0	1.50	3.52	15.21
4.3	16.62	13.0	1.47	4.88	15.0	1.35	4.13	17.44

L is obtained from Eq. (10-23), β_1 and β_2 from Figure 10-3, y_2/y_1 from Eq. (10-7), F_2 and F_3 from Eq. (10-10), and L_1+L_2 from Eq. (10-24).

An angle of 4.3° for the contraction is considered close enough. For more accurate results one should use Eq. (10-9) instead of Figure 10-3.

Expansions. The basic theory presented above can be used to delineate the flow field in expansions. Graphical methods for designing expansions are described by Ippen (1951). Expansions for preliminary studies can be designed by using the results of an analytical and experimental study conducted by Rouse, Bhoota, and Hsu (1951). Their results are presented in Figure 10-12. The expansion consists of a convex curve followed by a reverse curve. The convex curve is given by

$$\frac{B}{B_1} = \frac{1}{4} \left(\frac{x}{B_1 F_1} \right)^{3/2} + 1 \qquad (10-25)$$

The coordinates of the reverse curves can be obtained from Figure 10-12. Negative waves generated by the convex curve are compensated by positive waves formed by the reverse curve, and the flow is restored to uniformity at the end of the transition.

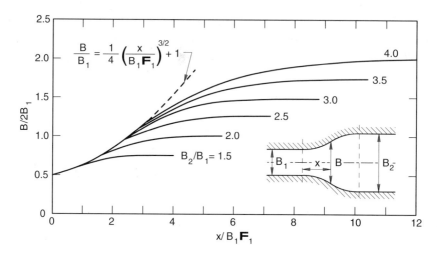

Figure 10-12. Dimensionless channel expansion curves.

PROBLEMS

10-1 Design a transition between a concrete-lined flume of rectangular cross section and a channel of trapezoidal section. The discharge in the channels is 15 m³/s. The flume is 6 m wide. The slope of the flume is 0.0004 and the Manning n is 0.012. The trapezoidal channel has side slopes of 1.5H:1V, a bottom width of 4.5 m, bottom slope of 0.0003, and Manning n of 0.02.

10-2 A gradual transition has been designed from a rectangular section to a trapezoidal section. The flow depth in the rectangular channel of 4.5-m width is 1.2 m. The trapezoidal channel has a bottom width of 3 m with sides sloping at 2H:1V. The depth in the trapezoidal channel is 1.8 m. The discharge in the channels is 6 m³/s. Determine the total change in bottom elevation from the rectangular section to the trapezoidal section. Neglect surface resistance.

10-3 Uniform flow is established in a steep, wide, rectangular concrete channel. An infinitesimal oblique wave originates from a construction joint in one of the channel walls. The oblique wave makes an angle of 25° from the channel wall. If the slope of the channel is 0.02 and the Manning n is 0.016, determine the discharge per unit width in the channel.

10-4 A rapid flow of 10 m³/s/m encounters a gradually expanding wall through a total deflection of 20°. The initial flow depth is 1.2 m. Determine (a) depth and velocity along the wall at the end of the deflection and (ii) disturbance angle of the initial and final wavelets.

10-5 A rectangular channel 3 m wide carries water at a depth of 1.0 m and a velocity of 15 m/s. The channel is turned abruptly through an angle of 30°. In order to prevent wave disturbance in the downstream channel, how far should the inner corner of the channel be offset longitudinally from the outer corner and what will be the width of the downstream channel?

10-6 Design a straight-walled contraction for a rectangular channel for the following conditions: upstream width $B_1 = 15$ m, downstream width $B_3 = 10$ m, and upstream Froude number $F_1 = 6$.

10-7 A supercritical flow defined by $F_1 = 4$ and $y_1 = 0.3$ m passes through a linearly contracting section of deflection angle θ, which causes a rise in depth by a factor of 2. The initial width of the channel is 4 m. The length of the contraction is defined by the intersection of the oblique wavefronts on the centerline of the channel. Beyond this point the bottom of the channel drops abruptly to a much lower level. Determine (a) the length of the contraction, (b) the width of the terminal section, and (c) the Froude number of the deflected stream.

10-8 Determine B_3 in Figure 10-11b for $F_1 = 6$, $\theta = 10°$ and $B_1 = 15$ m.

10-9 A high-velocity stream with a depth of 1.5 m and a Froude number of 4.0 emerges from a series of sluice gates between 1-m wide piers. Compute the flow depth immediately downstream of the pier on the assumption that the gate width is very large.

10-10 Consider flow in the channel included in Figure 10-7. The velocity and depth in the channel upstream from point A are 10 m/s and 0.6 m, respectively. The angle θ is 10° and the channel width upstream from point A is 6 m. Determine the depth and velocity at a point on the inclined wall 15 m downstream of point A.

10-11 A supercritical flow encounters a sudden positive wall deflection of 15° in a channel that carries a flow of 5 m³/s/m. What is the angle of the transverse wavefront if the depth upstream from the wavefront is 0.6 m? What is the downstream depth?

10-12 An abrupt negative wall deflection of 15° creates a gradual drop in water surface from an initial depth of 1.5 m to a final depth of 0.6 m in supercritical flow. Determine the unit discharge.

10-13 The downstream channel of Problem 10-8 is enlarged to a width of 15 m some distance downstream. Design an efficient expansion transition section.

10-2 Flow in Bends

Bends occur frequently in rivers. Bends are also provided in artificial channels to satisfy constraints in the alignment. The flow in bends is nonuniform due to normal acceleration. The flow in erodible-channel bends is more complex due to scour and deposition within the bend; such flows are beyond the scope of this book. Subcritical and supercritical flows are discussed separately as oblique wavefronts become significant in the latter flow.

Subcritical Flow

Consider flow around a bend, as shown in Figure 10-13. The streamlines within the bend region are curved in the xz plane; consequently there exists flow acceleration along the lateral direction. The pressure distribution is hydrostatic in the y-direction as the fluid acceleration a_y in the y-direction is zero and is obtained from Eq. (1-1) as

$$\frac{\partial p}{\partial y} = -\gamma \cos \theta$$

A similar equation in the z-direction can be written as

$$\frac{\partial p}{\partial z} = -\rho a_z$$

where a_z is the fluid acceleration in the z-direction. Integration of the two equations yields

$$p(y, z) = -\gamma y \cos \theta - \rho \int_{z_r}^{z} a_z \, dz + A$$

where z_r is the z-coordinate of the point of intersection of the free surface with the right bank (see Figure 10-13) and A is the constant of integration that can be obtained from the boundary condition that the pressure is zero at $y = y_{sr}$ and $z = z_r$ as

$$A = \gamma y_{sr} \cos \theta$$

where y_{sr} is the y-coordinate of the point of intersection of the free surface with the right bank. Substitution for A yields the following equation for the pressure distribution:

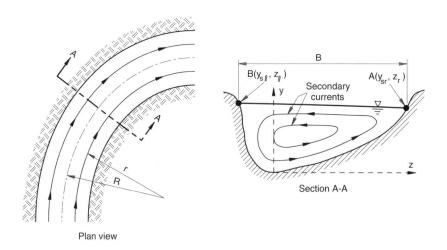

Figure 10-13. Flow in a bend.

$$p(y,z) = \gamma \cos\theta (y_s - y) - \rho \int_{z_r}^{z} a_z \, dz \qquad (10\text{-}26)$$

It can be shown (Rouse, 1946) that

$$a_z = \frac{u^2}{r} \qquad (10\text{-}27)$$

where r is the local radius of curvature of the streamlines in Figure 10-13. The evaluation of the pressure correction term requires the knowledge of the velocity distribution. For $B \ll r$, it can be assumed that $u \cong V$ and $r \cong R$, and Eq. (10-26) for $\cos\theta = 1$ can be written as

$$p(y,z) = \rho g \{ y_s(z_\ell) - y \} - \frac{\rho V^2}{R} (z - z_\ell) \qquad (10\text{-}28)$$

where R is the radius of curvature of the centerline of the channel. The difference in water-surface elevation between point B on the left bank and point A on the right bank can be obtained from Eq. (10-28) using the condition that the pressure at point B where $y = y_{s\ell}$ and $z = z_\ell$ is zero, i.e.,

$$y_s(z_\ell) - y_s(z_r) = \frac{V^2}{gR}(z_r - z_\ell) = \frac{V^2 B}{gR} \qquad (10\text{-}29)$$

where $B = (z_r - z_\ell)$ = surface width. The free surface is a plane for this case. A better approximation is that the streamlines are concentric circles and the velocity distribution is the same as in free-vortex flow, i.e., ur = constant (Prob. 10-14). The water surface is parabolic in this case. The outside wall of the channel in a bend must be made high enough to accommodate the increase in water level due to the bend.

Two other features of flow in bend should be mentioned. One is the head loss due to bend in addition to the normal head loss due to friction. The head loss due to bend is generally small and can be neglected. The second feature is the presence of the secondary flow in the bend. The explanation for the secondary currents is as follows. The velocity u is not uniform over a vertical section; it is small near the bed and is high near the free surface. Consequently the normal acceleration is larger near the free surface than the bottom. On the other hand the transverse pressure gradient is almost constant on a vertical section, because the pressure distribution is hydrostatic along the vertical direction. As a result the transverse pressure gradient near the bottom is greater than the transverse inertia term, and this imbalance sets up an inward flow near the bed. The continuity equation demands an outward flow near the free surface. The inward flow near the bed and the outward flow near the surface result in secondary currents in the bend. In erodible channels this secondary current is responsible for scour along the outer bank.

Supercritical Flow

There is an additional complication due to the presence of cross waves in supercritical flow in a bend. The cross-wave pattern for supercritical flow in a curved channel of constant width B and centerline radius r_c is shown in Figure 10-14. Positive and negative disturbances emanate from the outer and inner walls, respectively. Consequently the water surface rises along the outer wall and drops along the inner wall. The first positive disturbance AB and the first negative disturbance A'B start from the tangent points A and A', respectively. These two disturbances meet at point B. The flow is undisturbed upstream from ABA'. Beyond the point B the two initial disturbances AB and A'B do not propagate along straight lines as they are affected by the disturbances emanating from the inner and outer walls. As the positive disturbances cross the negative disturbances, the deflections are added. The first negative disturbance reaches the outer wall at point C where the water surface is a maximum. After point C the negative disturbances from the inner wall cause the water surface along the outer wall to drop. The water surface along the inner wall begins to drop at point A' and reach a minimum at point D where the positive disturbances from the outer wall begin affecting the flow. The water surface starts rising after point D. The first negative disturbance is reflected negatively at point C, and the first positive disturbance is reflected positively at point D. They continue to be reflected back and forth across the channel, causing the water surface to have a series of maxima and minima along the wall, approximately at angles θ_0, $2\theta_0$, $3\theta_0$, and so on, as shown in Figure 10-14. The angle θ_0 can be determined by approximating the distance AC by AC'. From the geometry

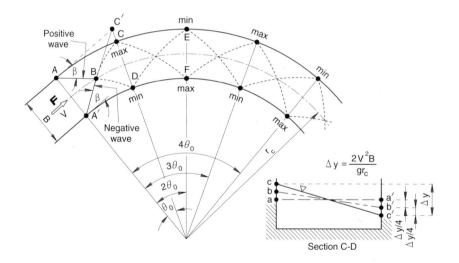

Figure 10-14. Cross-wave pattern in a curved channel.

$$AC' = \frac{B}{\tan \beta} = \left(r_c + \frac{B}{2} \right) \tan \theta_o$$

or

$$\tan \theta_0 = \frac{B}{(r_c + B/2)\tan \beta} \tag{10-30}$$

where $\beta = \sin^{-1}(1/F)$. The maximum and minimum surface elevations can be determined with Eqs. (10-21) and (10-22). The distance AC along the wall represents a half wavelength of water-surface oscillations. Thus the wavelength is equal to $2B/\tan\beta$.

Ippen and Knapp (1938) found that the maximum difference Δy in water-surface elevation between outer and inner walls for supercritical flow was about the twice the difference for subcritical flow, as shown for section CD in Figure 10-14, i.e.,

$$\Delta y = \frac{2V^2 B}{gr_c}$$

The lines a-a', b-b', and c-c' respectively represent the water surface in a straight channel, in a curved channel carrying subcritical flow, and in a curved channel carrying supercritical flow. In other words one-half of the rise in the water surface is due to wave disturbances and the other half is due to the centrifugal acceleration. The water surface in section EF in Figure 10-14 is level, as the effect of wave disturbances is cancelled by that due to centrifugal acceleration.

The pattern of cross waves will continue downstream from the plain circular bend unless it is modified to suppress the wave disturbances. The following methods for suppressing cross waves have been suggested by Knapp (1951):

A. Easement curves. The bend consists of three circular curves, as shown in Figure 10-15. The circular main curve of radius r_c and central angle θ_c is preceded and followed by another circular transition curve of radius $2 r_c$ and central angle θ_t that is given by

$$\theta_t = \tan^{-1} \frac{B}{(2r_c + B/2) \tan \beta} \tag{10-31}$$

This method is aimed at producing a positive counterwave with a phase shift of one-half of the wavelength and a height of one-half of the maximum disturbance for a plain circular curve. The water surface begins to rise from point A and to drop from point A'. The difference in water surface level between points B and B' is $V^2 B/(gr_c)$. The sidewalls beyond point A (A') are not involved in turning the flow but merely confine the flow as in the straight sections of the channel. The difference in water surface elevation is maintained throughout the entire length of the main curve. The downstream circular transition curve produces the interference pattern required to eliminate the disturbance in the downstream tangent. The total

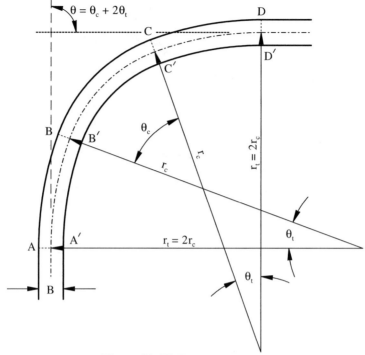

Figure 10-15. Easement curves.

total angle of the bend is $\theta_c + 2\theta_t$.

B. Banking of the channel bottom. Consider flow in a circular channel with vertical sidewalls and with the bottom sloping at an angle ϕ (Figure 10-16a) given by

$$\phi = \tan^{-1} \frac{V^2}{gr_c} \tag{10-32}$$

where r_c is the radius of the circular curve. The banking of the bottom supplies the lateral force necessary to change the direction of the flow. This force is equal to the centrifugal force due to velocity V and radius r_c. The water surface in the lateral direction is parallel to the channel bottom, as shown in Figure 10-16a. The sidewalls in this case are not involved in turning the flow and consequently do not generate wave disturbances. The banking is obtained either by raising the bottom on the outer wall or by lowering it on the inside wall or by a combination of the two. The banking is introduced, as well as eliminated, gradually by providing a spiral transition before and after the main curve, as shown in Figure 10-16b. The transition spiral curve is given by

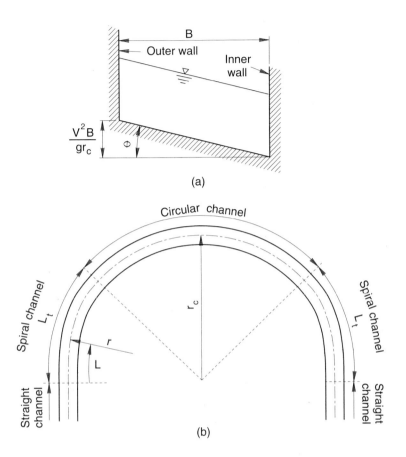

Figure 10-16. Banking of the channel bottom.

$$rL = r_c L_t \tag{10-33}$$

where r is the local radius of curvature at a distance L from the beginning of the transition and L_t is the length of the transition given by

$$L_t = 15 \frac{V^2 B}{g r_c} \tag{10-34}$$

C. Diagonal sills. Diagonal sills on the channel bottom near the entrance to and the exit of the bend are found to produce the necessary interference pattern to suppress the cross waves; the details are given by Knapp (1951).

PROBLEMS

10-14   Derive an expression for the difference between the water-surface elevation at the
two banks of a river bend of radius R assuming that the streamlines are concentric
and the velocity distribution in the bend is given by

$$ur = C$$

where C is a constant and r is the radius of any streamline.

10-15   A trapezoidal channel with a bottom width of 5 m and side slopes of 1H:1V has a
horizontal bend of radius 30 m. The channel carries a discharge of 30 m³/s and the
water depth at the inside of the bend is 2 m. Determine the water depth at the
outside of the bend.

10-16   A 90° bend is to be designed for a 10-m wide, rectangular channel carrying 180 m³/s
of water at a depth of 1.2 m. If the depth is not to exceed 2 m at any location, what
is the minimum bend radius required?

10-17   High-velocity flow that exists in a rectangular channel of 10 m width must turn by
90°. The depth of flow is 1.2 m and the velocity is 12 m/s. Design an easement
curve so that depth anywhere in the curve does not exceed 1.8 m.

10-18   Solve Problem 10-17 introducing a banking of the channel bottom. Design spiral
transitions.

10-3 Hydraulic Jump

The concept of the hydraulic jump was introduced in Chapter 3. A hydraulic jump
forms when a supercritical flow changes into a subcritical flow. The change in the
flow regime occurs with a sudden rise in water surface. Considerable turbulence,
energy loss, and air entrainment are produced in the hydraulic jump. A hydraulic
jump is used for mixing chemicals in water supply systems, for dissipating energy
below artificial channel controls, and as an aeration device to increase the
dissolved oxygen in water. Basic characteristics of the hydraulic jump in horizontal
rectangular channels are presented here; these characteristics are used in designing
stilling basins.

Energy Loss in the Jump

The loss of energy in the jump as given by Eq. (3-24) is

$$\Delta E_j = \frac{(y_2 - y_1)^3}{4 y_1 y_2} \qquad (3\text{-}24)$$

It can be shown that the relative energy loss in the jump, $\Delta E_j/E_1$, can be expressed
as (Prob. 10-19)

$$\frac{\Delta E_j}{E_1} = \frac{8F_1^4 + 20F_1^2 - \left(8F_1^2 + 1\right)^{3/2} - 1}{8F_1^2 \left(2 + F_1^2\right)} \tag{10-35}$$

Equation (10-35) is compared with the experimental data of the U. S. Bureau of Reclamation (1958) in Figure 10-17. The agreement between them is fairly good except for $F_1 < 2$. For a flow at the approach Froude number of 2.0, which corresponds to a relatively thick jet at low velocity entering the jump, the relative energy loss is about 7 percent. On the other extreme, for a flow at the approach Froude number of 19 produced by a relatively thin jet at high velocity entering the jump, the relative energy loss is about 85 percent. A significant amount of energy is lost in the jump, particularly at high Froude numbers.

Types of Jumps

Hydraulic jump in horizontal rectangular channels can be classified into five categories (Bradley and Peterka, 1957) based on the approach Froude number, F_1, as shown in Figure 10-18:

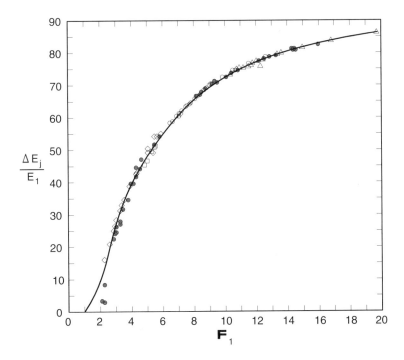

Figure 10-17. Loss of energy in jump on horizontal floor.

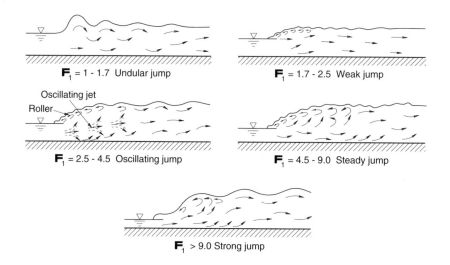

Figure 10-18. Types of hydraulic jump.

Undular jump ($1 < F_1 < 1.7$): The water surface shows undulations. The energy loss in the undular jump is insignificant.

Weak jump ($1.7 < F_1 < 2.5$): A series of surface rollers develops on the surface of the hydraulic jump. The water surface downstream of the jump is smooth. The energy loss in the weak jump is small.

Oscillating jump ($2.5 < F_1 < 4.5$): The incoming jet oscillates in a random manner between the bed and the surface producing surface waves that persist a considerable distance downstream and can cause bank erosion.

Steady jump ($4.5 < F_1 < 9.0$): The jump is well stabilized. The downstream extremity of the surface roller and the point at which the high-velocity jet tends to leave the floor occur in practically the same vertical plane. The energy dissipation in the steady jump ranges from 45 percent at $F_1 = 4.5$ to 70 percent at $F_1 = 9.0$.

Strong jump ($9.0 < F_1$): The water surface is very rough. The point at which the high-velocity jet tends to leave the floor lies upstream of the downstream extremity of the surface roller. The jump action is rough but effective.

Length of the Jump

The length of the jump is an important parameter that governs the length of the stilling basin. The length of the jump is defined as the distance between the front of the jump to a vertical section passing through the downstream extremity of the surface roller. The experimental results are presented in Figure 10-19. A considerable scatter in the experimental data is to be expected because of the difficulty in locating the ends of the jump. The length is practically constant at about $6y_2$ over the range $4.5 < F_1 < 13$, where y_2 is the depth downstream of the jump.

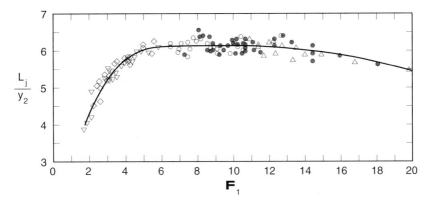

Figure 10-19. Length of jump on horizontal floor (data from different sources).

Surface Profile of the Jump

Knowledge of surface profile of the jump is useful in determining the height of the sidewalls and the pressure on the floor of the stilling basin. The experimental data from different sources (Rajaratnam and Subramanya, 1968) are presented in Figure 10-20. Data show a unique relationship between $y/[0.75(y_2 - y_1)]$ and x/X, where X is defined in Figure 10-20. The length X is empirically related to y_1 and F_1 as

$$\frac{X}{y_1} = 5.08F_1 - 7.82 \tag{10-36}$$

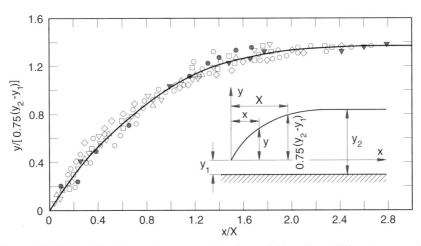

Figure 10-20. Profile of jump in rectangular channel (data from different sources).

Hager (1992) developed the following relationship between the flow depth y and the distance x from the beginning of the jump:

$$\frac{y - y_1}{y_2 - y_1} = \tanh \frac{x}{L_r} \tag{10-37}$$

where L_r is the length of the roller and is given by

$$\frac{L_r}{y_1} = -1.2 + 160 \tanh \frac{F_1}{20} \tag{10-38}$$

Control of the Jump

The location of a jump on an horizontal floor is very sensitive to the depth downstream of the jump. A slight variation in the downstream depth causes the jump to migrate substantial distances unless the jump location is stabilized by installing accessory devices such as baffle piers and sills on the floor, as shown in Figure 10-21. The momentum equation is applied to a hydraulic jump under the influence of baffle piers and can be written for a unit width of the channel as

$$\frac{y_1^2}{2} - \frac{y_p^2}{2} - F_p = \rho q \left(\frac{q}{y_p} - \frac{q}{y_1} \right) \tag{10-39}$$

in which F_b = drag force exerted by the baffle piers per unit width of the channel, y_1 = depth in section 1, and y_b = depth in section 2 with the baffle piers between sections 1 and 2. The first two terms on the left-hand side of Eq. (10-39) are the hydrostatic pressure forces on the vertical sections at 1 and 2. The experimental results of several studies analyzed by Ranga Raju et al. (1980) show that for a given geometry of the baffle piers the drag force F_b is a function of the distance the baffle piers from the toe of the jump, and the drag force increases with decreasing distance. The two forces that retard the flow and produce the momentum change are the hydrostatic pressure force in section 2 and the drag force on the baffle piers. While the former force decreases with decreasing downstream depth, the latter force increases because the distance between the baffle piers and the toe of the jump decreases due to downward migration of the jump. Consequently the jump migrates a small distance due to small decrease in downstream depth. A procedure to design baffle piers and sills is outlined by Ranga Raju et al. (1980).

Stilling Basins

A hydraulic-jump stilling basin is a short length of the paved channel downstream of a hydraulic structure where the energy dissipation within the hydraulic jump occurs so that the outgoing stream can safely be conducted to the channel below. The important design features of a stilling basin are those that promote the

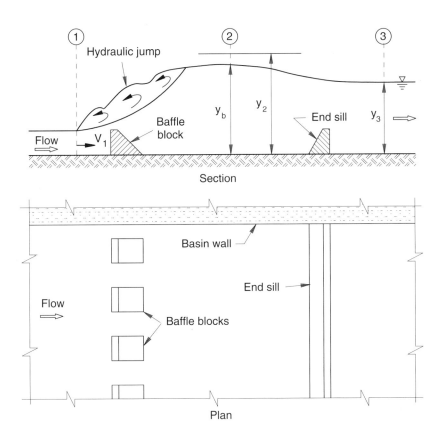

Figure 10-21. Control of jump by baffle piers.

formation of the jump within the basin, stabilize the jump, and make the basin as short as possible. The objectives are achieved by installing chute blocks, baffle piers, and end sills in stilling basins. Two stilling basins designed by the U. S. Bureau of Reclamation (1958) are presented in Figure 10-22. Basin III with the chute blocks, the baffle piers, and a solid end sill is recommended for small dams and outlet works. Basin II with the chute blocks and a dentated sill is generally used for spillways on high dams, where the velocity at the toe of the jump exceeds about 18 m/s. Note that Basin II does not include the baffle piers, as they are damaged in high-velocity flows. The damage is caused by cavitation.

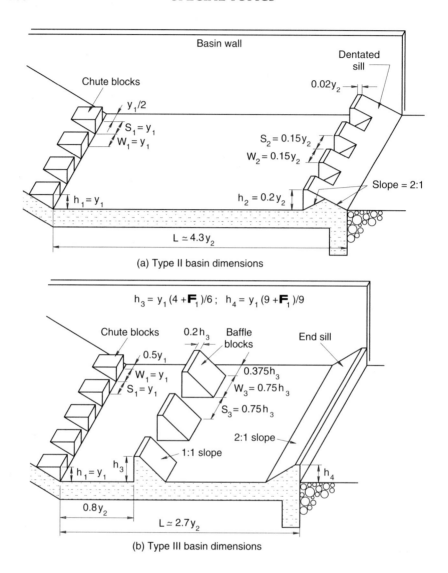

Figure 10-22. USBR Type II and Type III basins.

PROBLEMS

10-19 Using the basic equations for hydraulic jump in a rectangular channel, derive Eq. (10-35).

10-20 Water discharges at the rate of 300 m³/s over a spillway 12 m wide into a stilling basin of the same width. The lake level behind the spillway is 50 m and the elevation of the water surface in the river downstream is 25 m. Neglecting friction losses in the flow down the spillway, find the invert elevation of the stilling basin for a hydraulic jump to form within the basin. Also design a suitable stilling basin.

10-21 If it is found impossible to set the invert level in Problem 10-20 lower than 35 m, what must be the stilling basin width so that the jump will form within the basin.

10-22 Explain the effect of baffle piers in a stilling basin on the stability of a hydraulic jump.

10-4 Flow through Culverts

A culvert is a short conduit through an embankment constructed across a natural channel and is used to carry flow from one side of the embankment to the other, as shown in Figure 10-23. The culvert can flow full or partly full. The flow profiles in a culvert flowing partly full are the same as shown in Figure 4-10 for a mild culvert and in Figure 4-11 for a steep culvert. Culverts can be used for measuring

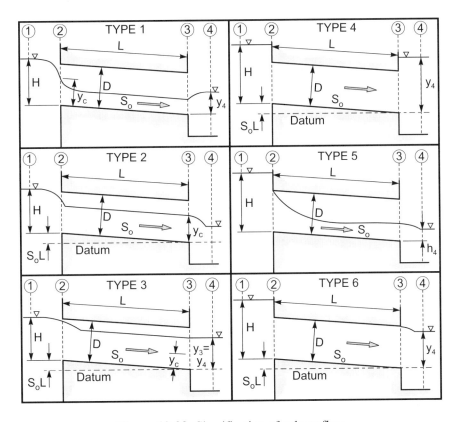

Figure 10-23. Classification of culvert flow.

discharge (Bodhaine, 1968). The determination of the discharge through the culvert is not simple because the flow is controlled by many factors including the inlet geometry, the slope, size and roughness of the conduit, and the headwater and tailwater levels. For purposes of discussion, four sections are identified in Figure 10-23. Sections 2 and 3 are situated at the inlet and the outlet of the conduit. Sections 1 and 4 are located at about one culvert opening width upstream of the culvert entrance and downstream of the culvert outlet, respectively. The following variables are identified: H = flow depth in section 1 above the invert of the culvert inlet, D = maximum vertical dimension of the cross section of the culvert, L = length of the culvert, S_0 = slope of the culvert, and y_4 = tailwater depth above the invert of the culvert outlet.

For convenience in computing discharge, culvert flow is classified into six types on the basis of the location of the control section, the type of control, and the headwater and tailwater elevations. The characteristics of the six types of flow are summarized in Table 10-2 and the flow profiles are shown in Figure 10-23.

Table 10-2. Characteristics of Culvert Flows

Type	Inlet Situation	Conduit Flow	Control Section Location	Control Section Type	Culvert Slope
1	Unsubmerged	Partly full	Inlet	CDC	Steep
2	Unsubmerged	Partly full	Outlet	CDC	Mild
3	Unsubmerged	Partly full	Outlet	ACC[*]	Mild
4	Submerged	Full	Outlet	ACC[*]	Any
5	Submerged	Partly full	Inlet	ACC[**]	Any
6	Submerged	Full	Inlet	ACC[#]	Any

[*]Tailwater flow condition; [**]Similar to sluice gate control;
[#] "Drowned" control.

Discharge Equations

Discharge equations for each type of flow are developed by application of the continuity and energy equations between the approach section and the control section.

Type 1 Flow. The flow in the culvert is a free-surface flow, and the culvert slope is steep. The control is a critical-depth control and is located in the inlet (section 2). The inlet is unsubmerged. The difference in headwater level and the water-surface level at the critical depth in the inlet in Figure 4-9a is $(1 + k_e)V_c^2/2g$ and is equal to about 0.5 y_c. For the inlet to remain unsumberged, H/D should, therefore, be less than about 1.5. The depth in section 2 is critical as long as the depth in the outlet (section 3) is less than y_u, where y_u is defined in Figure 4-11 and can be obtained from Eq. (4-26). According to Bodhaine (1968), $y_u \cong y_c + S_0L$; therefore the depth in section 2 is critical if $y_4 < y_c + S_0L$. Note that the flow at depth y_u is subcritical, and therefore $y_3 = y_4$. The energy equation between sections 1 and 2 can be written as

$$H + \alpha_1 \frac{V_1^2}{2g} = y_c + \frac{V_c^2}{2g} + \Delta h_e + \Delta h_{1-2}$$

where Δh_{1-2} = energy loss due to friction between sections 1 and 2 = $\overline{S_f} \, L_{1-2}$ = $\sqrt{S_{f1} S_{f2} L_{1-2}}$ = $L_{1-2}(Q^2/K_1 K_c)$; K = conveyance, defined in Eq. (2-16); L_{1-2} = distance between section 1 and 2; Δh_e = head loss due to entrance contraction = $[(1/C^2) - 1](V_c^2/2g)$, where C is the discharge coefficient. From the above equation the velocity in section 2 is

$$V_c = C \sqrt{2g\left(H + \alpha_1 \frac{V_1^2}{2g} - y_c - \Delta h_{1-2}\right)}$$

From the continuity equation the discharge is given by

$$Q = V_c A_c = C A_c \sqrt{2g\left(H + \alpha_1 \frac{V_1^2}{2g} - y_c - \Delta h_{1-2}\right)} \qquad (10\text{-}40)$$

The discharge coefficient that accounts for the contraction in the flow area is a function of the inlet geometry (Bodhaine, 1968). The critical depth in section 2 can be obtained from Eq. (4-24).

Type 2 Flow. The flow in the culvert is a free-surface flow, and the culvert slope is mild. The control is a critical-depth control and is located in the outlet (section 3). The inlet is unsubmerged. The difference in headwater level and the water-surface level at the normal depth in the inlet in Figure 4-9b is $(1 + k_e)V_n^2/2g$ and is equal to about $0.2y_n$. For the inlet to remain unsumberged, H/D should, therefore, be less than about 1.2. The depth in section 3 is critical as long as the tailwater depth is less than y_c. The energy equation between sections 1 and 3 can be written as

$$H + \alpha_1 \frac{V_1^2}{2g} + S_0 L = y_c + \frac{V_c^2}{2g} + \Delta h_e + \Delta h_{1-2} + \Delta h_{2-3}$$

From the above equation the velocity in section 3 is

$$V_c = C \sqrt{2g\left(H + S_0 L + \alpha_1 \frac{V_1^2}{2g} - y_c - \Delta h_{1-2} - \Delta h_{2-3}\right)}$$

From the continuity equation the discharge is given by

$$Q = V_c A_c = C A_c \sqrt{2g\left(H + S_0 L + \alpha_1 \frac{V_1^2}{2g} - y_c - \Delta h_{1-2} - \Delta h_{2-3}\right)} \qquad (10\text{-}41)$$

where $\Delta h_{2\text{-}3} = L\ (Q^2/K_2K_3)$. The critical depth in section 3 can be obtained from Eq. (4-24). The depth in section 2 can be obtained from the energy equation between sections 1 and 2 and can be approximated by

$$H = y_2 + \frac{V_2^2}{2gC^2} \tag{10-42}$$

Type 3 Flow. The flow in the culvert is a free-surface flow, and the culvert slope is mild. The control is the tailwater depth ($y_4 > y_c$) and is located in section 4. For the inlet to remain unsumberged, H/D should, therefore, be less than about 1.2. The depth in section 3 is equal to that in section 4. The outlet is unsubmerged if $y_4/D < 1$. Similar to type 2 flow, the discharge in type 3 flow is given by

$$Q = V_3 A_3 = CA_3 \sqrt{2g\left(H + S_0 L + \alpha_1 \frac{V_1^2}{2g} - y_3 - \Delta h_{1\text{-}2} - \Delta h_{2\text{-}3}\right)} \tag{10-43}$$

Type 4 Flow. The culvert flows full and the discharge can be computed from the energy equation between sections 1 and 4. Neglecting the velocity head in section 1 and the friction losses between sections 1 and 2 and between sections 3 and 4, the energy equation can be written as

$$H + S_0 L = y_4 + \frac{V_4^2}{2g} + \Delta h_e + \Delta h_{2\text{-}3} + \Delta h_{ex}$$

where Δh_{ex} = loss due to sudden expansion between sections 3 and 4 = $[(V_3^2/2g) - (V_4^2/2g)]$. Note that $V_2 = V_3$. Substitution for $\Delta h_e = [(1/C^2) - 1](V_3^2/2g)]$ and $\Delta h_{2\text{-}3} = n^2 L V_3^2/R_0^{4/3}$ in the above equation, where n is the Manning roughness coefficient for the culvert and R_0 is the hydraulic radius of the cross section of the culvert, an expression for the discharge can be obtained as

$$Q = CA_0 \sqrt{\frac{2g(H + S_0 L - y_4)}{1 + 2gC^2 n^2 L/R_0^{4/3}}} \tag{10-44}$$

where A_0 = cross-sectional area of the culvert.

Type 5 Flow. The culvert flows partly full, but the inlet is submerged, which occurs for H/D values larger than about 1.2–1.5 depending upon the entrance shape. The tailwater level is below the crown of the culvert at the outlet. The flow cross section beyond the inlet contracts in a manner similar to a sluice gate. The discharge equation similar to Eq. (9-23) can be derived and is given by

$$Q = CA_0 \sqrt{2gH} \tag{10-45}$$

For a culvert at mild slope the water-surface profile downstream from the point of maximum contraction, called vena contracta, an M_3 curve develops. The depth in section 3 is a function of H, L, S_0, R_0, n, and the shape of the edge at the top of the entrance (square, rounded, or beveled). If the characteristics of the culvert are such that y_3 becomes larger than D, the culvert will be sealed both at the entrance and the outlet. The culvert then runs full as in type 6 flow. Such a culvert is termed as *hydraulically long*; otherwise the culvert is *hydraulically short*. Whether a culvert is hydraulically long or short can be determined from the experimental data presented by Bodhaine (1968).

Type 6 Flow. The culvert is hydraulically long and flows full. The tailwater does not submerge the outlet, i.e., $y_4 < D$. Neglecting the approach velocity head and the energy loss due to friction between sections 1 and 3 in the energy equation between section 1 and 3, the discharge equation is obtained as

$$Q = CA_0 \sqrt{2g\left(H + z - y_3 - \Delta h_{2-3}\right)} \tag{10-46}$$

The pressure head y_3 in Eq. (10-46) is not equal to D and lies between $D/2$ and D. Instead of using Eq. (10-46), Bodhaine (1968) developed a functional relationship from the experimental data to compute the discharge.

Coefficient of Discharge

Bodhaine (1968) developed, by laboratory study, empirical relations for coefficients of discharge, C, for types 1–6 flows; the reader should refer to this study to determine the discharge coefficients. The coefficients vary from 0.39 to 0.98, and they have been found to be a function of the degree of channel contraction and the geometry of the culvert entrance. The radius of rounding or degree of bevel of corrugated pipes is the critical dimension.

Example 10-5. Show that the discharge in a box culvert with the following data is 14.7 m³/s: $H = 2.44$ m, square section with $D = 2.44$ m, $L = 18$ m, $S_0 = 0.0034$, Manning $n = 0.015$, $y_4 = 1.22$ m, and $C = 0.95$.

(a) $H/D = 2.44/2.44 = 1 < 1.2$; unsubmerged case.

(b)   Determine the critical depth: $q = 14.7/2.44 = 6.02$ m³/s/m; $y_c = (q^2/g)^{1/3} = (6.02^2/9.81)^{1/3} = 1.55$ m.

(c)   Determine the normal depth: $nQ/b^{8/3} S_0^{1/2} = 0.015 \times 14.7/2.44^{8/3} \times 0.0034^{1/2} = 0.350$: From Table D-1, $y_n/b = 0.774$; $y_n = 0.774 \times 2.44 = 1.82$ m.

(d)   Since $y_n > y_c$ and $y_4 < y_c$, it is type 2 flow.

(e)   Assume $y_2 = y_n$; $A_2 = 2.44 \times 1.82 = 4.44$ m²; $P_2 = 2.44 + 2 \times 1.82 = 6.08$ m; $R_2 = 4.44/6.08 = 0.73$ m; $K_2 = 4.44 \times 0.73^{2/3}/0.015 = 240.4$.
       $y_3 = y_c = 1.55$ m; $A_3 = 2.44 \times 1.55 = 3.78$ m²; $P_3 = 2.44 + 2 \times 1.55 = 5.54$ m; $R_3 = 3.78/5.54 = 0.58$ m; $K_3 = 3.78 \times 0.58^{2/3}/0.015 = 175.7$.
       $\Delta h_{2-3} = L(Q^2/K_2 K_3) = 18 \times 14.7^2/240.4 \times 175.7 = 0.09$ m.

(f)   From Eq. (10-41) upon neglecting the approach velocity and the head loss in the approach section

$$Q = CA_c \sqrt{2g\left(H + S_0 L - y_c - \Delta h_{2-3}\right)}$$
$$= 0.95(2.44 \times 1.55)\sqrt{2 \times 9.81\left(2.44 + 0.0034 \times 18 - 1.55 - 0.09\right)}$$
$$= 14.8 \text{ m}^3/\text{s}.$$

OK.

PROBLEMS

10-23 Determine the discharge through a box culvert with the following data: $H = 2$ m, square section with $D = 1.9$ m, $S_0 = 0.01$, $L = 20$ m, Manning $n = 0.015$, and $y_4 = 1.7$ m.

10-24 Solve Problem 10-23 if $y_4 = 2.1$ m.

10-25 Solve Problem 10-23 if $H = 3$ m.

10-26 Solve Problem 10-23 if the culvert is circular.

10-5 Surges in Power Canals

The start-up and shutdown times of hydraulic turbines used for electric power generation are rather small. During start-up and shutdown of a turbine there are sudden changes in flow conditions in the upstream and downstream power channels, which in turn develop surges in the channels. Two typical cases of the analysis of positive surges are given below.

Meeting of Two Surges

When two surges travelling towards each other meet, the result is formation of two new surges travelling away from each other, as shown in Figure 10-24. There are four unknown quantities, namely, y_3, V_3, V_{wd}, and V_{wu}. These four unknowns can be obtained by a simultaneous solution of the continuity and momentum equations across each of the two new surges. The solution can be simplified by assuming that the heights of the surges are small. For a surge shown in Figure 3-1, let $\Delta y = y_1 - y_2$, $\Delta V = V_1 - V_2$, and $\Delta y/y_1 \ll 1$, i.e., $y_1 \cong y_2 \cong y$. For small surges Eq. (3-7) can be written as

$$\Delta V = \pm \Delta y \sqrt{\frac{g}{y}} = \pm 2\Delta\left(\sqrt{gy}\right)$$

or

$$\Delta\left(V \pm 2\sqrt{gy}\right) = 0$$

or

$$V \pm 2c = \text{constant} \qquad (10\text{-}47)$$

where $c = \sqrt{gy}$. In Eq. (10-47) the positive and negative signs are valid for a surge moving upstream and downstream, respectively. The solution is presented in

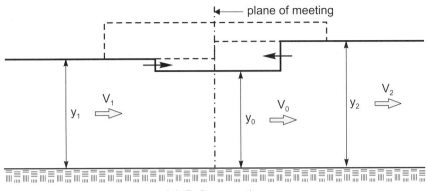

(a) Before meeting

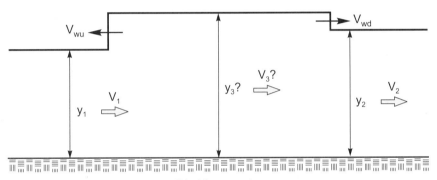

(b) After meeting

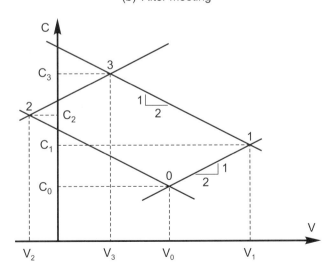

(c) Graphical solution

Figure 10-24. Meeting of two surges .

graphical form in Figure 10-24c. In the V-c plane, Eq. (10-47) is a family of straight lines with slopes of $\pm$ ½. The two unknowns V_3 and c_3 can be obtained from the following two equations. Equation (10-47) across the new surges moving upstream and downstream can be written respectively as

$$V_3 + 2c_3 = V_1 + 2c_1 \qquad (10\text{-}48)$$

$$V_3 - 2c_3 = V_2 - 2c_2 \qquad (10\text{-}49)$$

Surge Due to Sudden Load Rejection

It was shown in Section 7-7 that due to sudden closure of a sluice gate a positive surge moves upstream. The energy slope and the bottom slope of the channel were neglected in the analysis, and the height and velocity of the surge were assumed constant. A sudden rejection of load in a power canal also results in a surge that travels upstream. For establishing the height of the sidewalls of the canal, one needs to determine the maximum stage of water that could be developed as a result of a sudden decrease of discharge in the power canal to zero. As power canals could be fairly long, the effect of the bed slope could no longer be neglected. Due to bed slope the height and velocity of the surge change as it advances upstream. The water surface downstream from the surge front is approximately level, as shown in Figure 10-25, because velocity in that region is almost zero. The effect of resistance in that region also is negligible due to the same reasoning.

Let the surge advance a distance L_i in time t_i as in Figure 10-25. The canal cross section is assumed rectangular. The volume of water conveyed through the canal in time t_i filled the rectangular region ABCD and the triangular region CDE. Hence from the continuity equation

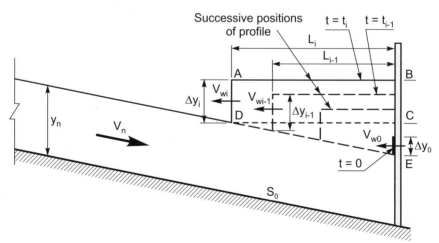

Figure 10-25. Positive surge moving upstream in a sloping channel.

$$V_n y_n t_i = L_i \Delta y_i + \frac{S_0 L_i^2}{2} \tag{10-50}$$

where V_n and y_n are the initial normal velocity and depth during the steady flow in the canal and the other symbols are defined in Figure 10-25. Equation (3-9), which is based on the continuity and momentum equations across a surge, can be written for the surge in Figure 10-25 as

$$V_{wi} + V_n = \sqrt{\frac{g}{2y_n}(y_n + \Delta y_i)(2y_n + \Delta y_i)} = \sqrt{gy_n\left[1 + \frac{3}{2}\frac{\Delta y_i}{y_n} + \frac{1}{2}\left(\frac{\Delta y_i}{y_n}\right)^2\right]}$$

where V_{wi} is the surge velocity. Assuming that $\Delta y_i \ll y_n$, the last term under the radical sign in the above equation can be neglected. The above equation then can be written as

$$V_{wi} + V_n = \sqrt{gy_n}\left(1 + \frac{3}{4}\frac{\Delta y_i}{y_n}\right) \tag{10-51}$$

Equations (10-50) and (10-51) can be solved for V_{wi} and Δy_i. From the known flow condition at time t_{i-1}, the flow conditions at time t_i can be determined as follows:

1. Compute the average surge velocity from $\overline{V}_{wi} = (L_i - L_{i-1})/(t_i - t_{i-1})$.
2. Compute V_{wi} from $\overline{V}_{wi} = (V_{wi} + V_{wi-1})/2$.
3. Compute Δy_i from Eqs. (10-50) and (10-51); if the two values are not the same, try another value of t_i.

Equation (10-50) for $t_i = 0$ can be written as

$$V_n y_n = V_{w0}\Delta y_0 \tag{10-52}$$

where V_{w0} and Δy_0 are the surge velocity and depth at time $t_i = 0$. The two unknowns can be obtained from Eqs. (10-51) and (10-52).

PROBLEMS

10-27 A positive surge 0.20 m high is moving in still water 15 m deep towards the dead end of a channel. Determine the height and the velocity of the reflected surge after the original surge hits the dead end.

10-28 Two positive surges 1.0 and 0.6 m in height move in opposite directions towards each other in a frictionless horizontal channel where the water is initially stationary at a depth of 9 m. Determine the flow conditions after the surges meet.

10-29 A rectangular channel delivers 30 m³/s to a hydropower station. The characteristics of the channel are width = 10 m, Manning $n = 0.015$, bottom slope = 0.002, and channel length = 2 km. Find the maximum height of surge reached at the upstream end of the channel on sudden complete shutdown from normal flow.

10-6 Roll Waves

A steady discharge of water in an open channel of sufficient length and slope can become unstable and may not result in a uniform flow. Instead a series of gravity waves, called *roll waves*, may develop in the chute channel. In the initial phase of development the waves have small amplitudes and the water surface is continuous as in Figure 10-26. As these small-amplitude waves propagate downstream, they

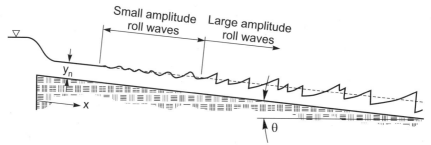

Figure 10-26. Development of roll waves.

eventually break and form a series of surges at regular intervals. The depth reaches a maximum just upstream, and a minimum just downstream, of the surge front (Figure 10-27). Roll waves in chute structure are objectionable because they may overtop the chute walls and cause surging in energy dissipators. Roll waves generally develop in chute channels that are longer than about 60 m and have bottom slope flatter than about 20° (Thorsky et al., 1967). The maximum wave height that can be expected is twice the normal depth for the slope.

Theoretical considerations for determining the instability of uniform flow are presented by Liggett (1975). Based on the analysis of Escoffier and Boyd (1962), the condition of stability is given by

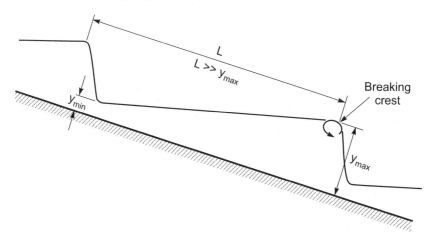

Figure 10-27. Schematic of a roll wave.

$$-1 < V < 1 \tag{10-53}$$

where V is called the Vedernikov number and is defined as

$$V = \kappa \varphi F \tag{10-54}$$

where κ = exponent of the hydraulic radius in the uniform flow equation (2-4) ($\kappa = 2$ for laminar flow; $\kappa = \frac{1}{2}$ or $\frac{2}{3}$ for turbulent flow using the relation of Chézy or Manning, respectively) and φ = channel shape factor defined as

$$\varphi = \left(1 - R \frac{dP}{dA}\right) \tag{10-55}$$

It can be shown that $\varphi = 1$ for very wide channels and $\varphi = 0$ for very narrow channels. If $V > 1$, the flow becomes unstable and the perturbation will amplify, leading to the formation of the roll waves. This condition for unstable flow is a necessary condition but is not a sufficient condition. Sufficient chute length is also necessary for flow to become unstable. For predicting the occurrence of roll waves in chute channels, Montuori (1963) incorporated the chute length in a design chart (Figure 10-28) that he developed for predicting the occurrence of the roll waves in chute channels. The equation for the curves in Figure 10-28 is

$$\frac{1 + \dfrac{2}{3} \dfrac{\varphi}{V}}{1 - V} \ln \varepsilon = \frac{g S_0 L}{V_n^2} \tag{10-56}$$

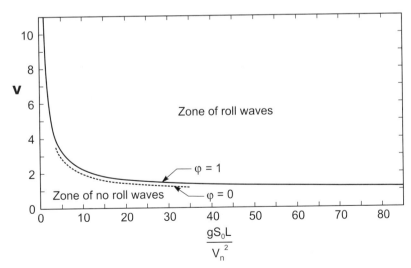

Figure 10-28. Prediction of occurrence of roll waves.

where L is the length of the channel and $\varepsilon = 10^{-4}$ is a calibration constant obtained from the experimental data. Note that the flow in Figure 10-28 is unstable only if $V > 1$.

Example 10-6. Is the flow in a wide rectangular channel with $y_n = 0.30$ m, $V_n = 10.0$ m/s, $S_0 = 0.10$ stable? The length of the channel is 150 m.

For a wide rectangular channel $\varphi = 1$, and $\kappa = \frac{2}{3}$ if the Manning equation is assumed applicable:

$$F = V_n/\sqrt{gy_n} = 10/\sqrt{9.81 \times 0.30} = 5.83$$

$$V = \kappa\varphi F = \frac{2}{3} \times 1 \times 5.83 = 3.89$$

$$gS_0L/V_n^2 = 9.81 \times 0.10 \times 150/10^2 = 1.47$$

The flow point ($V = 3.89$, $gS_0L/V_n^2 = 1.47$) lies in the stable zone in Figure 10-28. The channel is not long enough to develop roll waves.

PROBLEMS

10-30 What should be the minimum length of the channel in Example 10-6 for the roll waves to develop?

10-31 A rectangular channel 5 m wide and 250 m long carries a discharge of 25 m³/s. The slope of the channel is 0.15 and the Manning n is 0.014. Are the roll waves expected in the channel?

REFERENCES

Abbott, M. B. (1979). *Computational Hydraulics; Elements of the Theory of Free Surface Flows.* Pitman Publishing, London.

Abou-Seida, M. M. and Quraishi, A. A. (1976). A Flow Equation for Submerged Rectangular Weirs. *Proc. Inst. Civil Eng.,* 61, No.2, 685–696.

ASCE. Task Force (1963). Report on Friction Factors on Open Channels. *JHYD, ASCE,* 89, No. 2, 97–143.

Barnes, H. H. (1967). Roughness Characteristics of Natural Channels. U.S. Geological Survey, Water Supply Paper No. 1849, U.S. Government Press.

Basha, H. A. (1995). Routing Equations for Detention Reservoirs. *JHE, ASCE,* 121, No. 12, 885–888.

Blalock, M. E. and Sturm, T. W. (1981). Minimum Specific Energy in Compound Open Channel. *JHYD, ASCE,* 107, No. 6, 699–717.

Bodhaine, G. L. (1968). Measurement of Peak Discharge at Culverts by Indirect Methods. In *Techniques of Water Resources Investigations of the United States Geological Survey,* Book 3, Chapter A3. U.S. Geological Survey, Washington, DC.

Bradley, J. N. and Peterka, A. J. (1957). The Hydraulic Design of Stilling Basin: Hydraulic Jumps on a Horizontal Apron. *JHYD, ASCE,* 83, No. 5, 1–24.

Chaudhry, M. H. (1993). *Open-Channel Flow.* Prentice Hall, Englewood Cliffs, NJ.

Chaudhry, M. H. and Bhallamudi, S. M. (1988). Critical Depth in Compound Channels. *J. Hyd. Res., IAHR,* 26, No. 4, 377–495.

Chow, V. T., (1955). Integrating the Equation of gradually Varied flow, *JHYD, ASCE,* 81, No. 6, 1–32.

Chow, V. T. (1959). *Open Channel Hydraulics.* McGraw-Hill Book Company, New York.

Cunge, J. A. (1969). On the Subject of a Flood Propagation Computation Method (Muskingum Method). *J. Hyd. Res., IAHR,* 7, No. 2, 205–230.

Cunge, J. A., Holly Jr., F. M., and Verwey, A. (1980). *Practical Aspects of Computational River Hydraulics.* Pitman Publishing , London.

Dailey, J. W. and Harleman, D. R. F. (1966). *Fluid Dynamics.* Addison-Wesley, Reading, MA.

de Marchi, G. (1934). Saggio di Teoria del Funzionamento degli Stramazzi Laterali. *L'Energia Elletrica,* 11, No. 11, 849–860.

Diskin, M. H. (1961). End Depth at a Drop in Trapezoidal Channels. *JHYD, ASCE,* 87, No. 4, 11–32.

Dressler, R. F. (1954). Comparison of Theories and Experiments for the Hydraulic Dam-Break Wave. *Int. Assoc. Sci. Hydrol.,* 3, 319–328.

Eagleson, P. S. (1970). *Dynamic Hydrology.* McGraw-Hill Book Company, New York.

Escoffier, F. F. and Boyd, M. B. (1962). Stability Aspects of Flow in Open Channels. *JHYD, ASCE,* 88, No. 6, 145–166.

Ezra, A. A. (1954). A Direct Step Method for Computing Water-Surface Profiles. *Trans., ASCE*, 119, 453–462.

Farell, C. (1966). Discussion of Critical Analysis of Open-Channel Resistance by Hunter Rouse. *JHYD, ASCE*, 92, No. 2, 395–399.

French, R. H. (1985). *Open-Channel Hydraulics*. McGraw-Hill Book Company, New York.

Govinda Rao, N. S. and Muralidhar, D. (1963). Discharge Characteristics of Weirs of Finite Crest Width. *La Houille Blanche*, 5, 537–545.

Hager, W. H. (1992). *Energy Dissipators and the Hydraulic Jump*. Kluwer Academic Publishers, Dordrecht.

Hayami, S. (1951). On the Propagation of Flood Waves. Bulletin No.1, Disaster Prevention Research Institute, Kyoto University, Japan, December.

Henderson, F. M. (1966). *Open Channel Flow*. Macmillan Company, New York.

Henderson, F. M. and Wooding, R. A. (1964). Overland Flow and Groundwater Flow from a Steady Rainfall of Finite Duration. *J. Geophys. Res.*, 69, 1531–1540.

Henry, H. R. (1950). Discussion of Diffusion of Submerged Jets. *Trans, ASCE*, 115, 687–689.

Hildebrand, F. B. (1966). *Advanced Calculus for Applications*. Prentice-Hall, Englewood Cliffs, NJ.

Hinds, J. (1928). The Hydraulic Design of Flume and Syphon Transitions. *Trans., ASCE*, 92, 1423–1459.

Ippen, A. T. (1951). Mechanics of Supercritical Flow. *Trans., ASCE*, 116, 268–295.

Ippen, A. T. and Knapp, R. T. (1938). Experimental Investigations of Flow in Curved Channels. Report to Los Angeles County Flood Control District, reproduced by U. S. Engineering Office, Los Angeles.

Jain, S. C. (1971). Discussion of Effect of Channel Shape on Gradually Varied Flow Profiles. *JHYD, ASCE*, 97, No. 9, 1562–1565.

Jain, S. C. (1976). Discussion of Numerical Errors in Water Profile Computation. *JHYD, ASCE*, 102, No. 11, 1701–1702.

Jain, S. C. (1993). Nonunique Water-Surface Profiles in Open Channels. *JHE, ASCE*, 119, No. 12, 1427–1434.

Jain, S. C. and Fischer, E. E. (1982). Uniform Flow over a Skew Side-Weir. *JIDD, ASCE*, 108, No. 2, 163–166.

Kandaswamy, P. K. and Rouse, H. (1957). Characteristics of Flow Over Terminal Weirs and Sills. *JHYD, ASCE*, 83, No. 4, 1345-1 – 1345-13.

Kazemipour, A. K. and Aplet, C. J, (1982). New Data on Shape Effects in Smooth Rectangular Channels. *J. Hyd. Res., IAHR*, 20, No. 3, 225–233.

Kline, S. J., Morkovin, M. V., Sovran, G., and Cockrell, D. J. (1968). Computation of Turbulent Boundary Layers. *In Proceedings of the 1968 AFOSR-IFP-Stanford Conference*, Vol. 1, Stanford University., Stanford, CA.

Knapp, R. T. (1951). Design of Channel Curve for Supercritical Flow. *Trans., ASCE*, 116, 296–325.

Liggett, J. (1975). *Stability in Unsteady Flow in Open Channels*. K. Mahmood et al (Ed.). Water Research Publ., Fort Collins, CO.

Lighthill, M. J. and Whitham, G. B. (1955). On Kinematic Waves: I – Flood Movement in Long Rivers. *Proc. Roy. Soc. (London)*, A229, 281– 316.

Martin Vide, J. P. (1992). Open Channel Surges and Roll Waves from Momentum Principles. *J. Hyd. Res., IAHR*, 30, No. 2, 183–196.

McBeans, E. A., and Perkins, F. (1975), Numerical Errors in Water Profile Computation. *JHYD, ASCE*, 101, No. 11, 1389–1403.

Montuori, C. (1963). Discussion of Stability Aspects of Open Channel Flow by F. F. Escoffier. *JHYD, ASCE*, 89, No. 4, 264–273.

Montuori, C. (1968), Brusca Immisione di Una Corrente Ipercritica a Tergo di Altra Preesistente. *L'Energia Elettrica*, 45, No. 3, 174–187.

Montuori, C. (1993). Discussion of Open Channel Ssurges and Roll Waves from Momentum Principle. *J. Hyd. Res., IAHR*, 31, No. 1, 139–143.

Naudascher, E. (1984). Scale Effects in Gate Model Tests. In *Symposium on Scale Effects in Modelling Hydraulic Structures*, H. Kobus (Ed.). Institut fur Wasserbau, Universitat Stuttgart, Germany.

Parshall, R. L. (1926). The Improved Venturi Flume. *Trans., ASCE*, 89, 841–851.

Quintela, A. C. (1982). Discussion of Minimum Specific Energy in Compound. *JHYD, ASCE*, 108, No. 5, 729–731.

Rajaratnam, N. and Muralidhar, D. (1964a). End Depth for Exponential Channels. *JIDD, ASCE*, 90, No.2, 17–39.

Rajaratnam, N. and Muralidhar, D. (1964b). End Depth for Circular Channels. *JHYD, ASCE*, 90, IR1, 99–119.

Rajaratnam, N. and Subramanya, K. (1968). Profile of the Hydraulic Jump. *JHYD, ASCE*, 94, No. 3, 663–673.

Ranga Raju, K. G., Mittal, M. K., Verma, M. S., and Ganeshan, V. (1980). Analysis of Flow over Baffle Blocks and End Sills. *J. Hyd. Res., IAHR*, 18, No. 3, 227–242.

Rehbock, T. (1929). Discussion of Precise Weir Measurements. *Trans., ASCE*, 93, 1143–1162.

Reynolds, W. C. (1968). A Morphology of the Prediction Methods. In Proceedings of the *1968 AFOSR-IFP-Stanford Conference*, Vol. 1. Stanford University, Stanford, CA.

Roberson, J. A., Cassidy, J. J., and Chaudhry, M. H. (1988). *Hydraulic Engineering*. Houghton Mifflin Company, Boston, MA.

Rouse, H. (1936). Discharge Characteristics of the Free Overfall. *Civil Eng., ASCE*, 6, No. 4, 257–260.

Rouse, H. (1946). *Elementary Mechanics of Fluids*. John Wiley & Sons, New York.

Rouse, H. (1965). Critical Analysis of Open-Channel Resistance. *JHYD, ASCE*, 91, No. 4, 1–25.

Rouse, H. (Ed.). (1950). *Engineering Hydraulics*. John Wiley & Sons, New York.

Rouse, H., Bhoota, B. V., and Hsu, E. Y. (1951). Design of Channel Expansion. *Tran., ASCE*, 116, 347–363.

Schlichting, H. (1968). *Boundary-Layer Theory*. McGraw-Hill Book Company, New York.

Schoklitsch, A. (1917). Uber Dambruchwellen. *Stitzungsberichte, Mathematisch-naturwissenschaftliche Klasse, Akademie der Wissenschaften.* 126, 1489–1514.

Strelkoff, T. (1969). One-Dimensional Equations of Open-Channel Flow. *JHYD, ASCE,* 95, HY3, 861–876.

Swamee, P. K. (1988). Generalized Rectangular Weir Equations. *JHE, ASCE,* 114, No. 8, 945–949.

Thorsky, G. N., Tilp, P. J., and Haggman, P. C. (1967). Slug flow in Steep Chutes. Report No. CB-2, U.S. Bureau of Reclamation, Denver, CO.

Toch, A. (1955). Discharge Characteristics of Tainter Gates. *Trans., ASCE,* 120, 290–300.

U.S. Army Corps of Engineers, (1960). Floods Resulting from Suddenly Breached Dams – Conditions of Minimum Resistance, Hydraulic Model Investigation. Misc. Paper No. 2-374, Report 1, Waterways Experimental Station, February.

U.S. Army Corps of Engineers. (1961). Floods Resulting from Suddenly Breached Dams – Conditions of High Resistance, Hydraulic Model Investigation. Misc. Paper No. 2-374, Report 2, Waterways Experimental Station, November.

U.S. Army Corps of Engineers. (1979). HEC-2: Water Surface Profiles Users Manual. Hydrologic Engineering Center, Davis, CA.

U.S. Bureau of Reclamation (USBR). (1948). Studies of Crest for Overfall Dams. Boulder Canyon Final Reports, Bulletin 3, Part VI.

U.S. Bureau of Reclamation (USBR). (1958). Hydraulic Design of Stilling Basins and Energy Dissipators. Engineering Monograph No. 25, Denver, CO.

U.S. Bureau of Reclamation (USBR). (1977). *Design of Small Dams.* U. S. Govt. Printing Office, Washington, DC.

Villemonte, J. R. (1947). Submerged Weir Discharge Studies. *Eng. News Record,* 139, 866–869.

Vittal, N. and Chiranjeevi, V. V. (1983). Open Channel Transitions: Rational Method of Design. *JHE, ASCE,* 109, No. 1, 99–115.

White, F. M. (1999). *Viscous Fluid Flow.* McGraw-Hill, New York.

Wooding, R. A. (1965). A Hydraulic Model for the Catchment-Stream Problem. I. Kinematic-Wave Theory. *J. Hydrol.,* 3, No. 3/4, 254–267.

Wooding, R. A. (1966). A Hydraulic Model for the Catchment-Stream Problem. III. Comparison with Runoff Observations. *J. Hydrol.,* 4, 21–37.

Yarnell, D. L. (1934). Bridge Piers as Channel Obstructions. U.S. Department of Agriculture, Tech. Bull. No. 442, November.

Yen, B. C. (1973). Open-Channel Flow Equations Revisited, *JEMD, ASCE,* 99, EM 5, 979–1009.

Yevjevich, V. M. (1959). Analytical Integration of the Differential Equation for Water Storage. *J. Res. Nat. Bur. of Stand.,* 63B, No. 1.

INDEX